Environmental Chemistry of the Elements

Environmental Chemistry of the Elements

H. J. M. BOWEN

Department of Chemistry
The University of Reading
England

ACADEMIC PRESS · 1979

LONDON · NEW YORK · TORONTO · SYDNEY · SAN FRANCISCO

A Subsidiary of Harcourt Brace Jovanovich, Publishers

ACADEMIC PRESS INC. (LONDON) LTD
24/28 Oval Road
London NW1 7DX

United States Edition published by
ACADEMIC PRESS INC.
111 Fifth Avenue
New York, New York 10003

British Library Cataloguing in Publication Data

Bowen, Humphry John Moule
Environmental chemistry of the elements.
1. Environmental chemistry
I. Title
574.5 QD31.2 79-50305
ISBN 0-12-120450-2

Filmset by Eta Services (Typesetters) Ltd., Beccles, Suffolk
Printed in Great Britain

Preface

Information about the distribution of the chemical elements in the environment and living organisms is scattered over a very wide range of literature, which I reviewed in "Trace Elements in Biochemistry" in 1966. Current knowledge in the field has grown so rapidly since then that every chapter of that book has had to be completely rewritten, and the work has been given a new title. The aim is still a critical but not an encyclopaedic coverage, and a real effort has been made to reduce the number of literature citations by referring to recent review articles. Since no single individual can have seen, let alone read, the entire output of world literature, the coverage must be incomplete. The author here extends his apologies to all those whose work has been missed; this applies particularly to papers published in Japanese and Russian.

Despite the wealth of published work, much remains to be learnt in this field. In environmental studies, data for the platinum metals are still inadequate, and relatively few studies of the lanthanides have been published. Very little is known about the chemical species present in soils and waters. We still lack concentration profiles of the elements in major soil types, and depth profiles of oceanic concentrations are few in number. Our quantitative knowledge of the carbon and nitrogen cycles has improved, but is too imprecise for predictive modelling. The elemental composition of remarkably few organisms is well established, despite the welcome production of biological reference materials. Data for the platinum metals, lanthanides and many of the elements for the second and third long periods are scanty or unreliable. Several new essential elements have been discovered, but their exact function is usually uncertain. The number of proteins known to contain metals is growing rapidly and big advances can be expected in this field. Steady progress has been made in the study of isotopic fractionation by organisms. Interest in pollution has shifted from radioactive fission products, which have declined since the moratorium on nuclear bomb trials, to heavy metals such as lead, cadmium and mercury. Demonstrable biological effects of any contaminant on the natural world are still relatively few, but intensive research has produced some links between elemental excesses and patterns of human disease. The effects of elemental deficiencies are more often found in farm animals and plants.

This book is aimed at post-graduate workers on academic problems of inorganic biochemistry. It should be of interest to environmental chemists as well as to biochemists, ecologists and health physicists. Specialists will no doubt find that their subject has been condensed and oversimplified, but they should recall the limitations of time and space.

I should like to thank my colleagues and research students who have helped to clarify my thoughts about this book, and have encouraged me to rewrite it.

June 1979

H. J. M. BOWEN

Contents

7 Uptake and Excretion of Elements by Organisms 109

8 Essentiality, Deficiences and Toxicities of the Elements 129

9 Chemical Forms and Functions of the Elements 151

10 Biological Fractionation of Isotopes 181

The Periodic Table of Elements

The elements are arranged in order of atomic number: the figures given refer to valency states which are known to occur in the environment. Uncertain valency states are given in parenthesis. Elements beyond **Bk** have not been found in the environment.

	H																		
	1																		
He	Li	Be	B	C	N	O	F												
0	1	2	3	0,2,4	3,5	2	1												
Ne	Na	Mg	Al	Si	P	S	Cl												
0	1	2	3	4	3,5	2,4,6	1,(3,5,)7												
Ar	K	Ca	Sc	Ti	V	Cr	Mn	Fe	Co	Ni	Cu	Zn	Ga	Ge	As	Se	Br		
0	1	2	3	4	(3–)5	3,6	2–4	2,3,(–5)	2,3	2	0–2	2	3	4	3,5	2,4,6	1,(5)		
Kr	Rb	Sr	Y	Zr	Nb	Mo	Tc	Ru	Rh	Pd	Ag	Cd	In	Sn	Sb	Te	I		
0	1	2	3	4	5	4–6	(4–7)	3,4	0,3	0,2	0,1	2	3	2,4	3,5	2,4,6	1,3,5,(7)		
Xe	Cs	Ba	La	Ce	Pr	Nd	Pm	Sm	Eu	Gd	Tb	Dy	Ho	Er	Tm	Yb	Lu	Hf	Ta
0	1	2	3	3,4	3	3	3	3	2,3	3	3	3	3	3	3	3	3	4	5
						W	Re	Os	Ir	Pt	Au	Hg	Tl	Pb	Bi	Po	At		
						6	(4–7)	(4–8)	0,3,4	0,2,4	0,1,3	0–2	1,3	2,4	3	(2,4,6)	1		
Rn	Fr	Ra	Ac	Th	Pa	U	Np	Pu	Am	Cm	Bk	Cf	Es	Fm	Md	No	Lw		
0	1	2	3	4	5	4,6	5	3–6	3	3	3	3	3	3	3	2	3		

Symbols and Units

SI UNITS (Basic units in **bold**)

Quantity	Unit of measurement	Symbol	Equivalent
length	**metre**	m	
time	**second**	s	
mass	**kilogramme**	kg	
electric current	**ampere**	A	
temperature	**kelvin**	K	
luminous intensity	**candela**	Cd	
chemical substance	**mole**	mol	
volume	litre	l	$0{\cdot}001\ m^3$
pressure	bar	bar	$10^5\ kg\ m^{-1}\ s^{-2}$
energy	joule	J	$1\ kg\ m^2\ s^{-2}$
electric charge	coulomb	C	1 As
potential difference	volt	V	$1\ JA^{-1}\ s^{-1}$
radioactive decay rate	becquerel	Bq	$1\ s^{-1}$
energy deposition	gray	Gy	$1\ J\ kg^{-1}$

CONVERSION FACTORS

1 m	10^{10} Ångstrom units = 3·281 feet
1 m^2	10·76 $feet^2$ = $2{\cdot}471 \times 10^{-4}$ acres
1 s	$1{\cdot}157 \times 10^{-5}$ days = $3{\cdot}168 \times 10^{-8}$ years
1 kg	2·205 pounds = 0·001 tonnes
1 J	$6{\cdot}24 \times 10^{18}$ eV = 0·239 calories
1 J/mol	0·01036 eV/atom
1 Bq	60 disintegrations/minute = 27·03 picocuries
1 Gy	100 rads

VALUES FOR PHYSICAL CONSTANTS

atomic mass unit	1/12 mass of $^{12}C = 9{\cdot}1095 \times 10^{-31}$ kg
Avogadro number N	$6{\cdot}022 \times 10^{23}$ mol^{-1}
Boltzmann constant k	$1{\cdot}3807 \times 10^{-23}$ J K^{-1}
Electronic charge	$1{\cdot}6022 \times 10^{-19}$ C
Electronic mass	$9{\cdot}1095 \times 10^{-31}$ kg
Faraday F	96 485 C mol^{-1}
gas constant R	8·3144 J K^{-1} mol^{-1}
Planck's constant h	$6{\cdot}626 \times 10^{-34}$ Js
velocity of light c	$2{\cdot}9979 \times 10^{8}$ m s^{-1}

VALUES FOR SOME GEOPHYSICAL QUANTITIES

Earth	
mean radius	6371 km
surface area	$5{\cdot}09 \times 10^{8}$ km^2
volume	$1{\cdot}083 \times 10^{21}$ m^3
mass	$5{\cdot}977 \times 10^{24}$ kg
density	5·517 g cm^{-3}
Ocean	
surface area	$3{\cdot}61 \times 10^{8}$ km^2
volume	$1{\cdot}372 \times 10^{18}$ m^3
mass	$1{\cdot}39 \times 10^{21}$ kg
density	0·99987 g cm^{-3} at STP
Atmosphere	
mass	$5{\cdot}3 \times 10^{18}$ kg
density	1·29 mg cm^{-3} at STP

ABBREVIATIONS

a	atto = $\times 10^{-18}$
A	ampere
ADP	adenosine 5′-(trihydrogen diphosphate)
ATP	adenosine 5′-(tetrahydrogen triphosphate)
B-metals	Ag, Au, Bi, Cd, Cu, Hg, Pb, Sn, Zn
BP	before present
Bq	Becquerel
c	centi = $\times 10^{-2}$

C	coulomb
°C	degree celsius ($t/°C = T/K - 273{\cdot}15$)
Ci	curie = $3{\cdot}7 \times 10^{10}$ Bq
CoA	coenzyme A
DM	dry matter
DNA	deoxyribonucleic acid
e^-	an electron
eV	electron volt = $1{\cdot}602 \times 10^{-19}$ J
f	femto = $\times 10^{-15}$
F	faraday
FM	fresh matter
g	gramme
G	giga = $\times 10^9$
Gy	gray
h	Planck's constant
J	joule
k	kilo = $\times 10^3$
K	kelvin
l	litre
LD_{50}	lethal dose for 50%
m	metre *or* milli = $\times 10^{-3}$
M	mol/l *or* mega = $\times 10^6$
mol	mole
m/v	mass per volume
n	nano = $\times 10^{-9}$
N	Avogadro number
NAD	adenosine 5′-(trihydrogen diphosphate) 5′ → 5′ ester with 3-(aminocarbonyl)-1-β-D-ribofuranosylpyridinium hydroxide
$NADH_2$	ADP 5′ → 5′ ester with 1,4-dihydro-1-β-D-ribofuranosyl-3-pyridinecarboxamide
NADP	ADP 2′-dihydrogen phosphate 5′ → 5′ ester with 3-(aminocarbonyl)-1-β-D-ribofuranosylpyridinium hydroxide
$NADPH_2$	ADP 2′-dihydrogen phosphate 5′ → 5′ ester with 1,4-dihydro-1-β-D-ribofuranosyl-3-pyridinecarboxamide
p	pico = $\times 10^{-12}$
p.d.	potential difference
pH	$-\log[H^+]$
ppm	$\times 10^{-6}$ m/m or m/v
R	gas constant
RD_{50}	dose causing 50% of the population to respond
RNA	ribonucleic acid
s	second

STP	standard temperature and pressure (0 °C, 1·013 bar)
v/v	volume per volume
V	volt
X	any element
yr	year
α	alpha particle = $^{4}He^{2+}$
β	beta particle (electron from radioactive decay)
γ	gamma (electromagnetic) radiation from radioactive decay
δ	for two isotopes ^{A}X and ^{B}X, $\delta = 100[(^{A}X/^{B}X)_{sample} \times (^{B}X/^{A}X)_{std} - 1]\%$
μ	micro = $\times 10^{-6}$
ν	frequency

1 *The Atmosphere*

The concept of elementary cycles

The atmosphere and the ocean have long been regarded as static systems, and their elemental composition has been presented in most books as a simple list, with no indications of variability. We now believe that nearly all the components of these systems are in dynamic equilibrium, and that their concentrations may vary, even if analytical methods are insufficiently sensitive to measure this variation. A system in dynamic equilibrium is characterised by inputs, outputs and constant concentrations for each of its constituents. For constituent X, the total input rate must equal the total output rate R_X, and it is convenient to define a residence time T_X as follows

$$T_X = M_X/R_X$$

where M_X is the mass of substance X in the system.

A simple example concerns the water content of the atmosphere. This is normally equivalent to a rainfall of 2·5 cm over the whole planet, or $2{\cdot}9 \times 10^{15}$ kg of water. The mean rainfall of the Earth is 90 cm/year, so it is obvious that the atmosphere loses its water many times a year; the loss is made up by evaporation and transpiration. The residence time of water in the atmosphere must be

$$\frac{2{\cdot}5 \times 365}{90} = 10{\cdot}1 \text{ days}$$

and this is the most probable time period that any given water molecule will spend in the air.

Characteristics of the atmosphere (Wallace and Hobbs, 1977)

The Earth's atmosphere has a mass of $5{\cdot}3 \times 10^{18}$ kg and extends to at least 300 km altitude. Its density at ground level and STP is 0·00129 g cm^{-3}, but both pressure and density decrease uniformly with altitude; the temperature varies in a complex manner with altitude, and may range between 0–1000 K. Eighty per cent of the atmospheric mass resides in the troposphere, below about 10 km altitude, and this is the region of most interest to biochemists.

The stratosphere above the troposphere is mainly abiotic, but does contain substantial amounts of ozone which absorbs short-wave radiations from the sun. Among the planets, the Earth is unique in having an atmosphere containing large amounts of free oxygen. (For data on other planets, see Huntress, 1977.)

The commoner molecular species in the atmosphere are listed in Table 1.1, together with their concentrations, probable sources and sinks, and estimates of residence times.

TABLE 1.1 Sources, sinks, concentrations and residence times of (a) elements and (b) molecules in the atmosphere

Constituent	Concentration/ $\mu g\ m^{-3}$ STP	Sources	Sinks	Residence time/yr
Ar	$1{\cdot}6 \times 10^7$	^{40}K decay		$>10^7$
C	$1{\cdot}6 \times 10^5$			
Cl	1–2			
F	0·8			
H (dry air)	300			
He	920	U/Th decay	Outer space?	10^7
Kr	4100			$>10^7$
N	$9{\cdot}73 \times 10^8$			
Ne	$1{\cdot}6 \times 10^4$			$>10^7$
O	$2{\cdot}99 \times 10^8$			
S	3–80			
Xe	500			$>10^7$
$CClF_3$	0·8	Refrigerants and sprays	?	~50
CCl_2F_2	1·25	Refrigerants and sprays	?	~50
CCl_4	0·09–0·8	Microbes: industry	?	1
CH_3Br	0·013–0·06	Marine algae: fumigants	?	?
CH_3Cl	1.6	Marine biota?	?	?
CH_3I	0·012–0·06	Marine algae	Photolysis?	0·003
CH_4	1080	Swamp microbes	Soil?	2–5
CO	130	Marine microbes: cars	Soil: photochemical	1
COS	1·1–1·5	Photolysis?	?	20
CO_2	$5{\cdot}9 \times 10^5$	Respiration: combustion	Ocean: green plants	2–15
CS_2	0·24–1·3	Ocean	?	1
H_2	36–90	Swamp microbes	Soil: outer space?	5
H_2O	$\sim 3 \times 10^6$	Evapo-transpiration	Rain	0·027
O_2	$2{\cdot}99 \times 10^8$	Photosynthesis	Respiration: combustion	10^4
O_3	10–50 (−250)	Photochemical	Reductants	0·25
N_2	$9{\cdot}73 \times 10^8$	Microbes	Soil microbes	10^6
NH_3	0–15	Microbes	Soil: ocean	~0·003
N_2O	610	Microbes	Soil?	8–15
NO/NO_2	0–6	Microbes: combustion	Reductants	~0·01
SF_6	0·001	Industry?	?	?
SH_2	3–30	Swamp microbes: ocean	Oxidants	~0·002
SO_2	2–50	Combustion	Rain: oxidants	<0·01

References: see Junge (1972); Sandalls and Hatton (1977); Sandalls and Penkett (1977); Singh *et al.* (1977); Holland (1978)

Junge (1972) has pointed out that the atmospheric gases can be divided into a number of categories. These may be called inert gases, biocycle gases, and short-lived gases, with water as a very special case of the last category.

Inert gases

Neon, krypton and xenon have no obvious cycle, though some isotopes of Kr and Xe are stable products of uranium fission. Argon and helium are products of radioactive decay of ^{40}K and α-emitters respectively. Thus ^{40}Ar is by far the most abundant inert gas nuclide in the atmosphere. Helium, together with molecular hydrogen, would probably be more abundant but it is sufficiently light to escape from the Earth's gravitational field; hence it is constantly leaking into outer space at a very slow rate.

Biocycle gases

The two main components of the atmosphere, nitrogen and oxygen, are both believed to be largely if not entirely products of biological action. Nitrogen is an important constituent of other planetary atmospheres. It is produced by bio-degradation of organic compounds in large amounts, and is also taken up by specialised soil microbes: its residence time is very long.

Oxygen is found neither in volcanic gases nor in the atmospheres of other planets (except as a minor constituent). It might have been produced by photolysis of water in the upper atmosphere, but it seems likely that the primal atmosphere was anoxygenic. Oxygen began to build up in the atmosphere when green plants first became abundant, which is thought to have occurred about $1{\cdot}4 \times 10^9$ years ago, in the Pre-Cambrian era. It may still be increasing, but its residence time is so long that sufficiently accurate measurements to detect any changes cannot at present be made (Singer, 1970).

The other biocycle gases are hydrogen, carbon monoxide and dioxide, methane and nitrous oxide, which are all chemically inert at STP and have residence times of the order of years. Apart from CO_2, the evidence for their cycling is mostly recent and semiquantitative. Hydrogen is known to be produced by anoxygenic microbes and oxidised by aerobic microbes. It can be lost to outer space, or photo-oxidised to water by reactive species such as O_3 or O atoms in the stratosphere. Carbon monoxide, a human pollutant from car exhausts, is produced by photo-oxidation of methane and also by marine organisms, and is absorbed by soil microbes; it may also be oxidised to CO_2 by hydroxyl radicals in the stratosphere. A remarkable phenomenon is the difference in concentration in the two hemispheres – about 190 $\mu g/m^3$ north of the equator and 75 $\mu g/m^3$ south of it, which is believed to be due to lower rates of production in the Southern hemisphere (Weinstock and Chang, 1976).

Carbon dioxide concentrations in the atmosphere are buffered by the large reservoir of this gas dissolved in the sea. The two biological processes of photosynthesis and respiration appear to be in overall balance but it is possible to show daily fluctuations of CO_2 concentrations in the atmosphere (Wallace and Hobbs, 1977). Consequences of the recent injections of CO_2 from burning fossil fuels are considered in chapter 12. Carbon dioxide absorbs considerable amounts of solar infrared radiation. It is a constituent of the atmospheres of several planets and volcanic gases.

Methane is produced by anoxygenic bacteria e.g. in anoxic sediments, swamps and in the rumen of grazing mammals (Mah *et al.*, 1977). Large amounts are trapped as natural gas ($\sim 90\%$ CH_4) with petroleum deposits. It is not clear whether it is removed from the atmosphere by other microbes, or by oxidation to CO or CO_2. It is a major component of the atmosphere of Jupiter. Nitrous oxide appears to be a by-product of the nitrogen cycle; its sources have been identified but its sinks have not (Singer, 1970). It can be produced by the reduction of nitrates with iron(II) at room temperature between pH 6–8 as well as by bacterial reduction of nitrates (Moraghan and Buresh, 1977).

Water and short-lived gases

The atmosphere contains very variable quantities of water, whose sources and sinks are well known. Most of the gaseous water in the atmosphere gets there by evaporation from the ocean, but a substantial fraction is contributed by transpiration through plants. Water is the only atmospheric component which is liquid at STP, and its frequent deposition as rain cleans the atmosphere of much particulate matter and soluble trace gases. This has given rise to the problem of acid rain near industrial regions (Likens and Boormann, 1974). Water is also an important absorber of infrared solar radiation, and can be photo-dissociated by ultraviolet radiation to yield hydrogen atoms and hydroxyl radicals in the stratosphere.

$$H_2O \xrightarrow{h\nu} HO. + H.$$

Ozone is another short-lived component which is found mainly in the stratosphere, where its residence time is thought to be about 3 months. It is probably formed by the reactions (Crutzen, 1974; Logan *et al.*, 1978).

$$O_2 \xrightarrow{h\nu\ (<240\ \text{nm})} O + O$$

$$O + O_2 + X \rightarrow O_3 + X$$

(where X may be NO, Chameides and Stedman, 1976) and destroyed by e.g.

$$O_3 \xrightarrow{h\nu} O_2 + O$$

$$O_3 + O \xrightarrow{NO} 2O_2$$

It is also found as a pollutant near the ground, where its residence time may well be much shorter (Biersteker and Evendijk, 1976). Such a reactive gas can rapidly oxidise aerosols, organic matter or trace gases such as H_2, CO, SO_2, NO etc. Crutzen (1970) considers that ozone concentrations are controlled by nitric oxide in the atmosphere. Total atmospheric ozone is currently increasing (London and Kelley, 1974).

Ammonia is produced and consumed microbiologically, and is readily washed out by rain. It is involved in the production of ammonium sulphate aerosols. On other planets, such as Jupiter, it is an important gas and may have been so in the primal atmosphere of the Earth. Nitric oxide (NO) and its oxidation product nitrogen dioxide are also produced by microbes and by the stratosphere reaction $O + N_2O \rightarrow 2NO$. Both are common pollutants, formed wherever air is heated to a high temperature, and NO_2 is a reactive oxidant with similar properties to ozone.

Hydrogen sulphide is produced in coastal areas (and from volcanoes), but attempts to measure its concentration in the atmosphere are rarely successful. It is probably rapidly oxidised to sulphur dioxide – one suggested mechanism involves oxidation by hydroxyl radicals – and has a residence time of about 18 hours in the atmosphere (Sprung, 1976). Sulphur dioxide itself is an important pollutant with corrosive and biocidal properties (chapter 12). It is removed from the atmosphere by many processes, including wash-out by rain, absorption by limestone, and oxidation to sulphates. The latter process may take place either homogeneously or on aerosol particles, as both reactions occur at similar rates (Spedding, 1974). Other gaseous compounds of sulphur have been observed in the atmosphere, such as $(CH_3)_2S$ (Lovelock *et al.*, 1972) and $(CH_3)_2S_2$, COS and CS_2 (Banwart and Bremner, 1975).

Recently a number of halogen containing compounds have been observed in the atmosphere, and it is likely that chloroform, carbon tetrachloride and iodomethane are natural, fairly short-lived molecules with cycles of their own (see McConnell *et al.*, 1975, for review). As analytical techniques improve, it is likely that many further species may be detected. Mercury vapour is one example, and it has been observed after volcanic eruptions

(Siegel *et al.*, 1973). The occurrence of volatile derivatives of arsenic, tin and selenium is quite likely.

The mean rate of production of volcanic gases is about $7{\cdot}5 \times 10^{12}$ g/yr (Cadle, 1975); their composition is variable. Analyses due to Giggenbach (1975) are given below in mole per cent units. Water is usually the main component, while carbon dioxide and sometimes sulphur dioxide or hydrogen are relatively abundant. Minor constituents include CO, HCl, H_2S and SO_2, while CH_4 and HF are often present in traces.

Constituent	*mole % range*	*Constituent*	*mole % range*
CH_4	0·0003–1·1	HCl	0·02–2·6
CO	0·1–7·2	HF	0·001–0·02
CO_2	0·3–30	H_2S	0·01–4·5
H_2	0·001–30	SO_2	0·08–23
H_2O	23–99·3		

Note that all the short-lived components of the atmosphere exhibit great variability in the amount present, according to the proximity of their sources and sinks to the sampling point.

TABLE 1.2 Elemental content of air from various sources, in ng X/m^3 air at STP (columns 1–4 are clean air, columns 5–8 are contaminated air)

X	Shetland 1976	N. Norway 1974	N.W. Canada 1971	South Pole 1977	Europe median (range)	North America median (range)	Japan (2)	Hawaii or Etna volcano
Ag	<0·07			<0·0004	1 (0·2–7)	1 (<0·04–2·4)		30
Al	60	32	66	0·82	600 (160–2900)	2000 (150–3500)	1600–6200	20 k
As	0·6	1·9	0·3	0·007	16 (1·5–53)	15 (1·7–40)	1·5–190	5·5–850
Au	0·005			0·00004	(0·001–0·006)	(<0·003–0·3)	0·05	8
B					3·5	4		680
Ba		0·73		0·016	(6–90)	12 (0·2–28)		
Be					(0·9–4)	0·2 (0·1–0·3)		
Bi					0·3	(<1–3)		
Br	15	4·4	0·5	1·4	460 (13–2500)	240 (8–1200)	90–600	31–21 k
Ca	820	42	40	0·5	1400 (430–4300)	2800 (100–7000)	2 k–15 k	460–69 k
Cd	<0·8			<0·015	(0·5–620)	(<1–41)	25	8–92
Ce	0·1	0·06	0·24	0·0023	0·9 (0·2–4)	6 (0·02–20)	2–18	
Cl	4400	440	9	2·6	3000 (1400–5400)	1800 (9–6000)	500–11 k	1 k–990 k
Co	0·06	0·06	0·04	0·0005	(0·2–37)	3 (0·13–23)	1·5–4	4·5–27
Cr	0·7	0·7	0·6	0·005	25 (1–140)	60 (1–300)	20–70	45–67
Cs	0·04	0·02		0·0001	0·5 (0·06–1·5)	0·2 (0·07–0·3)	1·4	14
Cu	20	2·5	0·9	0·036	340 (8–4900)	280 (5–1100)		200–3 k
Dy					0·04			
Er					0·026			
Eu	0·004			0·000017	0·02 (0·005–0·08)	0·35 (0·01–1·7)		
F					1·5	(10–400)		high
Fe	90	48	71	0·84	1400 (130–5900)	3600 (260–14 k)	1 k–14 k	1 k–10 k
Ga				<0·0001	0·23	1		
Gd					0·045			
Ge					2·8			
Hf				0·00006	0·3	(0·0005–0·04)		

Table 1.2—*continued*

X	Shetland 1976	N. Norway 1974	N.W. Canada 1971	South Pole 1977	Europe median (range)	North America median (range)	Japan (2)	Hawaii or Etna volcano
Hg	<0·04	0·01	0·06		(<0·009–2·8)	(0·07–38)	1·6	18–250
Ho					0·013			
I	4	0·6	0·2	0·08	6 (3–15)	20 (0·4–60)	6–10	
In	<0·02	0·004	0·001	0·00005	0·2 (0·03–0·1)	0·09 (0·02–0·14)	1·2	
Ir					0·00027			
K	170	48	54	0·7	560 (140–1700)	8000 (200–40 k)	1400	94 k
La	<0·2	0·03	0·09	0·0005	1·5 (<0·1–10)	2·5 (0·5–6)	5	
Li					2·3			
Mg	480	77	16	1	430 (<160–1000)	3000 (60–11 k)	480–3 k	325
Mn	3	3	1·5	0·01	43 (9–210)	150 (6–900)	80–600	55–1300
Mo	<0·2				(<0·2–3·2)	(<1–10)		
Na	3300	550	18	3	1000 (350–1500)	1500 (120–5500)	1 k–7 k	110–87 k
Nb					0·12			
Nd					0·28	1		
Ni	<4	1·2	<2		25 (4–120)	90 (<1–120)	5–47	330
P					180	1100		
Pb	21		9–14	0·63	120 (55–340)	2700 (45–13 k)	200	28–1200
Pd					0·0007			
Pr					0·065			
Pt					<0·01			
Rb	<1			0·002	3 (<1–6·6)	1·1	0·6–20	
Rh					0·27			
Ru					0·27			
S				50	3200	8000 (2 k–13 k)		300 k
Sb	0·4	0·33	0·13	0·0008	8 (0·6–32)	12 (0·08–55)	2–63	45
Sc	0·015	0·005	0·044	0·00012	0·15 (0·03–0·7)	1·3 (0·1–3)	0·8	1–2
Se	0·5	0·26	0·043	0·0056	3 (0·15–11)	5 (0·06–30)	7·5	9–2100
Si		100			(1 k–6·3 k)	(5 k–63 k)		2700
Sm	<0·01	0·003	0·013	0·00009	0·08 (0·02–0·26)	0·4 (0·2–1)		
Sn					(1·5–800)	(<10–70)		
Sr				<0·5	2·2	(<7–50)	5–50	
Ta						0·011		
Tb					0·01			
Te					0·18			
Th	0·02	0·011	0·052	0·00014	0·16 (0·03–1)	(0·05–1·3)	1·3	
Ti	10	2·6	5	0·1	85 (22–210)	(<10–230)	140–1 k	
Tl					0·06	0·22		
Tm					0·004			
U					0·02	<0·5		
V	3	1·9	0·2	0·0015	30 (5–92)	400 (1–2000)	180	79
W	<0·5		0·016	0·0015	0·7 (0·35–1·5)	4 (0·03–6)	5–18	
Y					0·22		0·5–6	
Yb				<0·00005	(0·26–1·4)	(0·004–0·02)		
Zn	15	10	4	0·03	1200 (13–16 k)	500 (<10–1700)	100–6 k	10 k
Zr					0·7	5 (4–26)	3–25	

k, × 1000. Median values without ranges are often based on a single determination. Most of the data from Europe, North America and Japan came from large cities

References: Shetland data from Cawse, 1976; *N. Norway*, Rahn in Heindryckx and Dams, 1974; *N.W. Canada*, Rahn and Winchester, 1971; Volchok and Bogen, 1973; *South Pole*, Maenhaut and Zoller, 1977; *Europe*, Cadle, 1973, Dale *et al.*, 1973 (Glasgow), Hamilton, 1974 (London), Schramel *et al.*, 1974 (Munich), Dams *et al.*, 1975, Schutyser *et al.*, 1977 (Ghent), Cawse, 1976 (6 British sites); *North America*, Hwang, 1972 (U.S. cities), Dams *et al.*, 1970, Brar *et al.*, 1970 (3 Michigan sites), Zoller and Gordon, 1970 (Mass.), Brown and Vossen, 1970, Pillay and Thomas, 1971 (2 New York sites), Gray *et al.*, 1972 (Missouri), John *et al.*, 1973 (California), Aras *et al.*, 1973 (Maryland), Kuykendall *et al.*, 1974 (Ohio), Struempler, 1975 (Nebraska), Bate *et al.*, 1976 (Tennessee); *Japan*, Mamuro et al., 1970, Kato *et al.*, 1976; *Hawaii*, Cadle *et al.*, 1973; *Etna*, Buat-Ménard and Arnold, 1978

Aerosols in the atmosphere

A large amount of data has been collected recently on the elemental composition of atmospheric aerosols, almost all from the Northern hemisphere. Some of this is given in Tables 1.2 (as ng/m^3 air) and 1·3 (as ng/g aerosol). Tropospheric aerosols have been described by Junge (1963).

TABLE 1.3 Elemental composition of dust from three air samples in mg X/kg dust

X	Milan	London	Ocean	X	Milan	London	Ocean
Ag	13	0·1–11		Mn	1300	2–380	18–1300
Al	17 k	54–8 k	8 k–105 k	Mo	30	0·2–12	
As	85	0·5–80		N	24 k		
Au	3	<0·05		Na	9600	460–6800	3 k–67 k
B		0·3–25		Nb	0·1–2		
Ba	970	0·8–110	490	Nd		3	
Be		0·3–8		Ni	1500	0·4–230	91
Bi		0·01–4		O	287 k		
Br	700	7–2100		Os		0·001	
C	13 k			P	<100	13–460	
Ca	39 k	0·6 k–20 k		Pb	5600	30–7700	460
Cd	29	1–47		Pd	0·027	0·001	
Ce	140	0·03–5	8–160	Pr		0·02–0·8	
Cl	13 k	120–10 k		Pt	<0·1	0·001	
Co	17	0·06–31	9–50	Rb		0·2–350	
Cr	200	4–180	85 (–110)	Re		0·001	
Cs	2·1	0·03–28		Rh		0·001	
Cu	700	8–1100	160	Ru		0·1–2	
Dy	2.3	0.35		S	40 k	150–11 k	
Er		0·2		Sb	260	0·2–91	
Eu	0·45	0·2	0·5–3	Sc	2·9	0·02–0·8	13–42
F	2400	0·2–28	330–870	Se	12	1–9	
Fe	32 k	100–14 k	10 k–89 k	Si	61 k	120–18 k	
Ga	10	0·1–61	21	Sm	7·5	0·35	
Gd		3		Sn		0·3–160	30
Ge		1–28		Sr		1–63	100
H	17 k			Ta		0·09–1	
Hf	1	0·1		Tb		0·06	
Hg	22	1–14	7–84	Te		0·1–1	
Ho		0·1		Ti	2500	4–580	2·5 k–16 k
I	14	0·3–3		Tl		0·07–6	
In	0·4	0·1–4		Tm		0·035	
Ir	0·0012	0·001		U	1·1	0·01–0·1	
K	6900	350–87 k	7000	V	2000	6–540	0·5–140
La	75	0·01–3		W	17	1–3	
Li		0·05–10		Y		0·2–2	
Lu	0·2	0·03–0·3		Yb		0·25	
Mg	7·8 k	0·5 k–3·6 k	5 k–43 k	Zn	7100	9–2500	680
				Zr		0·1–9	

k, × 1000

Milan data from Dams *et al.* (1975), Schutyser *et al.* (1977); *London*, Hamilton (1974)—actually from Sutton, Surrey; *Ocean*, Bressan *et al.* (1974), Chester and Stoner (1973)

They have a size range of 5 nm–20 μm, and can be subdivided into small, charged Aitken particles with diameter <100 nm and large particles with diameters from 0·1 to 20 μm. The latter constitute the bulk of the mass of the aerosol and are responsible for natural haze and fog. Aerosols are not uniformly distributed, as they vary with both location and height – the concentration is highest near the ground (Cadle, 1973).

Aerosol particles originate from a variety of sources which can be summarised as

1. Salt particles, whose composition may differ from that of evaporated sea water.
2. Soil particles, containing large amounts of Al, Si and Ti; some of these may be volcanic.
3. Organic particles, evaporated from forest vegetation together with pollen and spores, at a rate of about 4×10^{10} kg/yr (Rasmussen and Went, 1965; Beauford *et al.*, 1975 and 1977).
4. Soot particles, often containing much carbon and unusual amounts of certain elements such as As, Cr, Cu, Pb, V and Zn.
5. Industrial haze particles, mainly ammonium sulphate with some ammonium nitrate.

Such aerosols have a residence time in the atmosphere of 10–30 days; they are washed out more or less efficiently by rain, and some of the heavier particles fall to earth by gravity (dry deposition). They constitute a neglected, but sometimes important, source of elements for organisms and must be considered in any quantitative studies of elemental cycling.

In Tables 1.2 and 1.3 we have made some attempt to divide aerosol samples into those from 'clean' areas and those from polluted areas but there is a continuous gradation between these. Clean areas (N.W. Canada, Tennessee, N. Norway, South Pole, much of the ocean) are characterised by low total solids, and the air has the following characteristics:

Al < 100; As < 2; Br < 10; Ca < 100; Cr < 1; Fe < 250; K < 200; Mn < 10; Sb < 0·5; V < 5; Zn < 15 ng/m^3.

Clean continental areas also have Cl < 15, Na < 120 ng/m^3, but this is not true for oceanic sites such as Norway and Shetland. More data for the Southern hemisphere, which is cleaner than the North, are badly needed. Differences between the hemispheres are shown by the data of Egorov *et al.* (1971) in Table 1.4. Further analyses of oceanic air are given by Duce *et al.* (1976).

One of the difficulties in studying a mass of data such as that in Table 1.2 is that we do not know the relative proportions of the five main types of aerosol particle in each sample. Cawse (1974, 1976) normalised his data by

TABLE 1.4 Elemental composition of oceanic air from the two hemispheres (Egorov *et al.*, 1971)

X	ng X/m^3; N. hemisphere	ng X/m^3; S. hemisphere
Al	121	12
Cd	2	0·14
Cr	7·2	0·23
Cu	12	12
Fe	180	7
Mn	7·9	0·24
Ni	2·9	0·35
Sn	3·5	0·27
Pb	4·4	1

Data are mean values from numerous measurements at sea

assuming that scandium concentrations have not been affected by human activity. He then calculated enrichment factors E for each element X as follows:

$$E = \frac{[X]_{air}\,[Sc]_{soil}}{[X]_{soil}\,[Sc]_{air}}$$

Table 1.5 compares Cawses enrichment factors with enrichment factors calculated from the data of Hamilton (1974). The latter has been normalised to titanium instead of scandium, using the mean soil composition from Table 4.4.

Other workers have attempted to calculate the percentage of their aerosols derived from different sources (Miller *et al.*, 1972; John *et al.*, 1973; Heindryckx and Dams, 1974; Hardy *et al.*, 1976). Thus John *et al.* found Cr, Cu, Hg, Sb, Se and Zn enriched in San Francisco air, relative to local soils.

Provisional conclusions are: 1. That aerosols throughout the Northern hemisphere in the latitude of Europe, Japan and North America have a broadly similar composition, with minor variations near industrial regions

TABLE 1.5 Enrichment factors for aerosols from Britain

Range of E	Assuming E(Ti) = 1 (Hamilton, 1974)	Assuming E(Sc) = 1 (Cawse, 1976)
<1–5	Al, Ba, Ca, F, Fe, Ga, La–Lu, Mn, Nb, Rb, Sc, Si, Sr, Th, U, Y, Zr	Al, Ce, Cr, Cs, Eu, Fe, K, La, Mn, Rb
5–50	B, Bi, Co, Cr, Cs, In, K, Li, Mg, Mo, Na, P, Se, Sn, Tl	Ca, Co, Mg, Mo, Ni, Sb, V
50–500	As, Au, Cd, Cu, Ge, I, Sb, V, Zn	Ag, As, Au, Cu, I, In, W, Zn
500–>5000	Ag, Be, Br, Cl, Hg, Pb, S	Br, Cd, Pb, Se

(e.g. high Ni and Co near Glamorgan (Cawse, 1974). 2. That aerosols collected near the sea coast have elevated Cl, Mg and Na concentrations. 3. That many elements in aerosols are derived largely from the local soil (cf. Table 4.4). 4. That certain elements are indicators of local contamination (see chapter 12), e.g. Br and Pb for leaded petrol residues; As, Cd, Cu and Zn for industrial activity and burning of coal; V for certain types of fuel oil. All elements are elevated in urban and industrial aerosols. The high enrichments for I and Se in British aerosols await explanation, though iodine is known to be released from the ocean photochemically (Miyake and Tsunagai, 1963). Se is also enriched in air from the South Pole, together with Br, Sb and Pb (Maenhaut and Zoller, 1977). In aerosols from Nigeria, soil-derived elements such as Al, Ce, Mg, Sc and Ti are much more abundant than they are in European aerosols, while industry-derived elements such as As, Br, Pb, Sb and Zn are much rarer (Beavington and Cawse, 1978).

Salmon *et al.* (1977) have investigated trends in aerosol behaviour at one station in Britain over the period 1957–1974. They noted the following:

(*a*) Seasonal variations, especially for As, Br, Hg, Pb, Sb, Se, V and Zn, possibly due to increased burning of coal in the winter (Fig. 1.1 for As in air).

(*b*) Medium-term fluctuations, very 'noisy', with a period of a few years.

(*c*) Long-term trends, with declining concentrations of most elements following the Clean Air Act of 1956. The air constituents which have increased during this period are Br, Na, NH_4^+, SO_4^{2-}, NO_3^- and possibly I. Of these, Br almost certainly comes from car exhausts and SO_2 from burning coal and oil.

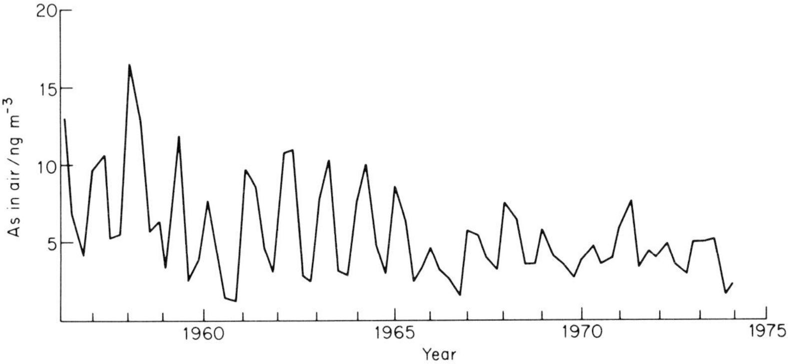

FIG. 1.1 Arsenic in the air near Harwell, England. Note (*a*) overall decline between 1957 and 1974, and (*b*) seasonal fluctuations with higher concentrations in winter. Data from Salmon *et al.* (1977)

An interesting discovery is that some air particulates are inhomogeneous. For example, Cr, Mn, Pb and V are enriched on the surface of fly-ash particulates (Linton *et al.*, 1977).

'Fossilised' aerosol deposits have been investigated in layers of the Greenland ice cap, which can be dated. High enrichment factors (relative to Al) have been found for Cd, Pb, S and Zn (but not for Fe, Mn, Na, Si or V) by Herron *et al.* (1977) even in dusts deposited a century or more ago, and it is suggested that these elements were supplied by volcanic emissions. It is also possible that these elements are cycled through the atmosphere as volatile organic compounds synthesised by anoxic bacteria (Wood, 1974). Very few data are available for either the total volume or the composition of volcanic gases. Data for Etna and Hawaii are given in Table 1.2 (Cadle *et al.*, 1973). This suggests that volcanic gases are not very different from industrial atmospheres but are particularly rich in boron, mercury and sulphur, and perhaps also in Ag, As, Au, Br, Cl, Cs, F, Ni and Se. For example, Hg is found at concentrations of 0·8–18 $\mu g/m^3$ (Siegel and Siegel, 1975).

Cycling of elements through the atmosphere

In view of the extremely inhomogeneous distribution of dust and metallic elements in the atmosphere, attempts to estimate the annual atmospheric input of metals to the ocean can be little more than order of magnitude estimates. Hence the figures given in Table 1.6 cannot be regarded as well founded.

TABLE 1.6 Estimates of atmospheric input to ocean in kg/year

X	Input/kg yr^{-1}	References
Cd	10^8	Kustin and McLeod (1977)
Cr	2×10^7	
Cu	2×10^8; $3{\cdot}6 \times 10^7$	Kustin and McLeod (1977); Boyle *et al.* (1977)
Hg	8×10^7	
I	$\sim 5 \times 10^8$	Miyake and Tsunogai (1963)
Ni	3×10^7	Kustin and McLeod (1977)
Pb	3×10^8	
V	2×10^7	
Sn	2×10^7	

Lantzy and Mackenzie (1979) estimated the annual fluxes of dust from soils and volcanoes to be 5×10^{14} g/yr and $1{\cdot}5 \times 10^{14}$ g/yr respectively. These estimates suggest that the ocean is a source of volatile compounds of As, Hg and Se.

2 *The Hydrosphere*

An inventory of the Earth's hydrosphere is given in Table 2.1.

TABLE 2.1 Inventory of the hydrosphere (after Strahler, 1972)

Reservoir	Mass of water/kg	Residence time/yr
Oceans	$1{\cdot}39 \times 10^{21}$	37 600
Ice	$2{\cdot}92 \times 10^{19}$	15 000
Ground water	$8{\cdot}3 \times 10^{18}$	
Lakes	$2{\cdot}3 \times 10^{17}$	6·2
Atmosphere	$1{\cdot}3 \times 10^{16}$	0·028 (see ch. 1)
Rivers	$1{\cdot}25 \times 10^{15}$	0·0337

Measurements of river discharge, rainfall and evaporation for the Earth suggest that rivers return $3{\cdot}7 \times 10^{16}$ kg water/yr to the sea (Holeman 1968; Turekian, 1969). Rainfall is $3{\cdot}47 \times 10^{17}$ kg/yr over the sea (area $3{\cdot}6 \times 10^{8}$ km^2) and $9{\cdot}9 \times 10^{16}$ kg/yr over land (area $1{\cdot}49 \times 10^{8}$ km^2), while evaporation is $3{\cdot}83 \times 10^{17}$ kg/yr from the sea and $6{\cdot}3 \times 10^{16}$ kg/yr from the land. Thus rivers return their own volume of water to the sea many times per year, and the mean residence time of a water molecule in rivers is about 12 days. On the other hand a water molecule in the sea has a long residence time of about 38 000 years. The ocean is fairly well mixed, with a mixing time of about 1600 years (Broecker, 1974).

Geologists agree that the process of dissolution of rocks by rivers, followed by transport to the sea, sedimentation, and recycling of water by evaporation have been going on for a very long time, so that a state of pseudo-equilibrium has been reached. Although this is not true thermodynamic equilibrium, it is likely that the ocean has not changed much in volume or content during the last 500 myr.

While the cycle of water is fairly well understood, there are technical difficulties in measuring the dissolved components. This is because river water transports such a wide range of sizes of components, from small molecules and ions to colloids, silt, sand, pebbles and even boulders. It is conventional to define dissolved matter in natural waters as that material which is not removed by an 0·45 μm filter which collects suspended matter.

Hence dissolved matter includes both molecular species and colloids, which may be further separated by ultra-centrifugation. There are not enough measurements of the ratio of suspended to dissolved matter to quote a reliable world average. Some recent measurements and estimates are quoted in Table 2.2.

All bodies of liquid water are self-purifying and lose elements to sediments by precipitation, biological action and adsorption on colloids. Small amounts of volatile elements such as iodine may also be lost to the atmosphere.

TABLE 2.2 Ratio of suspended/dissolved matter in some rivers

River	Dissolved matter/mg l^{-1}	Suspended/ Dissolved ratio	Reference
Amazon	53	2·5	Gibbs (1967)
Nile	119	3·7	Livingstone (1963)
Mississippi	166	1·1	Livingstone (1963)
U.S.A. rivers	?	2·8	Judson and Ritter (1964)
S.E. Asian rivers		4·2	Livingstone (1963)
World	120	4·0	Holeman (1968); Garrels and Mackenzie (1971)

Rivers

Useful physical data for rivers are not always known; recent source books giving flow-rates etc. include Morisawa, 1968; Unesco, 1971 and Whitton, 1975. The Amazon has the greatest discharge, equivalent to 20% of the world total, and has been fairly intensely studied, as have two other great rivers, the Nile and the Mississippi. Data for many of the great Asian rivers, such as the Amur, Ganges, Hwang-Ho, Mekong, Ob, Yangtse and Yenisey are scanty but these rivers are thought to carry 80% of the suspended matter transferred to the ocean each year (Holeman, 1968).

The elemental composition of the worlds rivers has been reviewed by Durum and Haffty (1963), Livingstone (1963), Turekian (1969, 1971) and Gibbs (1972), and much extra data is to be found in recent literature. Some of the earlier work gave much larger concentrations for some elements than does recent data, and technical problems confronting the freshwater analyst are still severe. Table 2.3 is an attempt to summarise the elemental composition of freshwater, but for most of the rarer elements the figures are provisional. River composition varies with climate, local geology and the presence of large cities or intensive agriculture, since rivers effectively integrate the soluble material from the basins they drain. In temperate regions the major species present are usually calcium and bicarbonate ions, while in the humid parts of the tropics the rivers are more dilute and contain more Al, Fe, Si and organic matter. Rivers draining arid regions usually contain much sodium, chloride and sulphate.

Since the average residence time of its water is of the order of days, river composition can change rather rapidly in response to external conditions. Regular annual fluctuations in phosphate and silicate content are characteristic features of temperate rivers where diatoms consume these elements in spring and release them later in the season. This example illustrates one of the most difficult problems for the river analyst, that of speciation. During a diatom bloom, the total phosphorus concentration does not change, but soluble phosphates are converted to organic phosphates within the living cells. It is possible to determine both inorganic and organic phosphates separately in this case, but more often it is not possible to determine the chemical species of an element whose total concentration is near the limit of detection. Some information on speciation may be obtained by the methods of Benes *et al.* (1976) or Florence and Batley (1977). It seems that while alkali metals, alkaline earth metals and halogens exist largely as simple ions in natural waters, most of the B-metals and transition metals are either complexed or adsorbed by colloidal matter to a greater or lesser extent. Mercury is unique in that it may exist as metal atoms, anions or organic complexes (Hinkle and Grenshaw, 1975). Indications of some of the likely species are given in Table 2.3. It should be noted that river waters usually contain much larger amounts of organic matter, which can act as a ligand, than does sea water (Jackson, 1975). Organic complexes of Al, Cu, Fe, Hg, Mn, Mo, Ni, Pb, Tl, V and Zn in fresh waters have been reviewed by Jenne (1977). A complicating factor is that hydrolysis reactions and formation of chelates may be quite slow processes. For example Hem (1968) comments that $Al(H_2O)_6{}^{3+}$ may take from 10 days to a year to polymerise sufficiently to form turbid solutions containing particles of about 0·1 μm diameter. Such reactions will not have time to reach equilibrium before the river water enters the ocean.

Ice

The amount of crushed rock delivered to the sea by melting ice has been estimated as 2×10^{12} kg/year, much less than the $2{\cdot}5 \times 10^{13}$ kg/year brought down by rivers (Garrels and Mackenzie, 1971). The rate of melting of icecaps is variable, but is probably of the order of 2×10^{15} kg/year, giving a residence time of $\sim$15 kyr.

TABLE 2.3 Elemental composition of freshwater

X	μg X/l Median	μg X/l Range	Species	Refs
Ag	0·3	0·01–3·5		C, D, K, L, La, R, S, T, U
Al	300	8–3500	Colloid	B, Ba, D, F, L, S, Su, T, V
Ar	600		Ar	D, L
As	0·5	0·2–230	Anion ± Me	An, D, L, La, Li, Ra, S, Su, T, U, W
Au	0·002	0·001–0·02	Anion?	Ba, Go, La, Li, S, W
B	15	7–500	$B(OH)_3$	Aj, C, D, L, Re, T
Ba	10	<3–150	Ba^{2+}	B, D, L, La, S, T, U, W
Be	0·3	0·01–1		C, D, L, M, U
Bi	0·02?			C, U
Br	14	0·05–55	Br^-	B, Ba, D, L, La, Li, S, T, W
C	11 k	6 k–19 k	HCO_3^-	D, L, T, V
Ca	15 k	2 k–120 k	Ca^{2+}	B, D, L, S
Cd	0·1	0·01–3	Org.	A, P, U
Ce	0·2	0·1–0·2	Colloid	B, Ba, S
Cl	7000	1 k–35 k†	Cl^-	B, L, R, V
Co	0·2	0·04–8		B, C, D, K, L, Li, P, S, T, U
Cr	1	0·1–6		B, Ba, D, K, L, La, Li, S, T, U
Cs	0·02	0·005–1	Cs^+	Ba, D, K, L, La, S, Sr, T
Cu	3	0·2–30	Org.	A, D, Hk, L, La, P, R, Re, Su, T, U
Dy		0·005		Hk
Eu	0·006	0·002–0·009		Ba, Hk, Li, S
F	100	50–2700*	F^-	D, L, Re, Su, T, V, Z
Fe	500	10–1400†	Colloid	B, Ba, L, P, R, S, Su, V
Ga	0·09			He
Hf	0·01	0·005–0·13		Ba, S
Hg	0·1	0·0001–2·8	Org. +	Ba, Bu, C, D, Dl, H, Ho, L, La, Li, S, T, Va
I	2	0·5–7	I^-	B, D, L, S, Su, T, W
K	2200	500–10 k	K^+	L, R, S, Sr, W
La	0·1	<0·05–0·8		B, Ba, Hk, La, Li, S, Si
Li	2	0·07–40	Li^+	C, D, L, Re, T, W
Lu	0·003	0·002–0·005		Ba
Mg	4000	400–6 k†	Mg^{2+}	L, R, S, V
Mn	8	0·02–130		B, Ba, D, Hk, Lk, La, P, R, Re, S, T, U
Mo	0·5	0·03–10	MoO_4^{2-}	Bt, D, K, L, Li, Su, T
N	50	2–1800	NO_3^-	L, Re, V
Na	6000	700–25 k†	Na^+	B, L, R, S, V
Nd	0·15	<0·06–0·25		Ba
Ni	0·5	0·02–27		A, D, L, P, R, Re, T, U
P	20	1–300	$PO_4H_2^-$	D, L, T, U

Table 2.3—*continued*

X	μg X/l Median	μg X/l Range	Species	Refs
Pb	3**	0·06–120		A, D, L, P, T, U
Ra	4×10^{-7}			L
Rb	1	0·6–9	Rb^+	B, D, K, L, La, S, Sr
Rn	$1{\cdot}7 \times 10^{-12}$		Rn	L
S	3700	200–40 k	SO_4^{2-}	L, R, Su, V
Sb	0·2	0·01–5	Sb(V)	K, La, Li, S, U
Sc	0·01	0·004–0·04		B, Ba, La, Li, S, Si, T
Se	0·2	0·02–1†	SeO_3^{2-}	Cu, K, La, Li
Si	7000	500–12 k†		L, R, Su, V
Sm	0·06?	0·01–0·12		Hk, S
Sn	0·009	0·004–0·09	Sn(IV) ± Me	C, Sb, U, Za
Sr	70	3–1000	Sr^{2+}	D, L, La, Re, S, Su, T
Ta	<0·002			Ba
Tb	0·003	0·001–0·005		Ba
Th	0·03	0·007–0·1		B, Ba, L, S, T
Ti	5	3–18		Ba, L, U
U	0·4	0·002–5		Be, D, F, G, L, La, Li, Re, T, W
V	0·5	0·01–20		Ba, C, D, F, L, Lk, S, Su, T
W	0·03	<0·02–0·1		La, U
Yb	0·01	0·005–0·2		Ba, S
Zn	15	0·2–100	Zn^{2+} Org.	A, B, Ba, D, Hk, L, La, Li, P, R, Re, S, Su, T, U, W
Zr	0·8	0·05–20		D, L, U

k, ×1000; Org., unknown organic complex. Medians are mostly subjective estimates. Evidence for species present is mostly weak

* F can occur at 14 000 μg/l in India (Paliwal, 1974)

† Higher values have been reported from some African rivers (Visser and Villeneuve, 1975)

** The prehistoric median was probably 0·5 μg/l (C. C. Patterson, 1971)

References: A, Abdullah and Royle (1974); Aj, Ahl and Jönsson (1972); An, Andreae (1977); B, Benes *et al.* (1976); Ba,Bate *et al.* (1976); Be, Bertine *et al.* (1970); Bt, Bertine and Turekian (1973); Bu, Burton and Leatherland (1971); C, Cowgill (1972); Cu, Cutter (1978); D, Durum and Haffty (1963); Dl, Dall'Aglio (1968); F, Fujinaga *et al.* (1973); G, Gladney *et al.* (1976); Go, Gosling *et al.* (1971); H, Hinkle and Grenshaw (1975); He, Heide and Ködderitzsch (1964); Hk, Hirose *et al.* (1978); Ho, Holden (1972); K, Kharkar *et al.* (1968); L, Livingstone (1963); La, Landström and Wenner (1965); Li, Lieser and Neitzert (1976); Lk, Linstedt and Kruger (1970); M, Meehan and Smythe (1967); P, Perhac (1972); R, Riemer and Toth (1968); Ra, Ray and Johnson (1972); Re, Reeder *et al.* (1972); S, Salbu *et al.* (1975); Sb, Schroeder *et al.* (1964); Si, Silker (1964); Sr, Sreekumaran *et al.* (1968); Su, Sugiwara (1968); T, Turekian (1969); U, Udodov and Parilov (1961); V, Visser and Villeneuve (1975); Va, Van der Sloot and Das (1974); W, Wedepohl (1969–74); Z, U.S. Geol. Surv. Water Supply Paper (1976) 2160; Za, Braman and Tompkins (1979)

Lakes and swamps

Lakes form convenient sized systems for estimating residence times, with inputs from streams, ground water, rainfall and back-diffusion from sediments. Outputs include streams, ground water, sediments and evaporation. Relatively few lakes have been studied in detail. Data for Lake Washington, Seattle, U.S.A. (Barnes and Schell, 1973; Crecelius, 1975) and Lake Michigan, U.S.A. (Klein, 1975) are summarised in Table 2.4. It can be seen that Br, Ca, Cl, Mg, Na and Sr have similar residence times to that of lake water, implying that they are lost mostly by outflow of water. On the other hand Al, Fe, Pb, Mn, Sc and Th have very short residence

TABLE 2.4 Elemental contents, outputs and residence times (T) for two lakes (A and B)

X	Mass in A/ kg	A output/ kg d^{-1}	T(A)/d	Mass in B/kton	B output/ kton yr^{-1}	T(B)/yr
Ag				1·5	0·021	700?
Al				132	420	0·31
As	5 k	12	415	4·9	0·16	30
Ba				180	6	30
Br				245	3	82
Ca				170 k	2·1 k	81
Cl				54 k	780	690
Co				0·88	0·1	9
Cr				8·3	0·72	11
Cs				0·069	0·017	4
Cu	13 k	23	560	25	0·51	49
Fe	152 k	13 k	12	93	210	0·44
Hg	110	0·32	340	0·13	0·0073	18
K				7·8 k	270	29
La				1	0·28	3·6
Mg				56 k	730	77
Mn				4·9	6·2	0·79
Na	$13{\cdot}6 \times 10^6$	14·9 k	913	25 k	300	84
Pb	2170	85	25			
Rb				4·9	0·65	7·5
S				29 k	510	57
Sb				1·1	0·019	58
Sc				0·015	0·052	0·29
Se				0·41	0·0091	45
Si				6·9 k	2·3 k	3
Sr				475	5·9	80
Th				0·012	0·053	0·23
V				1	0·39	2·6
Zn	74 k	134	550	78	2·9	27
H_2O	$2{\cdot}9 \times 10^{12}$	$3{\cdot}2 \times 10^9$	906	$4{\cdot}9 \times 10^9$	49×10^6	100

d, day; k, × 1000; ton, metric tonne
A, Lake Washington, U.S.A.: data from Barnes and Schell (1973) and Crecelius (1975)
B, Lake Michigan, U.S.A.: data recalculated from Klein (1975)

times compared with lake water, as they are rapidly lost by sedimentation. For most other elements, apart from silver where the data should be confirmed, the output rates due to outflow of water and sedimentation must be of the same order of magnitude.

It should be clear that lakes forming part of a river system function as gigantic sink traps, removing by sedimentation a number of elements such as Al, Fe and Si and also Ca (as $CaCO_3$). The chemical nature of the sediments will depend on local factors. The ultimate fate of lakes is to fill up with sediment and become swamps. Even then, a swamp can function as a sink trap removing nutrients. For example, the Upper Nile passes through one of the largest swamps in the world, the Sudd. Comparison of its composition before and after passing this swamp shows that chloride, sulphate and major cations are removed, while acidity ($[H^+]$) and ammonia are increased (Bishai, 1962).

Despite much research on the processes of sedimentation, the overall quantitative importance of the following processes in lakes is still uncertain.

1. Biological production of $CaCO_3$ (by molluscs etc.) or SiO_2 (by diatoms).
2. Absorption by hydrous oxides of Fe(III) and/or Mn(IV), formed in oxygenic waters.
3. Absorption by clay particles, which strongly bind large ions such as K^+ and NH_4^+.
4. Absorption by colloidal organic matter (Jackson, 1975).
5. Absorption by sulphides, produced in anoxygenic sediments by bacterial reduction. Annual cycling and variation of concentration with depth are described by Cowgill (1972), Andelmann (1973) and Bate *et al.* (1976).

Estuaries

The chemical processes occurring when fresh water meets salt water are obscure, largely because we rarely know the chemical nature of most dissolved elements. We might expect the following processes to occur in or near estuaries:

1. Deposition of suspended mud, and flocculation of colloidal hydroxides.
2. Desorption of some species from particulates because of competition with higher concentrations of ions in sea water.
3. Dissociation of some humic complexes in a medium of higher ionic strength.

In practice (1) obviously occurs, and both (2) and (3) have been shown to occur in model systems (Kharkar *et al.*, 1968; Van der Weijen, 1975). Careful studies by De Groot and Allersma (1975) have shown that there is little change in the chemical state of most metals in the estuary of an unpolluted tropical river, the Chao Phya, in Thailand. This contrasts with the situation for a polluted temperate river, the Rhine, where 80–90 % of the As, Cd, Cr, Cu, Hg, Pb and Zn, plus 50 % of the Ni, are desorbed from suspended matter in the estuary, and only La, Mn, Sc and Sm are unchanged. De Groot and Allersma ascribe the difference to the large amount of organic matter in the Rhine, compared with very little in the Chao Phya. The Rhine estuary is a site of intense biological activity, where many of the heavy metals are mobilised (i.e. enter true solution) during the oxidation of organic matter.

Estuaries can be classified as anoxic or oxygenic according to the nature of the sediment. Anoxic estuaries occur in deep fiords where microbial activity reduces sulphates to sulphides. We might expect such estuaries to trap thiophilic elements such as B-metals, but in practice only molybdenum has been shown to be trapped in this way (Gross, 1967).

Oxygenic estuaries are sites of intense biological activity with a high rate of cycling of carbon and organic matter. It is difficult to prove whether any given element X is trapped in estuarine muds or not, but Boyle *et al.* (1974) have suggested that good evidence for trapping is for the relation between [X] and [Cl^-] to be non-linear along the direction of flow.

Their data for the Merrimack river, Mass., U.S.A. suggests that iron is trapped, but silicon is not, in this estuary. Boyle *et al.*'s review of earlier work suggest that B, Ca, F, Mg, N, P, S and Si are probably not trapped by most estuaries, but that Fe is and Al, Cu, Mn, Ni and Zn may be (see also Boyle *et al.*, 1977). Cadmium appears to be trapped in the Severn estuary, Britain (Abdullah *et al.*, 1972) and mercury has been trapped in Minamata Bay, Japan (Hosohara *et al.*, 1961). The elements most likely to be trapped are those for which the mean concentration in river water greatly exceeds that in sea water, namely Al, Be, Ce, Fe, La, Mn, Sc and Zr. Much further work needs to be done on the mechanisms and quantitative significance of trapping of metals in estuaries, especially in connection with pollution studies (chapter 12). Burton and Liss (1976) give a good account of estuarine chemistry.

The ocean

Table 2.5 summarises data on elemental composition, probable species present, and residence time for the ocean. The dominant anions and cations present in sea water are Na^+, K^+, Mg^{2+}, Ca^{2+}, Cl^-, $SO_4{}^{2-}$ and $HCO_3{}^-$. The total salt concentration or salinity of sea water varies somewhat, but the ratio of the concentrations of major constituents to $[Cl^-]$ are remarkably constant, showing that the ocean is well-mixed. Data in Table 2.5 assume a chloride content of 19·35 g/l. However, many of the minor components, notably those with short residence times, do not have a constant concentration in the sea, but vary with depth and/or location. For example, Broecker (1974) has summarised the evidence that the concentrations of nitrate, phosphate and silica are markedly depleted in the upper layers of the ocean, from which they are effectively removed by plankton (Fig. 2.1). Bicarbonate, and to a detectable extent calcium, are somewhat depleted in surface water, again by biological precipitation. Chow and Goldberg (1960) suggest the barium content of surface water is about 25% that of deep water, so that barium sulphate may also be a component of bioprecipitation, but their data was not confirmed by Bolter *et al.* (1964). Substantial variations in the contents of Ag, Au, Cd, Co, Cu, Fe, Hg, Mn, Ni, Pb, Sb, Se and Zn have been measured in coastal waters and ascribed to river inputs from mineralised or industrial areas (e.g. Abdullah *et al.*, 1972; Preston, 1973; Schutz and Turekian, 1965; Gardner and Riley, 1974). Lead is notably high in surface ocean water compared to deep water owing to atmospheric input (Chow and Patterson, 1966). Profiles of Cd, Cu, Ni and Zn in the ocean have been measured by Bruland *et al.* (1979).

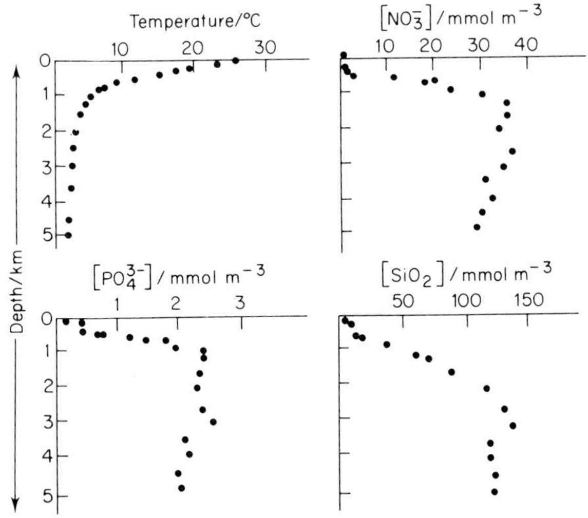

FIG. 2.1 Concentrations of nitrate, phosphate and silica in the Pacific Ocean as a function of depth. Data from 23 °S, 170 °W, collected by Y. Horibe (Broecker, 1974)

TABLE 2.5 Elemental composition of, and residence times, in sea water

X	Mean μg X/l	Range μg X/l	Res. time/yr	Theor. species	Ref.
Ag	0·04	0·03–2·7	5000	$AgCl_2^-$	Gr, Mp, St, Tu
Al	2	1–8·4	250	colloid, $Al(OH)_4^-$	F
Ar	450		28 000	Ar	
As	3·7	0·5–3·7	280 000	AsO_4H^{2-}, organic	An, Mp, Ra
Au	0·004	0·0005–0·027	75 000	$AuCl_2^-$	Mp, St
B	4440		11 M	$B(OH)_3$	
Ba	13	2–63	50 000	Ba^{2+}	Cg
Be	0·0056	0·0006	700?	$BeOH^+$	
Bi	0·02	0·015–0·02	38 000	BiO^+?	
Br	67 300		130 M	Br^-	
C	28 000			HCO_3^-, CO_3^{2-}	
Ca	412 000		1 M	Ca^{2+}	
Cd	0·11	<0·01–9·4	40 000	colloid, $CdCl_2$	A, Ab, Mp, Pr
Ce	0·0012		200	Ce^{3+}	
Cl	19·35 M		100 M	Cl^-	
Co	0·02	0·01–4·1	4000	Co^{2+}, $CoCO_3$	Gr, Mp, Ro, St, Tu
Cr	0·3	0·2–50	11 000	$Cr(OH)_3$, CrO_4^{2-}	Gr, Mp
Cs	0·3	0·15–0·42	560 000	Cs^+	Bo, Gr, Mp
Cu	0·25	0·05–12	3000	$CuOH^+$, $CuCO_3$	A, Ab, Bs, F, Mp, Pr, Sh
Dy	0·00091		6800	$DyOH^{2+}$	
Er	0·00087			$ErOH^{2+}$	
Eu	0·00013		800	$EuOH^{2+}$	
F	1300		500 000	F^-, MgF^+	
Fe	2	0·03–70	150	colloid, $Fe(OH)_2^+$	A, Mp, Pr
Ga	0·03		13 000	$Ga(OH)_4^-$	
Gd	0·0007			$GdOH^{2+}$	
Ge	0·05			$GeO_4H_3^-$	
H	110 M		38 000	H_2O	
He	0·007			He	
Hf	0·007		26 000	$Hf(OH)_4$	
Hg	0·03	0·01–0·22	11 000	$HgCl_4^{2-}$, organic	Bu, Ga, Mp, We
Ho	0·00022			$HoOH^{2+}$	
I	60	50–70	1·1 M	I^-, IO_3^-, CH_3I	Wb
In	0·00011				
K	399 000		6·8 M	K^+	
Kr	0·21			Kr	
La	0·0034		1300	La^{3+}	
Li	180	170–194	3·4 M	Li^+	
Lu	0·00015		2000	$LuOH^{2+}$	
Mg	1·29 M		12 M	Mg^{2+}	
Mn	0·2	0·03–21	9000	Mn^{2+}, $MnCl^+$, colloid	A, Pr, Sh
Mo	10	4–10	750 000	MoO_4^{2-}	Mp
N	640/15 500			$NH_3 + NO_3^-/N_2$	
Na	10·77 M		68 M	Na^+	
Nb	0·01	0·01–0·015			
Nd	0·0028		700	Nd^{3+}	
Ne	0·12			Ne	
Ni	0·56	0·13–43	440 000	Ni^{2+}, $NiCO_3$	A, Pr, Sh, St, Tu

Table 2.5 —*continued*

X	Mean µg X/l	Range µg X/l	Res. time/yr	Theor. species	Ref.
O	883 M		38 000	OH_2	
P	60	60–88	110 000	HPO_4^{2-}, $MgPO_4^-$	
Pa	$\leqslant 5\times10^{-8}$?	5×10^{-8}–2×10^{-10}			
Pb	0·03	0·03–13	380	$PbCO_3$, colloid	A, Ab, Cp, Pr
Po	$1{\cdot}5\times10^{-11}$	$(0{\cdot}7–2)\times10^{-11}$			Ch, Pa
Pr	0·00064			Pr^{3+}	
Ra	$8{\cdot}9\times10^{-8}$	$(3{\cdot}2–9)\times10^{-8}$	8400	Ra^{2+}	Br
Rb	120	67–195	4·5 M	Rb^+	Bo, Gr, Mp
Re	0·004	0·004–0·0084		ReO_4^-	Mb, Sc
Rn	6×10^{-13}		13 000	Rn	
Ru	0·0007				
S	905 000		9·2 M	SO_4^{2-}, $NaSO_4^-$	
Sb	0·24	0·18–5·6	45 000	$Sb(OH)_6^-$	Gr, Mp, St
Sc	0·0006	0·0006–0·12	2200	$Sc(OH)_3$	Gr, Mp
Se	0·2	0·052–0·2	38 000	$SeO_3^{2-}+SeO_4^{2-}$	St, Sw
Si	2200	2200–2900	12 000	$Si(OH)_4$, colloid	
Sm	0·00045		280	$SmOH^{2+}$	
Sn	0·004	0·002–0·81	17 000	$SnO_4H_3^-$, organic	Za
Sr	7900	7000–8500	4·2 M	Sr^{2+}	
Ta	0·002		38 000		
Tb	0·00014		1800	$TbOH^{2+}$	
Te				TeO_3H^-	
Th	0·001	0·0001–0·22	1200	$Th(OH)_4$	Cs
Ti	1		7500		
Tl	0·019				Ma
Tm	0·00017			$Tm\ OH^{++}$	
U	3·2	0·04–6	300 000	$UO_2(CO_3)_3^{4-}$	F, Gr, Ha, Mp, Va
V	2·5	0·9–2·5	200 000	$H_2VO_4^-$, HVO_4^{2-}	F
W	0·1	0·001–0·7	120 000	WO_4^{2-}	Mp
Xe	0·05			Xe	
Y	0·013			$Y(OH)_3$	
Yb	0·00082		3100	$YbOH^{2+}$	
Zn	4·9	0·2–48	12 000	Zn^{2+}, $ZnOH^+$	A, Ab, F, Gr, Mp, Pr, Sh
Zr	0·03		1400	$Zr(OH)_4$	

M = $\times 10^6$. Mean values and ranges, see Brewer (1975) or references. Residence times were calculated from $37\,600[X]_{sea}/[X]_{river}$ using $[X]_{river}$ values from Table 2.3. Theoretical species present, see Stumm and Brauner (1975)

References: A, Abdullah and Royle (1974); Ab, Abdullah *et al.* (1972); An, Andreae (1977); Bo, Bolter *et al.* (1964); Br, Broecker *et al.* (1976); Bs, Boyle *et al.* (1977); Bu, Burton and Leatherland (1971); Cg, Chow and Goldberg (1960); Ch, Cowen *et al.* (1977); Cp, Chow and Patterson (1966); Cs, Cherry and Shannon (1974); F, Fujinaga *et al.* (1973); Ga, Gardner and Riley (1974); Gr, Grancini *et al.* (1976); Ha, Hashimoto (1971); Ma, Matthews and Riley (1969); Mb, Matthews and Riley (1970); Mp, May *et al.* (1975); Pa, Parfenov (1974); Pr, Preston (1973); Ra, Ray and Johnson (1972); Ro, Robertson (1970); Sb, Sclater *et al.* (1976); Sc, Scadden (1969); Sh, Slowey and Hood (1971); St, Schutz and Turekian (1965); Sw, Sugimura *et al.* (1976); Tu, Turekian (1971); Va, Van der Sloot *et al.* (1975); Wb, Wong and Brewer (1976); We, Weiss and Crozier (1972); Za, Braman and Tompkins (1979)

Anoxic basins such as the Black Sea present a special case with marked variations of Co, Cu, Fe, Mn and Zn with depth (Brewer, 1975). These basins contain free hydrogen sulphide in concentrations from 0·03–100 mg/l (Richards, 1971).

Residence times in Table 2.5 have been estimated by assuming that input is wholly from rivers, and that a steady state has been established. If we neglect inputs from melting ice, rainfall, back-diffusion and submarine vulcanism, and assume that $3{\cdot}7 \times 10^{16}$ kg water is returned to the ocean each year by rivers, the residence time of any element X in the ocean is given by:

$$T_X = \frac{1{\cdot}39 \times 10^{21}[X]_{sea}}{3{\cdot}7 \times 10^{16}[X]_{river}} = \frac{37\,600[X]_{sea}}{[X]_{river}} \text{ years}$$

Such residence times are well established for the major components of sea water, but may be one or two orders of magnitude in error in the following cases:

1. When the analytical data for $[X]_{sea}$ or $[X]_{river}$ are inadequate.
2. When $[X]_{sea}$ is known to vary markedly with depth.
3. When T_X is found to be less than the mixing time of the ocean, about 1600 years.
4. When atmospheric inputs and outputs of X are substantial, e.g. for C, Hg, I, N, inert gases and possibly for some industrial contaminants such as Pb.
5. Where back-diffusion from marine sediments is substantial, as appears to be so for ^{226}Ra (Cherry and Shannon, 1974).

Residence times can also be calculated by assuming that $3{\cdot}2 \times 10^{12}$ kg of sediments are deposited from the ocean each year, and that these sediments have the same composition as mean crustal rock (Garrels and Mackenzie, 1971). This second estimate of residence time can be calculated from the expression

$$T_X = \frac{1{\cdot}39 \times 10^{21}[X]_{sea}}{3{\cdot}2 \times 10^{12}[X]_{rock}} = \frac{4{\cdot}33 \times 10^{8}[X]_{sea}}{[X]_{rock}} \text{ years}$$

using the elemental composition of crustal rock given in Table 3.3. They are not tabulated here.

The physical chemistry of sea water is the subject of several texts (Stumm and Morgan, 1970; Riley and Skirrow, 1975) but is by no means fully understood. For example, it is rarely possible to predict which insoluble form of any given element will be precipitated as sediment. Comparison of the composition of average river water with that of sea water shows that most elements have higher concentrations in the sea than in rivers, as is the case for sodium and chlorine. Elements of this kind will all have residence times longer than the residence time of water in the sea, i.e. greater than 38 kyr. The elements with higher concentrations in river water than in the sea can be subdivided according to the value of the ratio $R = [X]_{river}/[X]_{sea}$

$R \simeq 1$	Ar, Au, Ba, Bi, Cd, Hf, Sb, Se, Ta
$10 > R > 1$	Ag, Cr, Ga, Hg, Mn, Ra, Si, Sn, Ti, Zn
$100 > R > 10$	Be, Co, Cu, Eu, La, Lu, Nd, Sc, Tb, Th, Yb, Zr
$1000 > R > 100$	Al, Ce, Fe, Pb, Sm

These considerations suggest that sediments near the mouths of rivers should contain large amounts of aluminium and iron, together with other elements forming insoluble hydroxides, as is indeed the case.

The three main types of sediments found in the open ocean have been described in detail by Broecker (1974) and Chester and Aston (1976). These are:

1. Calcium carbonate: skeletons of marine organisms which inhabit the surface layers of the ocean, and which may redissolve when they sink into deep water which is undersaturated with calcite.
2. Silica: skeletons of other organisms, mostly diatoms, which may also partly or wholly redissolve in the deep ocean.
3. Red clays: aluminosilicates containing much iron, derived from continental rocks.

Other, rarer, types of sediments include marine phosphorites and manganese nodules, both of which appear to scavenge trace elements from sea water. The manner in which many important elements such as K or Mg are precipitated from the ocean as sediments is still obscure. Most of the rare elements in sea water are undersaturated with respect to the known minerals in sediments. Careful studies of the variation of concentration with depth are needed for many more of these elements (Holland, 1978).

The thermodynamic treatment of the stability of different chemical forms of an element in aqueous solution is conveniently represented by a Pourbaix diagram (Pourbaix, 1963). Pourbaix diagrams are graphs of pH against redox potential, and examples for Fe, I, Mn and S are given in Figs 2.2–2.5. It can be seen that the range of redox potentials in water is 1·25 V, but the absolute values of the upper and lower bounds vary linearly with pH. The upper bound corresponds to the oxidation of water to dioxygen, while the lower bound corresponds to reduction of water to dihydrogen. The range of pH is 4–9 for most natural waters, or 7·7–8·3 for sea water. One should bear in mind that these diagrams represent true equilibrium, which may not be obtained within the residence time of the water mass concerned. For example, dissolved organic matter in the presence of dioxygen is not in

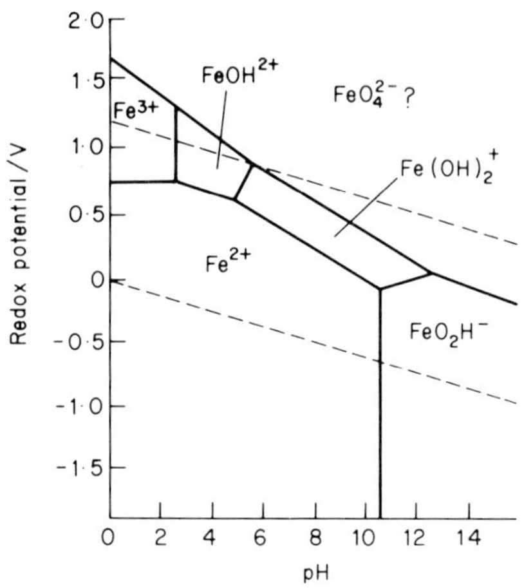

FIG. 2.2 Pourbaix diagram for iron in water. The region between the parallel dotted lines is the domain of stability of water itself

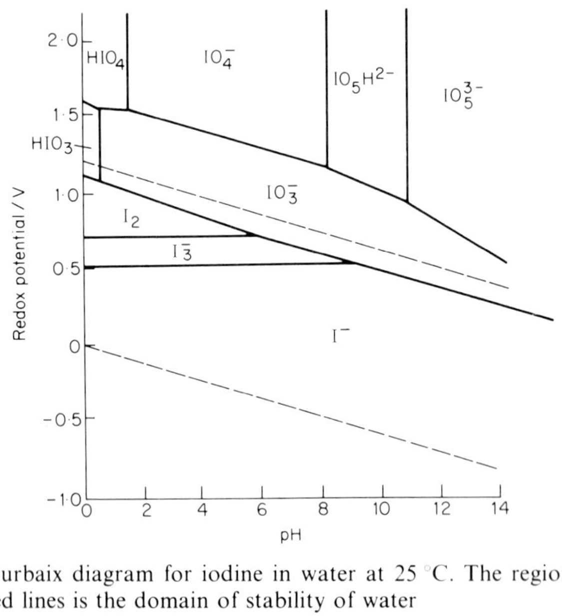

FIG. 2.3 Pourbaix diagram for iodine in water at 25 °C. The region between the parallel dotted lines is the domain of stability of water

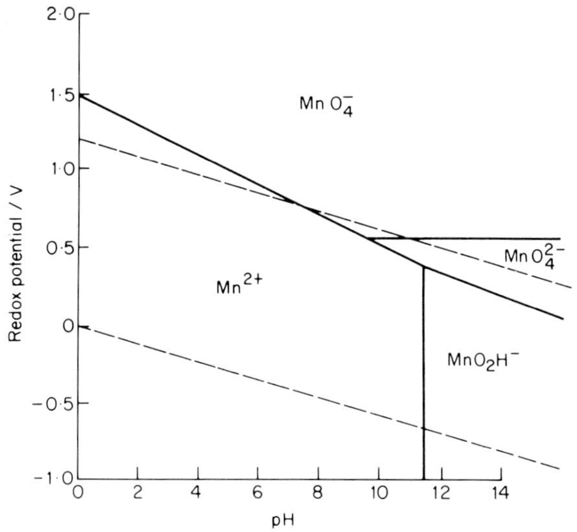

FIG. 2.4 Pourbaix diagram for manganese in water at 25 °C. The region between the parallel dotted lines is the domain of stability of water

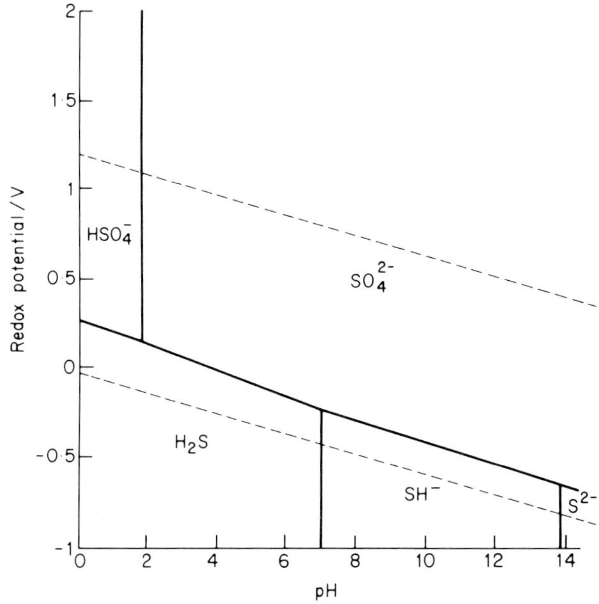

FIG. 2.5 Pourbaix diagram for sulphur in water at 25 °C. The region between the parallel dotted lines is the domain of stability of water

equilibrium; given time to react, it should form bicarbonate and nitrate ions. A major defect of Pourbaix diagrams is that they make no allowance for the presence of organic or colloidal ligands, which are either of unknown nature or for which no thermodynamic data are available.

The theory of chemical speciation in an ideal ocean containing neither organic matter nor colloidal material has been reviewed by Stumm and Brauner (1975), and the 'theoretical' species present are listed in Table 2.5. The main ligands for the elements are hydroxyl, chloride, carbonate and sulphate, distributed as follows:

Hydroxo/Oxo: Al, As, B, Be, C, Cr, Cu, Fe, Ga, I, La–Lu, Mo, N, P, Re, Sb, Sc, Se, Si, Sn, Te, Th, V, Y, Zn, Zr
Chloro: Ag, Au, Cd, Hg, Mn
Carbonato: Co, Cu, Ni, Pb, UO_2, Zn
Sulphato: Na

An alternative classification is into cationic, neutral and anionic species, thus:

Cations: Ba, Be, Bi, Ca, Ce, Co, Cs, Cu, F, Fe, K, La–Lu, Li, Mg, Mn, Na, Ni, Ra, Rb, S, Si, Sr, Zn
Neutral: B, Cd, Co, Cu, N, Ni, Pb, Sc, Si, Th, Y, Zn, Zr
Anions: Ag, Al, As, Au, B, C, Cr, F, Ga, Ge, Hg, I, Mo, N, P, Pb, Re, S, Sb, Se, Sn, Te, U, V, W

However it must be remembered that relatively few of these species have been conclusively identified in real ocean water (Florence and Batley, 1977). Other species may occur, namely low molecular weight organic complexes and larger colloidal complexes.

Low molecular weight organic complexes exist in river waters, and since the ocean contains 0·5–1 mg organic C/l, they may also occur in the sea. Evidence for unknown organic complexes of Au, Cd, Cu, Pb and Zn in water is given by Florence and Batley (1977) and Baker (1978). Cobalt certainly occurs in the ocean as the corrin complex Vitamin B_{12} (Riley and Skirrow, 1975). Under anoxic conditions, bacteria can use vitamin B_{12} to methylate a number of elements (Ag, As, Au, Cd, Hg, I, Pb, Pd, Pt, Se, Sn, Te and Tl) to form more or less volatile methyl derivatives (Fagerström and Jernelov, 1972; Wood *et al.*, 1977). This may account for the traces of methyl arsenic and methyl tin derivatives found in sea water (Andreae, 1977; Shaikh and Tallman, 1978; Braman and Tompkins, 1979).

Colloidal organic complexes resembling humates also exist in sea water, and may form ill-defined complexes with elements such as Cd, Cu, Hg, Pb, Sb and Zn (Fagerström and Jernelov, 1972; Florence and Batley, 1977). Theis and Singer (1973) have shown that humates can stabilise Fe(II) in solution and prevent its oxidation. Colloidal hydroxides of Al, Fe(III), Mn(IV) and Si are known to form at the pH of sea water, and to be potent scavengers of polyvalent elements (Jenne, 1968). The formation of some of these polymers is critically dependent on redox potential. When there is less than 1 mg O_2/l in sea water, iron and manganese are reduced to their soluble bivalent ions, and any ions absorbed on colloidal hydroxides of these elements are released. In general, sea water is well oxygenated and contains negligible amounts of reducing ligands such as sulphide ions.

A limited amount of work has been carried out on the valency states of elements in the sea. Iodine occurs mostly as iodate in deep water, but in the surface ocean up to 25% may be present as iodide (Wong and Brewer, 1976, 1977). The naturally occurring forms of As and Sb are predominantly pentavalent (Gohda, 1975; Valente, 1978). Chromium is thought to occur as both Cr(III) and Cr(VI) (Cranston and Murray, 1978), while thallium occurs mostly as Tl(III) (Batley and Florence, 1975). Both Se(IV) and Se(VI) co-exist in the deep ocean in approximately equal concentrations, but surface sea water contains mostly Se(VI) (Sugimura *et al.*, 1976; Cutter, 1978). Further work is needed to determine the valency states of Au, Ce, Hg and Te in the ocean.

Carbon, in the form of carbon dioxide, bicarbonates and carbonates, occupies a unique position in buffering the ocean against changes of pH. Broecker (1974) has shown how complex the situation is by measuring the variation in concentrations of these three species with depth, and pointed out that the deep Pacific is slightly more acid than is the deep Atlantic. Free CO_2 is present in very low concentrations in the sea, but there is twice as much per litre in deep water as in surface water. Bicarbonate and total carbon are also more concentrated in deep water than in surface water, but the carbonate ion is three times more abundant in surface water than in deep water. This agrees with the observation that biological precipitation of calcium carbonate takes place in surface water. Deep water is measurably more acid than is surface water, and is therefore able to dissolve calcite debris falling into it from above.

3 *The Elemental Geochemistry of Rocks*

The material in this chapter is discussed in detail in geological and geochemical texts, such as those of Mason (1966), Garrels and Mackenzie (1971) and Pettijohn (1977). Only an outline can be given here.

The Earth is made up of a number of concentric shells of distinct composition, as indicated in Fig. 3.1 and Table 3.1. The radii of the inner shells have been obtained by shock wave studies. Both temperature and pressure increase with depth, so that most of the material in the Earth exists under pressures exceeding 500 kbar, which are greater than those which can be sustained experimentally. Hence it is not surprising that the density of the Earth increases markedly with depth.

The core of the Earth is believed to be largely iron, with a similar composition to that of iron meteorites, although it has never been sampled. It probably contains both liquid and solid phases, since there is a sound-wave discontinuity at a depth of 5100 km. One estimate of its composition is 84% Fe, 11% Si and 5% Ni (Ringwood, 1966). Carbon, cobalt and the precious metals (Au, Ir, Os, Pd, Pt, Re, Rh and Ru) are thought to be

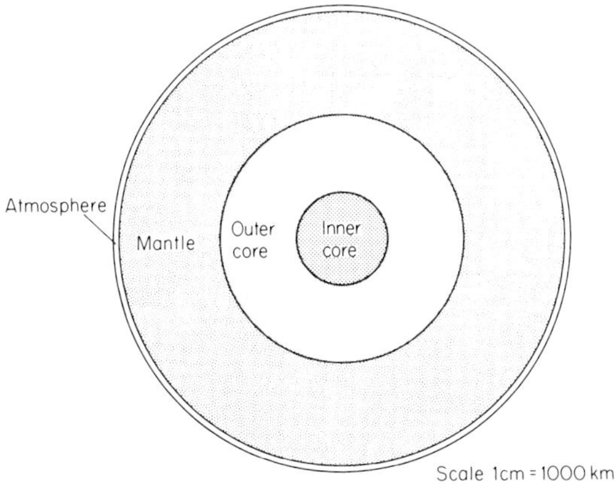

FIG. 3.1 Concentric shells making up the Earth. On this scale, both crust and hydrosphere are thinner than the lines used to demarcate the mantle

TABLE 3.1 Concentric shells making up the Earth

Shell	Mean thickness/km	Mass/10^{21} kg	Density/g cm^{-3}
Atmosphere	100	0·0053	0·00129 at STP
Hydrosphere	3·8	1·76	1·000 at STP
Crust	17	24	2·8
Mantle	2883	4016	4·5
Core	3471	1936	11·0
Whole Earth	6371	5975	5·52

enriched in the core, as may well be the following elements: As, Cu, Ga, Ge, Mo, Sb, Sn, Te and W.

The mantle is not readily accessible, but is thought to be closest to the surface beneath the mid-Atlantic and mid-Pacific ridges. Material exuded from Hawaiian volcanoes, with the composition of dunite, is believed to be a sample of the upper mantle. Dunite is an impure magnesium silicate containing about 13% of iron. Wedepohl (1968) has suggested that the mantle has the same composition as chondritic meteorites. It therefore seems likely that the crust differs from the mantle as follows with respect to its elemental composition:

1. The crust is richer in Group IA and IIA metals and in those elements which crystallise last from molten magma, i.e. Hf, La–Lu, Nb, P, Ta, Th, Ti, Tl, U, W, Y and Zr.
2. The crust is somewhat richer in silicon than the mantle, and also As, B, F, Ga, Hg, In, Sc and V.
3. The crust is somewhat poorer than the mantle in Mg and Fe.
4. The crust is much poorer than the mantle in Co, Cr, Ni and the precious metals.

The Earth's crust is readily accessible for analysis and is highly heterogeneous. It can be thought of as a series of large continental blocks, floating on, and depressing the surface of, the mantle. Crustal rocks can be broadly divided into igneous rocks, which have crystallised from molten magma, and sedimentary rocks, which have been formed by deposition from the ocean. Common sediments contain much more carbon, and much less sodium, than do igneous rocks.

Igneous rocks and magmatic crystallisation

The process of crystallisation of igneous rocks from molten magma is complex, but the broad outlines are as follows (Mason, 1966).

1. The dense mineral olivine (Mg_2SiO_4) crystallises first, and sinks to the bottom of the melt; the resulting rock is a basalt.

2. Insoluble iron salts, forming the minerals pyrrhotite (FeS), pyrite (FeS_2), magnetite (Fe_3O_4), chromite ($FeCr_2O_4$) and ilmenite ($FeTiO_3$) crystallise next, in relatively small amounts. There is not much sulphur in most magmas, but the FeS is able to carry down some part of the elements with insoluble sulphides, such as Co, Cu, Ni and Zn.

3. The bulk of the magma now solidifies, forming a continuous series of rocks whose contents of Al and Mg decrease, while those of Mn and Rb increase, as crystallisation proceeds (Fig. 3.2). The first rocks to crystallise

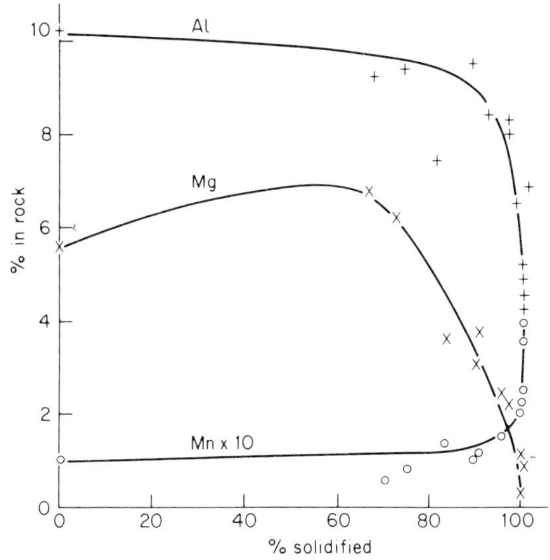

FIG. 3.2 Behaviour of Al, +, Mg, × Mn, ○ during crystallisation of magma. Data of Wager and Brown (1968).

here are called gabbros, followed by diorites and finally granites. All these rocks are mixtures of different minerals, of which the commonest are feldspars, amphiboles, pyroxenes, micas and quartz. Rarer minerals which crystallise here include apatite, fluorite, monazite, titanite, xenotime and zircon. The ideal formulae of these minerals are given in Table 3.2, but diadochic replacement is so common that it is rarely possible to give satisfactory chemical formulae of the commoner aluminosilicate minerals.

4. Finally the elements which have not been trapped in major or accessory minerals crystallise out separately, in small amounts, forming

TABLE 3.2 Common minerals formed in magnetic crystallisation (major constituents in capital letters)

Stage of crystallisation	Mineral	Ideal formula	Elements co-crystallised
1	OLIVINE	Mg_2SiO_4	Fe(II)
2	Pyrite	FeS_2	Ag, As, Cd, Co, Cu, Mo, Ni, P, Pd, Sn, Te
2	Pyrrhotite	FeS	
2	Magnetite	Fe_3O_4	
2	Chromite	$FeCr_2O_4$	
2	Ilmenite	$FeTiO_3$	
3	FELDSPAR	$MAlSi_3O_8$	Ba, Ca, K, Na, Rb, Sr
3	AMPHIBOLE	$L_{14}Si_{16}O_4(OH)_4$	Ca, F, Fe, Li, K, Mg, Mn, Na, Sr, Ti
3	PYROXENE	L_2SiO_6	K, Li, Mn, Na, Ti
3	MICA	$KMg_3(OH)_2AlSi_3O_{10}$	Ba, F, Fe, Li, Mn
3	QUARTZ	SiO_2	
3	Apatite	$Ca_5(OH)(PO_4)_3$	As, F, Pb, Se, V
3	Fluorite	CaF_2	
3	Monazite	$CePO_4$	La–Eu, Th
3	Titanite	$CaSiTiO_6H_2$	F
3	Xenotime	YPO_4	Gd–Lu
3	Zircon	$ZrSiO_4$	Hf
4	Miscellaneous pegmatites		Au, Be, Cs, Hf, La–Lu, Li, Nb, Sn, Ta, U, W, Y, Zr
5	Volatiles		Ar, C, H, He, Hg, N, O

L, Ca, Fe(II) or Mg; M, K or Na

pegmatites. These elements have unusual ionic radii which lead to their natural segregation.

5. At the same time volatile species such as carbon dioxide and steam are released to the atmosphere.

The elemental composition of igneous rocks

There are two reasons why it is difficult to give a reliable tabulation of the elemental composition of 'average crust'. Although the analytical data for a few standard rocks are now reasonably certain, there are not many complete analyses for large samples of crustal rocks. The main difficulty is that geologists cannot agree as to the relative proportions of the different sorts of igneous rocks in the crust. Taylor (1964) suggested that the crust consists of half basalt and half granite, which makes the calculation of its composition very simple. Wedepohl (1969) prefers a crust consisting of 44 % granite, 34 % granodiorite, 13 % gabbro and 8 % quartz diorite, while Sibley and Wilblad consider that 66 % granodiorite plus 34 % basalt is a better approximation to the truth.

Table 3.3 gives the elemental composition of average igneous rocks and rocks used as standard reference materials, as follows. Dunite DTS-1, probably resembling the mantle; Granite G-2 and average crustal granite; Basalt BCR-1 and average crustal basalt; Average crust (using Taylor's suggestion). Despite intensive efforts, there are still gaps to be filled in the analytical data, notably for rhenium and the platinum metals. In addition the amounts of the volatile elements Br, Cl, I, N and S in igneous rocks are poorly known. Anderson (1974) found 750 mg Cl/kg and 8100 mg S/kg as means for 30 Hawaiian basalts, which are very different from the average values given in Table 3.3.

The weathering of igneous rocks

Weathering includes all physical, chemical and biological actions which abrade or dissolve rocks. During weathering, all but a few extremely resistant minerals are chemically changed: exceptions include native gold, diamond, zircon, cassiterite and various gemstones. Physical weathering is important in regions with large variations in temperature, such as mountains and deserts, and in unstable regions such as cliffs and coasts. Alternate freezing and thawing splits rocks, and the grinding action of moving glaciers can produce 'rock flour'. In deserts, rocks are rounded and polished by the action of wind-blown sand, and the erosive action of the sea is evident on all rocky coasts.

The main agent of chemical weathering is water, especially when it contains small amounts of dissolved carbon dioxide which makes it more acidic. Sodium is readily leached from rocks by the continual action of water, while calcium, magnesium and potassium are dissolved somewhat more slowly. The other common rock constituents, aluminium, iron and silicon, are much less soluble but can ultimately be leached out by water. Where the percolating water is alkaline or neutral, silicon dissolves fastest, leaving behind laterites which consist of hydrous oxides of aluminium and/or iron. In acid regions, aluminium and iron usually leach out before silicon, forming sands of more or less pure silica; iron is especially readily leached under reducing conditions. The variation in solubilities of these three oxides with pH helps to account for these facts, though in practice many other constituents are present whose moderating effects on solubilities have not been tested. Krauskopf (1959) has given an excellent account of the factors known to affect the solubility of silica (see Fig. 3.3).

Biological weathering is rather less obvious, as a close cover of vegetation apparently protects rock surfaces from the effects of heavy rain. However, plant roots are able to split rocks by osmotic pressure, and the organic acids

TABLE 3.3 Elemental composition of some igneous rocks in mg X/kg

X	Dunite DTS-1	Basalt BCR-1	Granite G-2	Mean basalt	Mean granite	Mean crust
	(Flanagan, 1973)	(Flanagan, 1973)	(Flanagan, 1973)	(Taylor, 1964)	(Taylor, 1964)	(This work)
Ag	0·008	0·036	0·049	0·1 W	0·04 W	0·07
Al	12 700	72 000	81 500	87 600	77 000	82 000
Ar				<0·002 Y		1·2 So
As	0·03	0·7	0·25	1·5 W	1·5 W	1·5
Au	0·0008	0·00095	0·001	0·0005 C	0·0018 Sh	0·0011
B	<5	5	2	8 W	12 W	10
Ba	2·4	670	1900	250 W	730 W	500
Be		1·7	2·6	0·3 W	5 W	2·6
Bi	0·01	0·05	0·043	0·031 M	0·065 M	0·048
Br	0·2	0·15	0·3	0·44 W	0·3 W	0·37
C	220	330	220	610 W	360 W	480
Ca	1070	49 000	14 000	67 200	15 800	41 000
Cd	0·12	0·12	0·039	0·13 M	0·09 M	0·11
Ce	0·06	54	150	59 W	83 Ha	68
Cl	11	60	53	60	200 W	130
Co	130	38	5·5	35 He	1 T	20
Cr	4000	18	7	200	4	100?
Cs	0·006	0·95	1·4	1	5	3
Cu	7	18	12	90 W	13	50
Dy	0·003	6·3	2·6	6·9 He	5?	6
Er	<0·003	3·6	1·3	3·9 He	3·7 Ha	3·8
Eu	0·0009	1·9	1·5	2·2 He	2 Ha	2·1
F	15	470	1300	510 Sh	1400 Sh	950
Fe	60 000	94 000	19 000	56 000	27 000	41 000
Ga	0·2	20	23	17 W	19 W	18
Gd	<0·01	6·6	5	8 He	7·4 Ha	7·7
Ge	0·9	1·5	1·2	1·6 R	2 R	1·8
He				<0·001 Y		0·008
Hf	0·01	4·7	7·4	4·8 He	5·8 Sh	5·3
Hg	0·0087	0·011	0·039	0·012 W	0·08 W	0·05
Ho	0·003	1·2	0·4	1·5 He	1·3 Ha	1·4
I	0·11 W	0·16 W	<0·03	0·08 W	0·2 W	0·14
In	0·0025	0·095	0·034	0·058 R	0·04 R	0·049
Ir	0·001	0·000004	0·000002	0·00033 C	0·00002 Sh	0·000003?
K	100	14 000	37 000	8300	33 400	21 000
Kr				<0·000003 Y		0·00001?
La	0·04	26	96	21 He	43 Ha	32
Li	2	13	35	10	30	20
Lu	0·002	0·55	0·11	0·49 He	0·54 Ha	0·51
Mg	300 000	21 000	4600	45 000	1600	23 000
Mn	970	1400	260	1500	400	950
Mo	0·2	1·1	0·36	1	2	1·5
N	18 Sk	39 Sk	37 Sk	30 W	21 W	25
Na	52	24 000	30 000	19 400	27 700	23 000

Table 3.3—*continued*

X	Dunite DTS-1	Basalt BCR-1	Granite G-2	Mean basalt	Mean granite	Mean crust
	(Flanagan, 1973)	(Flanagan, 1973)	(Flanagan, 1973)	(Taylor, 1964)	(Taylor, 1964)	(This work)
Nb	<3	14	14	20	20	20
Nd	0·035 B	29	60	32 He	44 Ha	38
Ne				<0·00006 Y		0·00007
Ni	2300	16	5·1	150	0·5?	80?
O	431 000	455 000	484 000	467 000	492 000	474 000
Os	0·001	0·0001				0·0001?
P	9	1600	610	1400	700	1000
Pb	14	18	31	3 W	24 W	14
Pd	0·001	0·012	<0·0005	0·00084 C	0·0005?	0·0006?
Pr	0·006	7	19	7·8 Ha	11 Ha	9·5
Pt	0·003	0·002	<0·0005			0·001?
Ra	$1·3 \times 10^{-9}$	$5·6 \times 10^{-7}$	$7·1 \times 10^{-7}$			6×10^{-7}
Rb	0·053	47	170	37 W	150	90
Re	<0·0004	0·0008	<0·0007	0·00025 W	0·00056 W	0·0004
Rh	0·0009	0·0002				0·0002?
Ru	0·0025	0·001				0·001?
S	<10	390	24	250	270	260
Sb	0·46	0·69	0·1	0·2 W	0·2 W	0·2
Sc	3·6	33	3·7	25 He	6·9 W	16
Se	0·004 W	0·1 W	0·005 W	0·05	0·05	0·05
Si	189 000	255 000	323 000	240 000	336 000	277 000
Sm	0·004	6·6	7·3	7·3 He	8·5 Ha	7·9
Sn	1	2·6	1·5	1 W	3·5 W	2·2
Sr	0·35	330	480	465	285	370
Ta	<0·1	0·91	0·91	0·5	3·5	2
Tb	0·0003	1	0·54	1·2 He	1 W	1·1
Te	0·003 Sm	0·004 Sm	0·005 Sm			0·005?
Th	0·01	6	24	1·6 W	23 W	12
Ti	71	13 000	2800	9000	2300	5600
Tl	0·0005	0·3	1	0·08 M	1·1 M	0·6
Tm	0·001	0·6	0·3	0·42 Ha	0·54 Ha	0·48
U	0·004	1·7	2	0·43 W	4·4 W	2·4
V	10	400	35	250 W	72 W	160
W	0·04	0·4	0·1	0·36 Z	1·5 W	1
Xe				<0·000001 Y	0·000008	0·000002
Y	0·05	37	12	27 W	33 Ha	30
Yb	0·01	3·4	0·88	3·4 He	3·3 Ha	3·3
Zn	59	120	85	100 W	52 Sh	75
Zr	3	190	300	150	240 Sh	190

References: A, Aly *et al.* (1975); B, Brunfelt *et al.* (1974); C, Crocket *et al.* (1973); Ha, Haskin *et al.* (1968); He, Helmke and Haskin (1973); M, Marowsky and Wedepohl (1971); R, Rösler *et al.* (1968); Sh, Shawe (1976); Sk, Shukla *et al.* (1978); Sm, C. L. Smith *et al.* (1977); So, Soroiu and Cerci (1967); W, Wedepohl (1968) or in Wedepohl (1969–1974); Y, Dymond and Hogan (1973); Z, Helsen *et al.* (1978)

that they exude assist in the leaching processes. Some tropical species, such as bamboos, extract substantial amounts of silica, as well as the better-known plant nutrients, from the subsoil. Humic acids produced by bacterial oxidation of dead organic matter colour some rivers brown, and assist in the leaching of base-rich rocks.

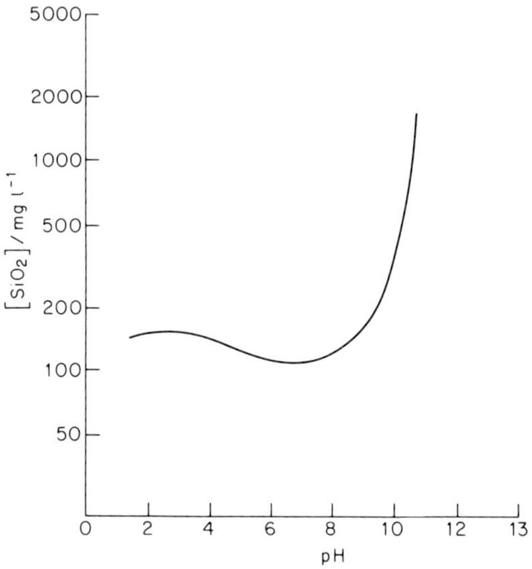

FIG. 3.3 Variation of solubility of silica with pH at about 25 °C. Data from Krauskopf (1959)

Sediments

The products of rock weathering which are deposited as sediments under the ocean can be classified under five main headings (Degens, 1965; Pettijohn, 1977).

1. *Resistates*, or unchanged minerals from primary rocks;
2. *Hydrolysates*, mostly clays from weathering of aluminosilicate minerals;
3. *Precipitates*, mostly calcium carbonate or silica, skeletons of marine plankton;
4. *Evaporates*, mostly salt or gypsum from drying salt lakes;
5. *Bioliths*, or resistant organic matter.

1. *Resistates* are mainly particles of quartz (SiO_2), corundum (Al_2O_3) or haematite (Fe_2O_3) which have been carried down to the sea as small

particles. In a few parts of the world there are notable concentrations of cassiterite (SnO_2 – S.E. Asia), rutile (TiO_2), ilmenite ($FeTiO_3$ – black sands of S.E. Asia), and monazite ($CePO_4$ – S. India, Brazil) whose particles are graded by the actions of rivers, winds or beach waves. Resistate deposits occur at or near the mouths of rivers around all the continents, but are absent from the deep ocean.

2. *Hydrolysates* are clay minerals of colloidal dimensions which are formed by weathering of aluminosilicates; they are abundant in soils (chapter 4) and while they may be formed in the sea, they are certainly transported in sufficient quantity by rivers and winds to account for their distribution in the ocean (Price, 1976). The commoner clay minerals in ocean sediments are kaolinite (ideal formula $Al_4Si_4O_{10}(OH)_8$) in the tropics, illite (ideal formula $K_2Al_4Si_8O_{20}(OH)_4$) in temperate latitudes, and montmorillonite (ideal formula $K_2Al_{10}Mg_2(Si_8O_{10}(OH)_4)_3$) in regions of volcanic activity. Both illite and montmorillonite strongly absorb potassium, but not sodium, into their lattices and may be the main sinks for potassium in the ocean. As colloids they also absorb ions with multiple charges with avidity.

The loss of elements during the weathering of igneous rocks to clay minerals is illustrated by the data of Patterson and Roberson (1961), which is summarised in Table 3.4. These workers analysed Hawaiian basalt and its weathering product kaolinite for major elements, with the following results:

TABLE 3.4 Changes in elemental composition when basalt weathers to kaolinite

X	g X/kg basalt	g X/kg kaolinite
Al	40	88
Ca	69	<0·7
Fe	102	207
K	5·1	0·2
Mg	72	5
Na	13·4	0·4
Si	210	118

Almost all the A-metal cations are lost during weathering, together with some of the silicon, leaving a residue rich in aluminium and iron.

3. *Precipitates* of calcium carbonate and silica are formed by organisms in the ocean and accumulate as sediments in the shallower parts of the sea floor. Chemically precipitated calcite is rare and only occurs in very warm and shallow bays of the sea. Manganese nodules are concretions of hydrous iron(III) and manganese(IV) oxides, which are abundant in some parts of

the ocean bed. They grow in diameter at rates of about 1 μm/yr, and absorb a number of rare elements. Other, much rarer, precipitates include francolite ($Ca_5(PO_4)_3F$) from the bones of fish, $BaCO_3$, $BaSO_4$, $FeCO_3$, $SrCO_3$, and $SrSO_4$. Iron(II) sulphide precipitates in anoxic seas, such as the deeper parts of the Black Sea, and absorbs other elements which form insoluble sulphides.

4. *Evaporates* from drying salt lakes or arms of the sea have been much studied because of their economic importance. When sea water is evaporated, the order of deposition by forming crystals is:

$$\text{Calcite } (CaCO_3) > \text{Gypsum } (CaSO_4 2H_2O) > \text{Salt } (NaCl) > Na_2SO_4 1OH_2O > \text{Salts of K and Mg}$$

In very dry areas, sodium carbonate and sodium nitrate, with traces of arsenates, borates and iodates, may form local deposits at the final stage of evaporation.

5. *Bioliths* include peat and petroleum, and are laid down in anoxic regions where specialised bacteria remove the oxygen atoms from dead organic matter. They sometimes absorb unusual elements (e.g. uranium in some peats; Mo, Ni and V in some petroleum) but are usually >95% organic matter.

The elemental composition of sediments has been used to investigate geochemical problems. Landström *et al.* (1967) have investigated the elemental composition of sediment cores from a Swedish lake which has been a branch of the sea at various periods in its history. The clays deposited from freshwater and sea water in this lake differ little except that marine clays showed considerable enrichment in boron, bromine and possibly selenium.

Gorham and Swaine (1965) have shown that oxidised and reduced muds from the freshwater Lake Windermere, U.K. differ very little in elemental composition. The reduced muds are enriched in sulphur, and to a small extent in Co, Ni, Pb and Sn, but not in Ag, Cu, Mo or Zn. On the other hand pyrite nodules, found in ancient sediments, are strongly enriched in thiophilic elements such as Ag, As, Au, Co, Cu, Pb, Sb, Se and Zn (Steinnes and Mukherjee, 1973).

Diagenesis and metamorphism

Diagenesis is the slow process of conversion of sediments to sedimentary rocks, by processes involving biological action, ion-exchange, recrystallisation and cementation. Characteristic features of diagenetic changes are decreases in water content, decreases in organic matter and physical hardening. Well-known examples include the following changes:

Resistates → Sandstones by cementation with silica (or rarely calcite).
Clays → Shales, by expulsion of water and chemical changes under pressure.
Calcium carbonate → Limestone or dolomite, by cementation and/or ion-exchange with Mg^{2+}.

$$\text{Gypsum} \begin{array}{l} \nearrow \text{Anhydrite by loss of water.} \\ \searrow \text{Calcite + sulphur by the action of anoxic bacteria.} \end{array}$$

Peat → Coal, with loss of hydrogen.
Petroleum → Asphalt, with loss of hydrogen.

Changes in elemental composition during diagenesis have seldom been studied, since the process is slow and cannot be hastened experimentally. The data in Table 3.5 illustrate the generalisation that elemental composition does not change very much during diagenesis. If one compares the elemental analyses of marine clays (Chester and Aston, 1976) with those for shales (Wedepohl, 1968), it can be seen that conversion appears to involve gains of Ba, Hg, Se, Sr and U, with losses of B, Br, Cu, Mn, Mo, Ni, Pb and rare earths. Few of the changes are of more than a factor of 5, and many may not be real; that is, more extensive analyses of representative samples may show them to be statistically insignificant. Similar remarks apply to the conversion of lignite to coal, for both of which ranges of elemental composition are given in Table 3.6. The conversion may involve losses of Ag, Al, Ca and Fe, but for most elements the similarity in elemental composition is striking.

The scanty evidence on elemental changes during the diagenesis of plant remains to coal has been reviewed by Nicholls (1968). He mentions As, Ge and Mo as being enriched in coal with respect to shales, and the data of Tables 3.5 and 3.6 suggest that B, S, Se and U are also enriched in this way. In order to obtain clues as to how this enrichment arose, it is necessary to make analyses across a transect of an entire coal seam, and to analyse both the original coal and its ash. Certain elements tend to concentrate at the margins of the seam, notably germanium, and often As, Cr, Mn, Mo, Pb, V and Zn; this suggests that these elements were absorbed from ground water during diagenesis. Nicholls considers that organic matter absorbs B, Ge, Mo and some Co and Ni, while sulphide mineral impurities absorb As, Co, Cu, Ni, Pb and Zn. Other elements (Cr, Ga, V) may be absorbed by clay mineral impurities. Ni and V are enriched in plant fossils which occur in coal (Meisch *et al.*, 1978).

TABLE 3.5 Elemental composition of major sediments in mg X/kg

X	Marine clays	Marine carbonates	Mean shale	Mean limestone	Mean sandstone	Mean sediment
	(Chester and Aston, 1976)		(Wedepohl 1968, 1969–1974)			
Ag	0·11	0·06 T	0·07	0·12	0·25	0·057
Al	94 000	42 000	88 000	9000	43 000	72 000
As	13	1	13	1	1	7·7
Au	0·0027 C	0·0016 W	0·0025	0·002	0·003	0·0024
B	230	55	130	20	30	100
Ba	2300	190	550	90	320	460
Be	2·6		3	<1	<1	2?
Bi	0·5 M		0·48 M	0·17 M	0·18 M	0·4
Br	130		24	6	1	19
C	4500	66 000	14 000	116 000	16 000	29 400
Ca	10 000	203 000	16 000	340 000	31 000	66 000
Cd	0·21	0·23	0·22 M	0·028 G	0·05	0·17
Ce	345	35	96	20	78	83
Cl	22 000 W		160	76	510	190
Co	74	7	19	0·1	0·3	14
Cr	90	11	90	11	35	72
Cs	6	0·4	5·5	0·5	0·5	4·2
Cu	25	30	39	5·5	30	33
Dy	6?	2·7	5·8	2·1	6·9	5·4
Er	3 W	1·5	4	1·7	4·9	3·8
Eu	1·8 W	0·6	1·2	0·4	2?	1·2
F	730	550 W	800	220	180	640
Fe	58 000	27 000	48 000	17 000	29 000	41 000
Ga	20	13	23	2·5	6	18
Gd	7·4 W	3·8	6	2·7	7·1	5·6
Ge	2	0·2	2	0·31	1·2	1·7
Hf	4·1	0·41	2·8	0·3	3·9	2·5
Hg	0·08 M	0·46	0·18 M	0·16 M	0·29 M	0·19
Ho	1·2 W	0·8	1·8	0·28	2?	1·6
I	30 W	31 W	19 Be	14 Be	0·1 Be	16?
In	0·07 W	0·02	0·057	0·009 R	0·01 R	0·044
Ir	0·0003 Ba, C					
K	27 000	12 000	24 500	3100	15 000	20 000
La	31 W	10	49	10	42	41
Li	57	5	76	7·5	38	56
Lu	0·8?	0·5	0·8	0·16	0·8?	0·7
Mg	23 000	14 000	16 000	5800	11 500	14 000
Mn	6700	1000	850	620	460	770
Mo	8 Bt	3?	2·6	0·16 Bt	0·2	2
N	160		600	78	120	470
Na	11 000	5900	5900	1300	10 400	5700

Table 3.5—*continued*

X	Marine clays	Marine carbonates	Mean shale	Mean limestone	Mean sandstone	Mean sediment
	(Chester and Aston, 1976)		(Wedepohl 1968, 1969–1974)			
Nb	14	4·6	18	0·3	0·05	13
Nd	35 W	14	41	9	38	32
Ni	225	30	68	7	9	52
O	477 000	494 000	483 000	494 000	492 000	486 000
Os	0·00023 Ba				0·000055	
P	610	650	700	700	440	670
Pb	80	17 W	23	5·7	10	19
Pd	0·0037 C					
Pr	9·6 W	3·3	11	2·5	11	9·7
Pt			0·0001?			0·0001?
Ra			11×10^{-7}	4×10^{-7}	7×10^{-7}	10^{-6}
Rb	110	10	160	52?	46	135
Re	0·001?	0·004?				
S			2400	1100	2300	2200
Sb	1	0·15	1·5	0·3	0·05	1·2
Sc	19	2	13	1	1	10
Se	0·17	0·17?	0·5 K	0·03 K	<0·01 K	0·42
Si	258 000	126 000	275 000	32 000	327 000	245 000
Sm	6·2 W	3·8	7	2·1	8·4	6·4
Sn	3·2 S	1·5 S	6	0·5	0·5	4·6
Sr	18	2000	300	610	20	320
Ta	1·2 L		2	0·05	0·05	1·5
Tb	1·1 W	0·6	1	0·44	2?	1
Te			<0·1			
Th	5		12	1·7	3·8	9·6
Ti	5000	2300	4600	300	3500	3800
Tl	0·9 M	0·16 W	1·2 M	0·14 M	0·36 M	0·95
Tm	0·56 W	0·1	0·6	0·16	1?	0·6
U	1		3·7	2·2	0·45	3·1
V	120	20	130	45	20	105
W	1·1 W	0·11 W	1·9	0·56	1·6	1·7
Y	32 W	42	41	23	54	40
Yb	2·9 W	1·5	3·9	1·6	4·4	3·6
Zn	165	35	120	20	30	95
Zr	150	20	160	20	220	150

References: Ba, Barker and Anders (1968); Be, Becker *et al.* (1972); Bt, Bertine and Turekian 1973); C, Crocket *et al.* (1973); G, Gong *et al.* (1977); K, Koljonen (1973); L, Landström *et al.* (1967); M, Marowsky and Wedepohl (1971); R, Rösler *et al.* (1968); S, Smith and Burton (1972); T, Turekian (1969); W, Wedepohl (1968, 1969–1974) (*see also* Turekian and Wedepohl (1961) and Haskin and Gehl (1962). For Ar, He, Kr, Ne and Xe in shale, see Canalas *et al.* (1968)

TABLE 3.6 Elemental composition of minor sedimentary rocks in mg X/kg

X	Manganese nodules	Phosphorites	Petroleum	Lignite	Mean coal (range)	Coal SRM 1632
Ag	6	0·02–3	0·0001 Bg	6·5	0·05 (0·01–3·7)	0·06
Al	31 000	9000	0·5 Bg	20 000	10 000 (3 k–30 k)	18 000
As	9–190	30	0·26	1–55	5 (0·3–93)	6·5
Au	0·0025		0·00044		0·0004	0·00085
B	290 W	< 50	0·002 Bg	70 (1–400)		
Ba	2000	100	0·1 Bg		200 (1–3000)	330
Be		0·5	< 0·003		1 (0·1–7)	
Bi	8	< 0·05–0·4			0·05 (0·02—4)	
Br	28–140		049		5 (1–50)	18
C		6000	860 000		800 000	
Ca	25 000	314 000	15	2 k–17 k	1500 (500–37 k)	4000
Cd	8	0·01–25	0·03		0·2 (< 0·01–22)	
Ce	720	100	0·01 Bg	6–35	12 (1–30)	20
Cl			39		500 (10–8000)	900
Co	3000	< 3–5	0·05	1–15	4 (1–90)	5·4
Cr	14	2–1000	0·05	10–200	10 (2–400)	20
Cs	0·5		0·004	1–20	0·3 (0·03–9)	1·4
Cu	2600	100	1·3	6–25	15 (3–180)	15
Dy	42	12		0·5–5	2·5 (0·2–5)	1·2
Er	19	9	0·001 Bg		0·5	
Eu	14	2·8	0·0009	0·06–0·8	0·5 (0·1–0·9)	0·33
F		31 000			80 (1–480)	
Fe	160 k	7700	25	2 k–15 k	8000 (500–43 k)	8500
Ga	10	< 10	0·01	2–8	5 (0·3–30)	5·8
Gd	6·4	15			1·2 (0·1–4)	3·6
Ge	1·3 W		0·001		4 (0·5–120)	70
H			84 000		50 000	
Hf	0·01–1				0·9 (0·1–4)	0·92
Hg	0·5	0·2	0·05		3 (0·01–21)	0·23
Ho	8	3·3			0·19 (0·1–0·4)	0·24
I	120–900	0·8–280 W	0·7		1 (0·1–14)	5
In	0·25	0·03			0·02 (0·01–0·6)	0·017
Ir	0·009				0·005?	0·0035
K	6400	4200	5	150–8 k	3000 (50–6500)	2900
La	160	56	0·005 Bg	2–20	5 (0·3–40)	10
Li					10? (0·1–160)	0·12
Lu		1·2		0·05–0·25	0·07 (0·01–0·4)	0·14
Mg	18 000	1800	0·1 Bg	1100–8100	2000 (100–3500)	1700
Mn	160 000	30	0·6	20–90	50 (3–900)	41
Mo	410	30	0·03		3 (0·3–30)	3·1
N			7 k–20 k	12 500	15 000	
Na	19 000	4500	4	40–550	400 (100–6000)	380
Nb	1–24				10 (5–40)	
Nd	190	57			9 (4–36)	8·7

Table 3.6—*continued*

X	Manganese nodules	Phosphorites	Petroleum	Lignite	Mean coal (range)	Coal SRM 1632
Ni	4900	<2–1000	10		10 (1–80)	15
O	476 000	429 000	6–44 k		64 k (1 k–280 k)	
P	2200	138 000	0·24		130 (6–4000)	
Pb	870	2–14			10 (2–370)	30
Pd	0·0055				0·005?	<0·005
Pr	50	15			1·8	
Pt					<0·07	0·27?
Ra					2×10^{-7} (10^{-9}–10^{-6})	
Rb	40		0·15		20 (1–150)	19
Rh					<0·002	
Ru					0·02?	0·018
S	750	7200	15 000 W		15 k (1 k–120 k)	
Sb	4–25	0·2–7	0·05	6–9	1 (0·1–9)	3·5
Sc	10	10	0·001	0·7–6	5 (0·5–30)	3·6
Se		1–300	0·3		3 (0·04–10)	2·9
Si	86 000	56 000			30 k (5 k–110 k)	32 000
Sm	90	11		0·9–1·8	1·2 (0·2–6)	1·6
Sn	2·7	0·18	0·01 Bg		2 (1–47)	1·4
Sr	825	1000	0·1 Bg		150 (20–1000)	160
Ta					0·2 (0·06–8)	0·24
Tb	10	2·1			0·22 (0·1–2)	0·3
Te	48				0·1? (<0·1–0·4)	
Th		1–16		0·8–8	2 (0·1–10)	2·2
Ti	6400	600	0·1 Bg	150–1100	500 (200–1800)	900
Tl	100	<0·03–1			0·2 (0·01–2)	
Tm	<2·4	1·1			0·06	
U	10	90	0·015	1–64	1 (0·005–200)	1·4
V	440 W	300	50	10–110	20 (2–130)	35
W	60				0·5 (0·1–4)	0·77
Y	310	300	0·001 Bg		7 (1·5–25)	
Yb	6·4	6·7		10–40?	0·37 (0·1–1·5)	0·69
Zn	710	300	30	30–300 W	50 (3–300)	31
Zr	650	30			50 (8–500)	

k, ×1000

Bg, Bertine and Goldberg (1971); W, Wedepohl (1969–1974)

Data sources: Manganese nodules: Glasby (1973, 1977), Chester and Aston (1976); Phosphorites: Goldberg *et al.* (1963), Gulbrandsen (1966), Langmyhr *et al.* (1977); Petroleum: Veal (1966), Shah *et al.* (1970), Bertine and Goldberg (1971), Zaghloul *et al.* (1973), Yen (1975), Hofstader (1976), Elofson *et al.* (1977); Lignite: Akcetin *et al.* (1973); Coal: Headlee and Hunter (1953), Joensuu (1971), Clark and Swaine (1962), Haskin *et al.* (1966), Nicholls (1968), Weaver (1973), Block and Dams (1973, 1974), Hamilton (1974), Ruch *et al.* (1977), Babu (1975); Coal SRM 1632: Ondov *et al.* (1975), Nadkarni (1975, 1977), Steinnes and Rowe (1976), Rowe and Steinnes (1977), Wangen and Gladney (1978)

Brätter *et al.* (1971) have analysed samples of coexisting calcite and dolomite and found small enrichments of many elements in dolomite. Approximate enrichment factors are:

0·9–1·2	Cs, K, La, Mn, Sr, Th
1·2–2·4	Al, Ce, Cr, Sc, U, V, Yb
2·4–4·8	Co, Eu, Fe, P, Tb
>4·8	Mg, Na

These results suggest that the calcite is forming dolomite by recrystallisation rather than by solid-state diffusion.

Metamorphism involves the conversion of sedimentary rocks to denser, metamorphic rocks at high temperatures and pressures, which occurs with deeply buried sediments. It involves solid state reactions and partial recrystallisation, but not total melting, so that fossils are distorted or destroyed, e.g.

Sandstone	→	quartzite
Shale	→	slate or schist
Limestone	→	marble
Coal	→	graphite

Although the changes in elemental composition during these changes have not been thoroughly studied, they are probably relatively minor ones.

If sedimentary rocks are remelted by volcanoes, they become liquid magma which will crystallise to form igneous rocks. Submarine vulcanism along mid-oceanic ridges has produced rocks rich in iron and manganese, and perhaps also in As, B, Ba, Cd, Cr, Hg, Pb and V (Boström *et al.*, 1969). In the Red Sea, submarine vulcanism is associated with pools of metal-rich brines (Degens and Ross, 1969). Quantitative estimation of the mass of submarine volcanic metamorphosed and remelted sediments is very difficult.

Mass and composition of sediments

There is no agreement as to the total mass or the relative proportions of different types of sediment. Garrels and Mackenzie (1971) suggest that 30 $\pm 10\%$ of the Earth's crust consists of sedimentary rocks, and that the masses of the main types in the crust are determined by the mass balance relation:

Igneous Rock	+ CO_2	+ HCl	→ Shale	+ Limestone	+ Sandstone	+ Water
1000 g	50 g	27 g	865 g	122 g	79 g	11 g

This gives the masses of sediments in the crust shown in Table 3.7. Other geologists give different estimates for these masses or their ratios (e.g. Wedepohl, 1968; Maynard, 1976; Pettijohn, 1977; Sibley and Wilblad, 1977).

TABLE 3.7 Masses of sedimentary rocks in the crust (Garrels and Mackenzie, 1971)

Type of sediment	Mass/10^{18} kg
Shale	1730
Submarine volcanic	860?
Limestone	350
Sandstone	260
Gypsum	40
Salt	21
Coal	3

Table 3.5 gives the elemental composition of average sediments, shales, limestones and sandstones, while Table 3.6 gives similar figures for manganese nodules, phosphorites, coal and crude oil. The mean sedimentary rock composition in Table 3.5 has been calculated as 74% shale + 15% limestone + 11% sandstone.

A notable feature is the close similarity in elemental composition of 'average' igneous rock and 'average' sedimentary rock. Sedimentary rocks are depleted in sodium, which remains behind in the ocean. They are somewhat enriched in calcium, and also in carbon, nitrogen and a number of 'volatile' elements which readily form bonds with carbon and may well be associated with the organic matter in sediments – As, B, Br, Cl, Hg, I, S, Sb and Se. These elements may arise, at least in part, from degassing of the mantle beneath the ocean. According to Price (1976), Br and I are exclusively contained in the organic fraction of sediments, and are lost from deep ocean sediments. The existence of metal-organic compounds in sediments is somewhat speculative, despite the correlation between the content of organic carbon and metals such as Cu, Mo, Ni, Pb, U and Zn. Nissenbaum and Swaine (1976) extracted organic humates from reduced marine sediments and found them to be markedly enriched in Ag, Cu, Ni, Pb, Sn and Zn, but it is difficult to be sure that these enrichments are not artefacts of the extraction process. Gulbrandsen (1966) considers that the following elements are associated with the 2% of organic matter in marine phosphorites; Ag, As, Cd, Cr, Mo, Ni, Sb, Se, V and Zn.

According to Chester and Aston (1976), certain elements are useful indicators of sedimentary history. Thus Al and Ti are indicators of terrestrial debris. Opal (SiO_2) is an indicator of bio-precipitation by diatoms, as is calcium carbonate for other types of organisms. Calcium carbonate is the purest sediment, impoverished in most trace elements apart from Mg, Sr and Y. Ba is a rare indicator of bio-precipitation of $BaSO_4$, while phosphorus is usually an indicator of francolite which is enriched in Ag, As, Cd, Cr, F, La, Mo, Nd, P, Sb, Se, U and Y. Manganese is an indicator of chemically precipitated manganese hydrous oxides, which are enriched in many elements, notably Co and Ni, but also Ag, As, Bi, Cd, I, Ir, La, Mo, P, Pb, Sb, Tl, V and Zn. However, both deep sea muds and the volcanic region near oceanic ridges are enriched in Co, Cu, Mn, Ni, Pb and Zn with respect to unpolluted near-shore sediments.

Outline of the geological history of the Earth

The chart below gives a brief summary of the approximate dates of some important events in the Earth's crust and atmosphere (Pettijohn, 1977), together with some key observations in the fossil record. The dates have been obtained by measuring the amounts of long-lived radioactive nuclides and their daughter products.

Date/ 10^9 yr BP	Earth's crust	Atmosphere	Fossil record
4·5	Formation of earth		
4	Differentiation of crust		
	Crust probably molten		
3·5			
	Crust solidifying	Forming: mostly	
	Oceans formed	$CO_2 + N_2$?	
	Oldest known rocks		Algae? in S. Africa (Engel *et al.*, (1968)
3			
2·5	Sediments mostly hydrous oxides of Mn and Fe; limestones rare		Oldest biogenetic limestones (S. Africa)
2	Fe/Mn declining in sediments?	Becoming oxygenic?	Oldest S bacteria (Shidlovsky, 1965)
			N fixation in Canada (Hu)
			Oldest fungi in Canada (Bt)
1·5			Calcareous algae frequent (Gl)
			Oldest eucaryotes; California (Hu)
		$O_2 \sim 2\%$	Oldest protozoa
1	Oldest sedimentary phosphates		
0·5	Oldest gypsum and rock salt		Oldest land plant; Britain (Hu)
0·4	Limestone, coal, oil common	CO_2 declines?	Plants and animals abundant
		O_2 19%	
0			

BP, before present; Bt, Barghoorn and Tyler (1965); Gl, Gravelle and Lelubre (1957); Hu, Hutchinson (1970)

4 *Soils*

The soil is the product of the effects of climate and living organisms on rocks, and its nature is determined by the composition of the parent rocks, the type of climate, and its age. Numerous classifications of soils have been proposed, none of which are perfect. A simplified version of the U.S. Department of Agriculture's classification of soils (1975) is given in Table 4.1. A map of world soil-types is given by Kellogg (1975).

The heterogeneous nature of most soils is shown by a series of layers of different colours and textures when a trench is dug. The upper layer, containing more or less decomposed plant debris, is called the A layer. Below this is the B layer, which consists of rock fragments more or less completely weathered to clays and hydrous oxides, mixed with organic matter. Below this again is the C layer of subsoil or parent rock, which is little modified by weathering or the addition of organic matter. These layers should be analysed separately if the composition of the soil is to be understood.

The heterogeneity of soil is equally obvious when each layer is examined by optical or electron microscopy, X-ray diffraction or electron microprobe methods. Individual particles from the same soil differ widely in chemical composition, solubility and ion-exchange properties, as well as in their pH and content of dioxygen. For example, a crumb of soil 3 mm or more in diameter may have an oxygenic outer layer and an anoxic core. This

TABLE 4.1 Classification of soils

Soil type	Main features	Layering	Where found
Entisol	Primary minerals present	Not obvious	Arctic tundra
Histosol	$>20\%$ organic matter	Not obvious	Temperate peat bogs
Inceptisol	Primary minerals present	A organic	Deciduous forest
Mollisol	Nutrient/organic-rich	A organic, B clay	Grassland
Spodosol	Strongly leached	Ferric oxide layer between B and C	Coniferous forest
Vertisol	$>35\%$ clay, nutrient-rich	Well-mixed	Temperate region
Aridisol	Dry: little organic matter	B clay	Desert
Alfisol	Nutrient-rich	A organic, B clay	Tropics
Ultisol	Nutrient-poor	B clay	Tropics (Mean T $>$ 8°C)
Oxisol	Nutrient-poor	B hydrous oxides	Tropics

microheterogeneity adds greatly to the difficulties of interpreting soil properties.

Gieseking (1975) has given a thorough description of the constituents of soils, which may be considered under six headings, as follows:

1. *Coarse inorganic particles*, with a diameter exceeding 200 nm, constitute 5–90% of the dry mass of most soils. They have a mean density of about 2·7 g cm^{-3} and can be subdivided into gravel, sand and silt by sieving.

2. *Colloidal inorganic particles*, with a diameter less than 200 nm, constitute 10–80% of the dry mass of most soils, and can be separated from coarser particles by filtering a suspension of soil in water. They have a mean density of about 2·4 g cm^{-3} and are responsible for the colour, texture and ion-exchange properties of most soils.

3. *Organic matter* constitutes 1–40(–90)% of most soils, and is very difficult to separate from colloidal inorganic matter, though its mean density of 0·7 g cm^{-3} is lower. Its amount can be estimated by the amount of CO_2 formed on combustion at 500 °C, or by titration with strong oxidants. It contributes to the colour, texture and ion-exchange properties of many soils.

4. *Living organisms* may constitute up to 0·2% of the fresh weight of most soils, but as the bulk of them are unicellular they cannot be separated quantitatively. Their density is in the range 0·9–1·1 g cm^{-3}. They are very important in the process of soil formation and the turnover of nutrients, and may control the pH and redox potential of soils.

5. *The soil solution*, which is held in capillary tubes in particles in fractions 2 and 3, can be removed by centrifugation and may constitute as much as 50% of the fresh mass of a soil. It is thought to be the nutrient medium sustaining both plant roots and micro-organisms.

6. *The soil atmosphere*, or gas held in the soil, can be removed by a vacuum pump. It has a density of 0·0013 g cm^{-3} and may constitute up to 50% of natural soils by volume.

These constituents will be discussed in a little more detail below.

Coarse inorganic particles

The principal minerals found here are the same as those found in rocks, but their relative proportions differ widely. Minerals which are resistant to weathering, such as quartz, zircon and tourmaline, are enriched in soils, while minerals which weather quickly, such as feldspars, micas, amphiboles and pyroxenes, are only found in young soils and are absent from mature soils. Table 3.4 showed the changes which occur when basalt weathers to kaolinite, and Table 4.2 gives more extensive data on the elemental composition changes when basalt weathers to soil (in Israel; Navrot and Singer, 1976). It can be seen that soluble cations of Ca, K, Mg, Na and Sr are mostly lost during weathering, Al and Fe appear to increase because they are not leached, while other elements (including Si) change relatively little. Similar results have been reported by many workers, e.g. Wolfenden (1965) for the weathering of andesites under tropical conditions in Borneo, and Simonson (1970) for U.S.A. soils.

TABLE 4.2 Elemental composition changes when basalt weathers to soil (Navrot and Singer, 1976)

Element X	mg X/kg basalt	mg X/kg soil
Al	60 000	104 000
Ca	61 000	2 600
Co	59	81
Cr	170	160
Cu	52	57
Fe	80 000	103 000
K	10 000	4600
Mg	43 000	4800
Mn	1200	1300
Na	25 000	5600
Ni	200	300
P	5900	3000
Si	240 000	250 000
Sr	1200	36
Ti	19 500	12 000
Zn	95	150

If primary minerals still exist in a soil, it is possible to predict the approximate abundance of some elements, but such predictions are unreliable for mature soils where the primary minerals have disappeared. For example, on Cretaceous limestone young soils may contain up to 80% of calcium carbonate from the original rock. Weathered soils on the same geological formation may contain less than 0·5% calcium carbonate and consist mainly of the hydrous oxides of iron and aluminium, with some clay and organic matter. Wakatsuki *et al.* (1977) have suggested that the ratio [Si]/([Ca]+[K]+[Mg]+[Na]) in the coarse inorganic fraction is a measure of the 'resistate maturity' of a soil: their paper should be consulted for details.

Colloidal inorganic particles

The secondary minerals in the colloid fraction from soils are of two types, clays and hydrous oxides. Clays are aluminosilicates with layer structures, many of which swell in water and function as cation-exchangers (Marshall, 1964). The hydrous oxides of Al, Fe and Mn also have limited ion-exchange properties, as well as acting as specific absorbents for phosphate and some of the rare elements (Norrish, 1975; Jenne, 1977).

Clay minerals occurring in nature are complex, non-stoichiometric compounds with the following general properties:

1. Density about $2{\cdot}4\ \mathrm{g\ cm^{-3}}$ (Bear, 1964).
2. All Si atoms are surrounded by a tetrahedron of four oxygen atoms, while Al atoms may be similarly located or have an octahedron of six oxygen atoms round them.
3. Hydroxyl groups in clays are always attached to Al, never to Si.
4. Almost any Si atom can be replaced by an Al atom plus a univalent cation such as K^+ to maintain electrical neutrality.
5. The aluminosilicate layers are either neutral or negatively charged, and are held together either by hydrogen bonds or by cations plus hydrogen bonds.
6. The cations holding the layers together can be exchanged for H^+ or other cations, so that natural clays contain a mixture of cations depending on their history.
7. The layers from different clay minerals can and do interleave themselves (Gieseking, 1975).
8. Many clay particles are coated with a surface layer of amorphous allophane, or hydrous oxides, or organic matter, which modifies their properties (Jenne, 1977).

The order of affinity of cations for clay minerals is determined mainly by their charge, and to a lesser extent by their size. For kaolinite, the order is $Th^{4+} > La^{3+} > Ca^{2+} > K^+$, but this order can be profoundly modified in the presence of organic ligands. For example, the citrate anion reverses the order of affinity of the above-mentioned ions for kaolinite (Mitchell, 1964). In nature, the commonest exchangeable cation in soils is usually Ca^{2+}, followed by Mg^{2+}, K^+ and Na^+. In very acid soils Al^{3+} and Mn^{2+} may become important exchangeable cations, as some of the hydrous oxides begin to dissolve at pH less than 4. In very alkaline soils, in deserts or salt marshes, Na^+ may become the dominant cation. If Na^+ constitutes more than 15% of the exchangeable cations, the soils are said to be sodic. Sodic soils have a pH $> 8{\cdot}5$, a characteristic texture, and support a specialised vegetation.

The cations K^+ and NH_4^+ are said to be fixed by some (not all) clay minerals, such as montmorillonite and vermiculite. When these clays are wetted, they swell and allow large ions such as K^+ (r = 133 pm) or NH_4^+ (r = 148 pm) to enter between the layers. On drying, the layers come together again and are held so tightly by electrostatic attraction that they no longer expand much on wetting. Other large ions such as Ba^{2+}, Rb^+, and Cs^+ can also be fixed in this way, but smaller ions such as Ca^{2+}, Mg^{2+} and Na^+ are not.

The anion-exchange properties of clay minerals are much less marked, as suitable sites probably only occur at the edges of the colloidal particles. Multiply charged ions such as phosphate (PO_4^{3-}) are much more strongly held than are singly charged ions such as Cl^- or NO_3^-[8]. The fluoride anion is exceptional in that it can replace the natural hydroxyl groups in clays, so that it can be fixed in a non-exchangeable form.

Fixation by chemical bonding can also occur with hydrous oxides, which are particularly common in mature tropical soils. The red-brown goethite (FeO_2H) and hydrated forms of iron(III) oxide are responsible for the colour of the majority of soils. Gibbsite ($Al(OH)_3$) and boehmite (AlO_2H) are colourless and very common. The hydrous oxides of manganese in soils are numerous, complex and still not well understood (Norrish, 1975). The nature of allophane, an amorphous mixture of hydrous oxides of Al and Si, is also unclear. In alkaline soils, hydrous calcium phosphates such as apatite may form. These materials all have a low capacity for anion and cation exchange, as well as more specific absorbent properties. For example, hydrous iron oxides strongly absorb As, Cr, Mo, P, Se and V in acid soils.

TABLE 4.3 Some properties of soil colloids (ion exchange capacities for a Kenyan soil; Sanchez, 1976)

Name	Ideal formula	Ions fixed	CEC/ equiv. kg^{-1}	AEC/ equiv. kg^{-1}
Montmorillonite	$K_2Al_{10}Mg_2(Si_8O_{10}(OH)_4)_3$	Ba, Cs, F, K, Rb	1·18	0·01
Vermiculite	$K_2(Al, Fe, Mg)_6Si_8O_{20}(OH)_4$	Ba, Cs, F, K, Rb	0·85	<0·01
Illite	$K_2(Al, Fe, Mg)_4Si_8O_{20}(OH)_4$	Ba, Cs, F, K, Rb	0·19	0·03
Halloysite	$Al_4Si_4O_{10}(OH)_84H_2O$	F	0·18	0·15
Kaolinite	$Al_4Si_4O_{10}(OH)_8$	F	0·04	0·02
Allophane	$(Al_2O_3)_p(SiO_2)_qnH_2O$		0·51	0·17
Gibbsite	$Al(OH)_3$	P	0·05	0·05
Goethite	FeO_2H	As, Cr, Mo, P, Se, V	0·04	0·04
Lithiophorite	$Mn_3O_4nH_2O$	Ba, Co, Ni, V		
Apatite	$Ca_5(PO_4)_3OH.H_2O$	As, Ba, F, Pb		
Humus	$C_{27-100}H_{29-80}O_{10-12}N$	Br, Cu, Hg, I, Pb, Sn, U	1·36	0·06

CEC, Cation exchange capacity; AEC, Anion exchange capacity

Hydrous manganese oxides are thought to carry most of the Co as Co(III), together with some Ba, Ni and lanthanides (Means *et al.*, 1978). In alkaline soils, calcium phosphate absorbs As, Ba, Cd, F, Pb and other elements (Norrish, 1975; Howard, 1977; Jenne, 1977).

Wakatsuki *et al.* (1977) have suggested two 'indices of maturity' for soils. Their 'hydrolyzate index' is $[Al]_{clay}/[Al]_{sand}$, while their 'oxidate index' is $([Fe]_{organic}+[Fe]_{goethite})/[Fe]_{sand}$ where sand represents our coarse inorganic fraction. It remains to be seen how useful these indices will prove.

Table 4.3 summarises data on ideal formulae, fixation properties and ion-exchange capacities for the colloidal material in soils. The ion-exchange capacities are representative examples measured at the natural pH of a soil from Kenya (Sanchez, 1976).

Organic matter

A very large number of organic compounds occur naturally in soils (Gieseking, 1975; Jackson, 1975), and here we have only the space to discuss how some of these interact with inorganic matter. We will consider three groups of compounds: (*a*) small molecules (*b*) fulvates and humates and (*c*) enzymes.

Small organic molecules are the products of partial decomposition of organic matter by bacteria and fungi, and have a short residence time in the soil. They include most of the common amino acids and also acetic, butyric, citric, formic, 2-ketogluconic, malic, oxalic, tartaric and many lichen acids. One effect of these acids is to lower the pH of the soil solution, which renders primary minerals more soluble. They may also chelate, and so render more soluble, elements such as Al, Cu and Fe: specific ionophores which chelate iron are described in chapter 7. Many other small molecules, such as inositol phosphate, fragments of nucleic acids and organic sulphur compounds may well interact with inorganic ions but this is not proved (Halstead and McKercher, 1975).

Fulvates and humates are weather-resistant organic polymers, which may result from the activities of fungi (Zunino and Martin, 1977). Fulvates are somewhat soluble in water, have a molar mass of about 1000 and a residence time in soil of 50–500 years. Humates are much larger polymers with a molar mass of about 150 000 and a residence time in soil of 300–1400 years (Campbell *et al.*, 1967). Typical compositions are: Fulvates; 50·9 % C, 3·3 % H, 0·7 % N, 44·8 % O, 0·3 % S: Humates; 56·4 % C, 5·5 % H, 4·1 % N, 32·9 % O, 1·1 % S (Dixon and Weed, 1977). Their main functional groups are –COOH (carboxylate) together with both alcoholic and phenolic hydroxyl groups. They attach themselves to other soil colloids and help to prevent clays from flocculating, so contributing to the crumb-like texture of soils. Their ion-exchange properties are illustrated by the data in Table 4.3. Their cation exchange capacity must be due to carboxylate groups, and is greatly reduced at pH below 4.5. The insolubility of humates may arise because they exist as calcium humates in most soils. By treatment with the sodium salt of an acid which precipitates calcium, such as the fluoride, oxalate or metaphosphate, much of the organic polymer can be dissolved as sodium humate. In fertile soils the ratio of clay to humus is often such that the ion exchange capacity of both colloidal fractions is of the same order of magnitude. Both fulvates and humates function as ligands for metals, and artificial acetylation, esterification and methylation implicates the carboxylate and phenolic hydroxyl groups in ligand formation (Schnitzer and Skinner, 1963, 1965). The solvent action of humates on common minerals has been studied by Baker (1973). The affinity for metals appears to follow the Irving-Williams order (Stevenson, 1977), but this tells us little about the sites involved; at least three different sites are used in the uptake of zinc by humates (Randhava and Broadbent, 1965). It is known that Hg(II), Cu, Pb and Sn(II) are exceptionally tightly bound by humates, while Mn(II) and the alkali and alkaline earth metals are at most loosely bound. Histosols have also been reported to concentrate Ge, Mo, Se, U and V, but the mechanism is unknown, and most of the iodine and bromine in soils appears to be in the organic fraction (Vinogradov, 1959). Ill-defined humates containing Al, Ca, Cu and Fe have recently been isolated from soils (Griffith and Schnitzer, 1976), and 22 elements have been determined in humates (Chen *et al.*, 1978).

Unfortunately the methods used to isolate humates and fulvates in this work involved pH changes from less than 1 to 12, and consequently there were opportunities for absorption and desorption of elements. The humates isolated had an ash content of 1·4–4%, while the fulvates had an ash content of 10–35%. The latter included up to 20–30% of Al, Na and Si, up to 1–4% of Cu, Fe, Mg and P and smaller quantities of Ag, B, Cr, Mn, Ni, Pb, Ti and V. This suggests that the fulvate fraction was more or less contaminated by clay minerals and perhaps also by iron phosphate, both of which can occur in amorphous states when the particle size is very small. It is not clear how much of the trace elements was absorbed by the fulvates and how much by the inorganic contaminants. One would expect most of the metals in humates and fulvates to be held in exchangeable form, much as they are in ion-exchange resins.

The existence of substantial quantities of free enzymes in soils has been clearly demonstrated. Kiss *et al.* (1976) conclude that the amounts of free amylase, cellulase, invertase and urease in soils greatly exceeds that in the microbial population. These enzymes continue to function when the living microbes have been inactivated by treatment with toluene or gamma radiation, and must play a substantial role in the cycling of organic nutrients in soils. Only when soils have been extensively polluted by heavy metals such as copper or zinc do these free enzymes lose their activity (Tyler, 1974).

Living organisms

The organisms inhabiting the soil can be classified as producers, predators or decomposers as described in the next chapter. Producers include minute, rootless algae and bryophytes which live on the surface of the soil, and seed plants whose roots may penetrate from 0·01–50 metres into the soil. Closed vegetation of seed plants absorbs about 10 g inorganic salts per square metre of soil per year. The residence time of most elements in vegetation is of the order of 0·5–3 years (Bowen, 1977a), and the salts are returned to the soil when the plant material is shed or dies.

The predators living in the soil include vast numbers of small arthropods, mostly mites and springtails, and smaller numbers of larger animals such as earthworms. Most of these obtain their energy by partially digesting organic matter and returning the residue to the soil.

Most of the decomposition of organic matter occurs as a result of the activities of micro-organisms, whose metabolism is inhibited by cold ($<5\,^{\circ}C$) or by drought (relative humidity $<85\,\%$). In well-aerated soils, most of the microbes are oxygenic, and the main process occurring is the oxidation of organic matter to carbon dioxide and water. In waterlogged or poorly aerated soils, where anoxic micro-organisms predominate, the oxidation of organic matter may be arrested at the stage where organic acids are produced, and humus may accumulate. Humus contains sufficient phenolic groups to be resistant to microbial attack. Under extremely anoxic conditions, bacteria may use sulphates as a source of oxygen, and the resultant hydrogen sulphide may react with and immobilise elements such as Cu, Fe, Mo or Zn as insoluble sulphides. Conversely oxygenic bacteria such as *Thiobacillus* are able to oxidise these sulphides to sulphates and so liberate the metals in a soluble form (Ehrlich, 1971). Bacteria and fungi are able to use highly insoluble phosphates such as apatite or iron(III) phosphate as a source of phosphate.

Microbial processes can change the valency state of iron and manganese in soils, with marked effects on their solubility (Aristovskaya and Zavarzin, 1971). Oxygenic algae, and especially bacteria such as *Thiobacillus ferrooxidans*, *Gallionella* and *Leptothrix* are able to oxidise Fe(II) to Fe(III), which usually ends up as goethite. The bacterial genus *Pedomicrobium* is probably responsible for the 'iron pan' or iron oxide layer in spodosols, which is produced by oxidation of downwardly migrating iron(II) fulvates to hydrous iron(III) oxide. Some anoxic bacteria can carry out the reverse reaction, giving a blue-green soil rich in iron(II), which is said to be gleyed. Gleying is accelerated by the presence of organic acids, which help to dissolve the goethite before it is reduced. In the case of manganese, oxygenic bacteria such as *Corynebacterium*, or fungi such as *Cladosporium*, can convert Mn(II) to Mn(IV) which forms insoluble hydrous oxides such as birnessite and lithiophorite. Waterlogging, adding fresh organic matter or shortage of dioxygen stimulate the growth of bacteria which can use the oxygen in Mn(IV) oxides, producing soluble salts of Mn(II) (Ehrlich, 1971). Other redox reactions known to occur in soils include As(III) $\leftrightharpoons$ As(V), Cr(VI) $\rightarrow$ Cr(III) (Bartlett and Kimble, 1976), I(I) $\rightarrow$ I(V), Mo(IV) $\rightarrow$ Mo(VI), Se(IV) $\rightarrow$ Se(VI), Se(IV) $\rightarrow$ Se(O), Te(IV) $\rightarrow$ Te(O) and V(V) $\rightarrow$ V(IV), and it is highly probable that Au, Sb and U undergo similar valency changes (Ehrlich, 1971). In strongly anoxic conditions it has long been known that microbes can produce volatile methyl derivatives of some non-metals, e.g. $HAs(CH_3)_2$: McBride and Wolfe (1971); $As(CH_3)_3$: Cox and Alexander (1973); $Se(CH_3)_2$ and $Te(CH_3)_2$: Ehrlich (1971), Alexander (1977). These observations have been extended by the researches of Wood *et al.* (1977), who showed that methyl cobalamin is probably the methylating agent involved (see chapter 9). Many more elements are now known to form volatile methyl derivatives in nature, including CH_3I: Lovelock *et al.* (1973); $Hg(CH_3)_2$: Wood *et al.* (1968); $Pb(CH_3)_4$: Wong *et al.* (1975); Sn methyls and perhaps also methyl derivatives of Au and Pt: Jernelov (1975).

The soil solution

Relatively few observations have been made on the composition of soil solutions, but those that have suggest that the range of composition is similar to that of river water, as given in Table 2.3 (see Bowen, 1966). The soil solution contains most of the chloride, sodium and nitrate, together with the trace amounts of borate, bromide and lithium, present in the soil. All these are highly mobile ions in soil, which are leached from the upper layers in regions of high rainfall, but rise to the surface forming white crusts of NaCl or $NaNO_3$ in arid regions. The movement of the soil solution through the soil after rain is responsible for the layered structure of most soils.

It might be thought that if all the different minerals present in a soil were known, the composition of the soil solution could be calculated from their solubilities. This is not so because the solubility of colloidal matter depends on particle size, notably for silica. Solubilities also depend on pH (Fig. 3.3). Thus the solubilities of hydrous iron(III) and manganese(IV) oxides are minute above pH 5, but are quite large at pH 4. Hydrous aluminium oxides are insoluble at pH 7 but dissolve in weak acids or bases. The solubility of calcite is critically dependent on the concentration of carbon dioxide in the water used to dissolve it; pure water at 16 °C dissolves 13 mg $CaCO_3$/l while a 1 % solution of CO_2 dissolves 211 mg $CaCO_3$/l.

Radiotracer experiments have shown that the soil is in dynamic equilibrium with its soil solution, and that the exchange time is of the order of seconds. Hence the concentration of a nutrient X in the soil solution should remain constant even when X is being removed by plant roots. The intensity factor for the nutrient supplying power of a soil is the logarithm of the concentration of the nutrient in the soil solution. The corresponding capacity factor is the total amount of exchangeable nutrient in the soil. Neither of these factors are easy to measure. The total exchangeable nutrient can be estimated by eluting a column of soil with excess ammonium acetate, and measuring the cations displaced by ammonium ions. More acid reagents, which are frequently employed, may extract from ten to a hundred times as much of the rarer elements, but dissolve unknown amounts of calcite, hydrous oxides and humates. For Ca^{2+}, Sr^{2+} and $PO_4{}^{3-}$, the experimental relation between concentrations in the soil and soil solution suggests that at least two different sources of these ions occur in the soil (Fried and Shapiro, 1961). Consequently empirical methods, mentioned in chapter 8, are used by agricultural scientists who wish to compare the amounts of nutrients in different soils.

The soil atmosphere

In well-drained soils the soil atmosphere consists of air of nearly normal composition, while in poorly-drained soils the dioxygen content declines and the content of carbon dioxide increases. In permanently frozen soils the dioxygen in normal air is entirely replaced by CO_2. 0·01 % of dioxygen by volume appears to be a critical concentration in the soil atmosphere. Above this concentration, the functional organisms are oxygenic, while below it, anoxic bacteria are dominant. Soils are a source of the gas nitrous oxide, a by-product of the activities of denitrifying bacteria (chapter 5); the concentration of this gas increases with depth in soils (Schütz *et al.*, 1970). They also act as a source of gaseous products of radioactive decay, such as He, ^{40}Ar and Rn (Megumi and Mamuro, 1974). Anoxic soils produce volatile derivatives of sulphur and selenium, and possibly other elements (Abu-Erreish *et al.*, 1968; Kodota and Ishida, 1972).

The composition of soils

Table 4.4 attempts to summarise the data for mean elemental composition and range of composition of soils. Data for soils near ore bodies (Ernst, 1974), heavily polluted soils and serpentine soils (notably rich in Cr, Mg and Ni but poor in Ca) have not been used. Insufficient data is available to compare the different types of soils defined in Table 4.1. Despite the limitations of the data in Table 4.4 caused by inadequate sampling, inaccurate analyses and possible contamination, the means agree fairly well with the mean values for igneous and sedimentary rocks (Tables 3.3 and 3.5). Detailed comparison of the tables shows that carbon and nitrogen are strongly enriched in soils with respect to rocks. As, Br, Cd, Hf, I, Pb, Sb, Se, Sn and Zr may be somewhat enriched in soils, while B, Ca, Cl, F, Hg, K, Mg, Na, Sr, Tl and U may be slightly depleted, but confirmation of these findings must await more and better data. There is scope for a great deal more work on the elemental analysis of soil profiles as well as soil fractions. Present-day techniques for fractionating soils are relatively drastic, and the fractions obtained may not be representative. One result is that attempts have been made to correlate elemental analyses with factors such as the content of clay or organic carbon, without any clear-cut results. The great heterogeneity of soil types would lead us to expect no uniform results from soil profile studies, though useful correlations might be made in individual soil types. Thus the enrichment of C, N, P and S commonly found in the upper layers of soils can be correlated with the large amount of organic matter there, at least in alfisols, inceptisols, mollisols and spodosols. Heavy metals such as Ag, As, Cu, Hg, Pb, Sb and Zn are also often found to be enriched in the upper layers of soils; the reason is not clear, but at least in the case of lead it appears to be a recent artefact due to pollution. Elements reported to be enriched in the lower layers of soils, where clay minerals and or hydrous oxides accumulate, include Al, Fe, Ga, Mg, Ni, Sc, Ti, V and Zr. We would expect B, Cl, Li and Na to be enriched at the surface of aridisols but depleted at the surface of soils where rainfall is abundant, as described above.

Rates of formation of soils

There is surprisingly little data on this topic. It is known that clay is not formed when rocks weather at less than 5 °C, and very little is formed below 10 °C. In temperate North America from 1–20 mg clay are formed per kg rock each year (Bear, 1964). Estimates of the rate of weathering of rock include a world mean of 27 g rock m^2/yr (Barth,1961); 42 g chalk m^2/yr under a rainfall of 0·6 m/yr near Cambridge, England (Perrin, 1965); and 80 g

TABLE 4.4 Medians and ranges of elemental composition of soils in mg X/kg soil

X	Median soil content (Range)	Ref.
Ag	0·05 (0·01–8)	A, C, Cj, G, P, S, U
Al	71 000 (10 k–300 k)	C, Cw, V, W
As	6 (0·1–40)	C, Ch, Cj, Cw, O, P, S, T, U, V, Z
Au	0·001–0·02?	We, X
B	20 (2–270)	A, C, Ch, S, U, V
Ba	500 (100–3000)	A, C, Ca, Ch, Cj, G, N, S, U, V, Y, Z
Be	0·3 (0·01–40)	C, G, Ms, S, U V
Bi	0·2 (0·1–13)	A, Cj, U
Br	10 (1–110)	Ch, Cw, L, U, V, Y
C (org.)	20 000 (7 k–500 k)	C, V
Ca	15 000 (700–500 k)	C, Ca, Th, V, W, Z
Cd	0·35 (0·01–2)	Cj, D, J, St, T, Th, U, V
Ce	50 (3–170)	C, Cw, N, U, V, Y, Z
Cl	100 (8–1800)	Cj, L, U, V, Y
Co	8 (0·05–65)	A, C, Ca, Ch, Cj, Cw, G, K, N, S, Th, U, V, Y, Z
Cr	70 (5–1500)	A, C, Ca, Cj, Cw, G, N, S, Th, U, V, Y, Z
Cs	4 (0·3–20)	Cj, Cw, N, S, U, Y, Z
Cu	30 (2–250)	A, C, Ca, Ch, G, P, S, T, Th, U
Dy	5 (2–12)	U, V
Er	2 (0·6–6)	U, V
Eu	1 (0·1–3·2)	K, N, U, V, X, Y, Z
F	200 (20–700)	C, Ch, S, U, V, We
Fe	40 000 (2 k–550 k)	C, Ca, Cj, Cw, K, N, S, Th, V, W, Z
Ga	20 (2–100)	A, C, Ca, G, S, U, V, Y
Gd	4 (2–6)	U, V
Ge	1 (0·1–50)	A, Sb, U, V
Hf	6 (0·5–34)	H, N, U, X, Y, Z
Hg	0·06 (0·01–0·5)	An, C, Cj, D, M, V
Ho	0·6 (0·4–2)	U, V
I	5 (0·1–25)	A, C, L, U, V, We
In	1 (0·7–3)?	Cj, Y
K	14 000 (80–37 k)	B, C, Ca, V, W, Z
La	40 (2–180)	C, G, K, N, S, U, V, Y, Z
Li	25 (3–350)	A, C, G, S, Sh, Th, U, V
Lu	0·4 (0·1–0·7)	N, U, V, Z
Mg	5000 (400–9000)	C, Ca, Cj, Th, V, W
Mn	1000 (20–10 k)	A, C, Ca, Cj, Cw, G, P, S, Th, V, W, Y, Z
Mo	1·2 (0·1–40)	A, Ch, Cj, G, Ku, S, U, V
N	2000 (200–5000)	V
Na	5000 (150–25 k)	C, Ca, Cj, Cw, V, W, Z
Nb	10 (6–300)	C, U
Nd	35 (4–63)	C, N, U, V, Z
Ni	50 (2–750)	A, C, Ca, Ch, Cj, Cw, G, S, T, Th, U, V
O	490 000	V
P	800 (35–5300)	C, G, V, W
Pb	35* (2–300)	A, C, Ca, Ch, Cj, Cw, G, P, S, T, Th, V

Table 4.4—*continued*

X	Median soil content (Range)	Ref.
Pr	7 (3–12)	U, V
Ra	0·8 μ (0·2 μ–5 μ)	Me, S, V
Rb	150 (20–1000)	A, G, N, S, U, V, Y, Z
S	700 (30–1600)	U, V
Sb	1 (0·2–10)	Cj, Cw, N, O, P, U, Wa, Y, Z
Sc	7 (0·5–55)	C, Cj, Cw, K, N, S, U, X, Y, Z
Se	0·4 (0·0112)	A, C, Ch, Cj, Cw, S, U, V
Si	330 000 (250 k–410 k)	C, V, W
Sm	4·5 (0·6–23)	K, N, U, V, W, Y, Z
Sn	4 (1–200)	A, Ch, Cj, P, S, U
Sr	250 4–2000)	A, C, Cj, G, S, Th, U, V
Ta	2 (0·4–6)	H, N, X
Tb	0·7 (0·1–1·6)	H, N, U, V, X, Z
Th	9 (1–35)	B, C, Cw, N, S, U, V, Y, Z
Ti	5000 (150–25 k)	A, C, Ch, G, S, V, W
Tl	0·2 (0·1–0·8)	Cj, U
Tm	0·6 (0·3–1·2)	U, V
U	2 (0·7–9)	B, C, S, U, V, Y, Z
V	90 (3–500)	A, C, Cj, Cw, G, S, U, V, Y
W	1·5 (0·5–83)	U, Y
Y	40 (10–250)	C, G, S, U
Yb	3 (0·04–12)	C, H, N, U, V, X, Y, Z
Zn	90 (1–900)	A, C, Ca, Cj, Cw, D, S, T, Th, U, Y, Z
Zr	400 (60–2000)	C, Cj, G, S, U, V, X

k, $\times 1000$; μ, $\times 10^{-6}$; org., organic

*12 prior to global contamination; up to 1000 in urban soils (Patterson, 1965)

References: A, Aubert and Pinta, 1977 (Compilation, world soils); An, Anderssen, 1967 (Hg in Swedish soils); B, Bunker and Bush, 1968 (Colorado soils); C, Connor and Shacklette, 1975 (35 N. American soils); Ca, Casagrande and Erchall, 1977 (2 peat soils); Ch, Chapman, 1966 (Compilation, miscellaneous soils); Cj, Chattopadhyay and Jervis, 1974 (1 Canadian soil); Cw, Cawse, 1974 (7 British soils); D, Davies, 1976 (12 British soils); G, Gorham and Swaine, 1965 (Scottish soils); H, Houtman, 1973 (1 Javanese soil); J, Jervis *et al.*, 1974 (Cd in Canadian soils); K, Kline and Brar, 1969 (12 world soils); Ku, Kubota, 1977 (Mo in N. American soils); L, Låg and Steinnes, 1976 (5 Norse soils); M, McKeague and Kloosterman, 1974 (Hg in Canadian soils); Me, Megumi and Mamuro, 1974 (Ra in Japanese soils); Ms, Meehan and Smythe, 1967 (Be in Australian soils); N, Naidenov and Travesi, 1977 (43 Bulgarian soils); O, Ohno and Yatazawa, 1970 (3 Japanese soils); P, Presant and Tupper, 1965 (Canadian soils); S, Swaine 1955 (Compilation, world soils); Sb, Schroeder and Balassa, 1967 (Ge in soils); Sh, Shacklette *et al.*, 1973 (Li in N. American soils); St, Steinnes, 1977a (14 Norse soils); T, Thornton and Webb, 1975 (37 British soils); Th, Tjell and Hovmand, 1978 (51 Danish soils); U, Ure *et al.*, 1978–9 (10 Scottish soils); V, Vinogradov, 1959 (Compilation, Russian soils); W, Wakatsuki *et al.*, 1977 (16 Japanese soils); Wa, Ward and Lakin, 1954 (Sb in N. American soils); We, Wedepohl (1969–1974) (Compilation, miscellaneous soils); X, Van der Klugt *et al.*, 1977 (1 Dutch soil); Y, Dantas and Ruf, 1975 (9 Brazilian soils); Z, Koons and Helmke, 1978 (4 Canadian soils)

gneiss m^2/yr under a rainfall of 1·2 m/yr in eastern U.S.A. (Johnson *et al.*, 1968). In recently glaciated areas, the age of a soil can sometimes be estimated with some precision. Temperate soils appear to have ages in the range 10^3–10^6 years, but no data are available for tropical soils, some of which are very deeply weathered and may be as much as 7×10^7 years old in Malaya.

Residence times of elements in soils

Using tracer techniques at Rothamsted, Jenkinson and Rayner (1977) have shown that the turnover of organic matter in arable soil is best fitted by a five-compartment model, whose concentrations and residence times are given below:

Name of compartment	C/kg m^{-2}	Residence time/yr
Decomposable plant litter	0·001	0·24
Resistant plant litter	0·047	3·3
Soil organisms	0·028	2·4
Physically stabilised organic matter	1·13	71
Chemically stabilised organic matter	1·22	2800

This agrees fairly well with carbon-14 dating of purified fulvates and humates which have ages of 50–1400 years (Campbell *et al.*, 1967) and with other estimates (Scharpenseel *et al.*, 1968; Rafter and Stout, 1970; Nye and Tinker, 1977).

Data on the residence times of other elements in soils have been reviewed by Bowen (1975b); the few that exist are little more than estimates based on rather imprecise figures for inputs and outputs. Readily leached elements, such as Ca, Cd, Mg and Na have residence times in the range 75–380 years, Hg has a residence time of 500–1000 years, while other elements such as As, Cu, Ni, Pb, Se and Zn have residence times of the order of 1000–3000 years in temperate soils. In tropical rain forests, where nutrient cycling is particularly rapid, Golley *et al.* (1969) estimate a residence time of about 40 years for all nutrient elements in the soil. Benninger *et al.* (1975), using data for ^{210}Pb, confirm that lead has a residence time of the order of kiloyears in soil. Much more data is needed, especially for undisturbed soils in regions far from industrial pollution, to amplify the picture. It seems that the processes of soil formation involve time scales of the order of millenia, and it is a sobering thought that the current dispersion of huge tonnages of lead into soils will be measurable for several thousand years into the future.

5 *The Biosphere and Elemental Cycles*

The *biosphere* is an ambiguous term which can either refer to the totality of living matter, or to those parts of the Earth's surface inhabited by organisms. The latter can be termed the *domain* of the biosphere. This domain extends from about 10 km above sea level to 10 km below it, and includes the soil, the ocean and the uppermost layer of marine sediments.

The biosphere shows far more morphological variety than does the geosphere. For example, there are only about 2000 named minerals in the lithosphere, but there are at least $3{\cdot}3 \times 10^5$ species of green plant, $9{\cdot}3 \times 10^5$ species of animal and 8×10^4 species of bacteria and fungi known to science. The total number of existing species (including an estimate of undescribed species) may well be as many as three million (Wigglesworth, 1964) and the number of extinct species must be of the same order.

The mass of any group of organisms, or of the whole biosphere, is called the standing crop or biomass, and can be measured in terms of fresh material (FM) or dry matter (DM). Fresh weights of most organisms are about four times their dry weights, and the latter are less variable. Sometimes biomasses are expressed in terms of mass per unit area, and to avoid ambiguity units must be clearly stated. There are many practical problems involved in the estimation of biomasses of small organisms in heterogeneous materials such as soils, and there is no satisfactory method of estimating the biomass of bacteria or fungi in soils containing much dead organic matter. According to current estimates, the lithosphere, or Earth's crust to a depth of 16 km, weighs $1{\cdot}5 \times 10^{22}$ kg, the hydrosphere weighs $1{\cdot}4 \times 10^{21}$ kg, the atmosphere $5{\cdot}1 \times 10^{18}$ kg and the whole biosphere $1{\cdot}8 \times 10^{15}$ kg DM. Since the Earth's surface area is $5{\cdot}1 \times 10^{14}$ m^2, there are 3·6 kg of dry living matter per square metre.

The relatively small mass of the biosphere belies its great importance in the cyclical processing of the elements. One group of organisms, the green plants, are able to absorb a fraction of solar radiation and store this as chemical energy, which is then used to absorb carbon dioxide from the atmosphere for growth and all other living processes. The chemical energy can be used to carry out chemical processes such as oxidation, reduction, precipitation and selective concentration of elements that would not occur under abiotic conditions. The results of biochemical processes operating over long periods of time can be seen in hills of chalk or limestone rock, coal seams, oil wells and in the oxygen of the atmosphere.

Several jargon terms have been used to describe the rate of carbon dioxide utilisation by green plants. The *production* of a given part of the biosphere is the rate of increase of biomass per year. It may be *gross production*, which is the total amount of dry matter formed by photosynthesis, or *net production*, corrected for losses due to respiration. *Productivity* is the production per unit area, and may also be either gross or net. The units for production and productivity should be clearly specified. For example, production may be quoted in kg DM/yr, or as kg C/yr, which differ by a factor of 2·2. Techniques for measuring production and productivities are relatively young, and reliable measurements have only been made in the last two decades.

The first estimate of production was made by Liebig (1840), who assumed that the land produced 0·5 kg DM/yr m^2 and that marine production was negligible. This leads to an estimated global production of $0{\cdot}34 \times 10^{14}$ kg DM/yr. Subsequent workers have realised that both terrestrial and marine production vary within wide limits, and better estimates can be made by subdividing the Earth into different vegetational regions and measuring the mean production in each of these regions. Table 5.1 and Fig. 5.1, taken

TABLE 5.1 Areas, biomasses, net production and productivities of vegetational regions of the Earth (slightly modified from Lieth and Whittaker, 1975)

Vegetation region	Area/ 10^{12} m^2	Plant biomass/ 10^{12} kg DM	Animal biomass 10^{12} kg DM	Biomass/ area/ kg DM m^{-2}	Production/ 10^{12} kg DM yr^{-1}	Productivity/ kg DM m^{-2} yr^{-1}
Tropical rain forest	17	765	0·33	45	37·4	2·2
Other tropical forest	7·5	260	0·09	35	12	1·6
Temperate evergreen forest	5	175	0·05	35	6·5	1·3
Temperate deciduous forest	7	210	0·11	30	8·4	1·2
Coniferous forest	12	240	0·06	20	9·6	0·8
Savannah	15	60	0·22	4	13·5	0·9
Scrub	8·5	50	0·04	6	6	0·7
Cultivated land	14	14	0·006	1	9·1	0·67
Temperate grassland	9	14	0·06	1·6	5·4	0·6
Marshland	2	30	0·02	15	6	3
Lakes	2	0·05	0·01	0·02	0·8	0·4
Tundra/mountain tops	8	5	0·003	0·6	1·1	0·14
Semi-deserts	18	13	0·008	0·7	1·6	0·09
Deserts/Glaciers	24	0·5	0·00002	0·02	0·07	0·003
Total land	149	1837	1·0	12·2	117·5	0·78
Open sea	332	1	0·8	0·003	41·5	0·125
Continental shelf	27	0·28	0·16	0·001	9·8	0·36
Estuaries	1·4	1·4	0·021	1	2·1	1·5
Coral reefs/Algal beds	0·6	1·2	0·012	2	1·6	2·5
Total sea	361	3·9	1·0	0·01	55·0	0·155
World total	510	1841	2	3·6	172·5	0·336

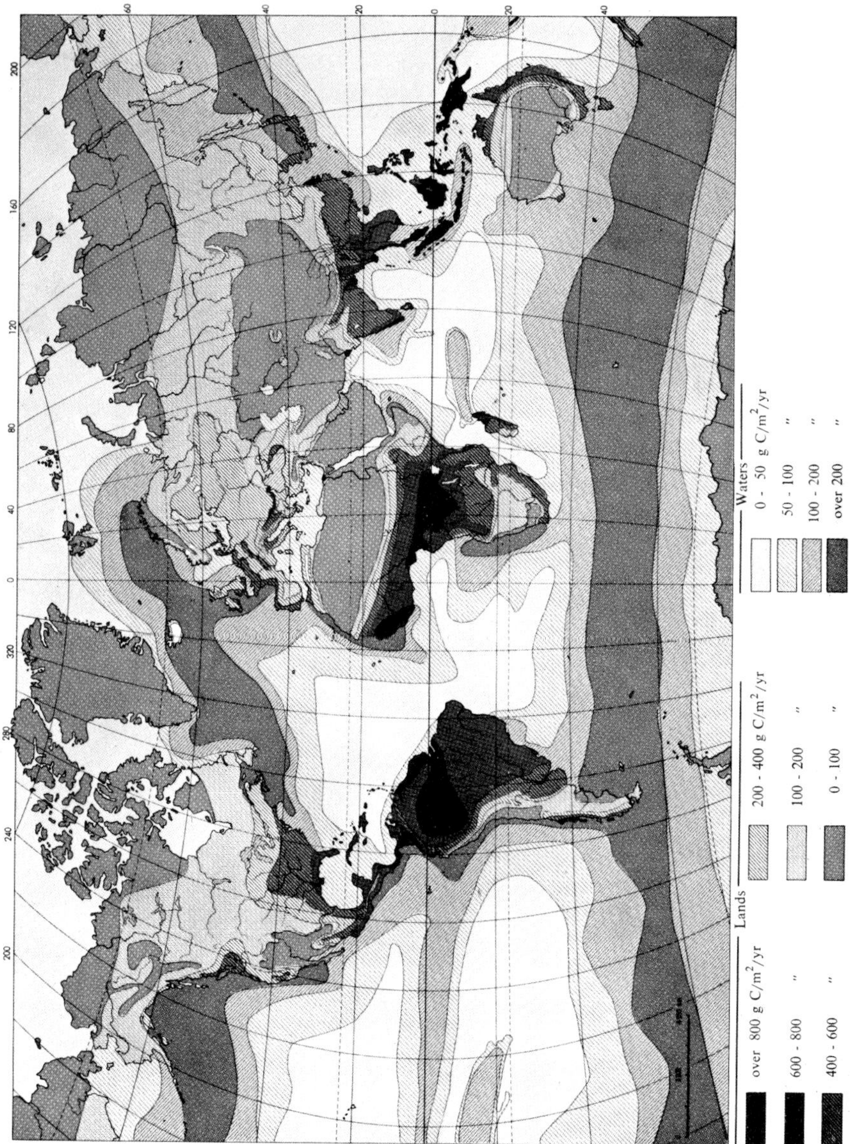

FIG. 5.1 Productivity contours of the World. From Lieth and Whittaker (1975)

from Lieth and Whittaker (1975), summarise data on areas, biomasses, net productivity and production from the major vegetational regions of the world. Short-term seasonal fluctuations have been averaged out, though they are important in the temperate zone and in areas with dry seasons. It must be emphasised that the data from which the mean values in Table 5.1 were calculated are not very extensive in time, and that other workers favour different assessments (Rodin *et al.*, 1974). The following points are not controversial:

1. The areas of some communities of great biological interest, such as rivers (area $2{\cdot}4 \times 10^{11}$ m^2), beaches (area $1{\cdot}4 \times 10^{10}$ m^2) and zones of upwelling in the sea (area 4×10^{11} m^2) are so small that their contributions to the global biomass and production may be neglected.
2. Terrestrial vegetation generally supports a far greater biomass per unit area than does oceanic vegetation, since the latter consists of unicellular plankton: about 99·5% of the world's biomass occurs on land.
3. The net productivities of land and ocean vegetation do not differ by such large factors. Highly productive vegetation, such as tropical forests, marshes and coral reefs, have similar net productivities.
4. Although the number of animal species exceeds the number of plant species, the biomass of animals is only about 0·1% of the total plant biomass.

Within the biosphere we can distinguish *communities* and *ecosystems*. A natural *community* is the totality of living material in a distinctive part of the biosphere, such as the Sahara desert or the Sargasso Sea. Communities may be transient, as in an arable field or a forest clearing, or climax, as in an undisturbed forest. Climax communities are extremely complex systems and may affect both the local climate and the type of soil formed. The term *ecosystem* includes both the community and its local environment, including the soil or water providing nutrients.

A community uses solar energy to carry the chemical elements through continuous cycles from dead to living material and back again. An outline of how matter and energy flow through the biosphere is shown in Fig. 5.2.

Odum (1975) has classified the components of communities according to their function.

Producers include all green plants, which contain chlorophyll and are able to convert water and carbon dioxide into living matter and oxygen. Their photosynthetic efficiencies are estimated to be:

Gross:	Land	1·6%;	Sea	0·2%;	*Total*	0·62%
Net:	Land	0·6%;	Sea	0·13%;	*Total*	0·27%

In other words, the average percentages of gross photosynthetic production lost as heat by plant respiration are as follows (Whittaker, 1975):

Land 63 %; Sea 33 %; *Total* 57 %

Predators include the entire animal kingdom, which depend on producers for their food, and can be subdivided into herbivores and carnivores. They consume about 10 % of the net primary production on land, and up to 40 % of that in the ocean. The conversion of plant tissue into animal tissue has an energy efficiency of 8–18 %, the remaining energy being lost as heat.

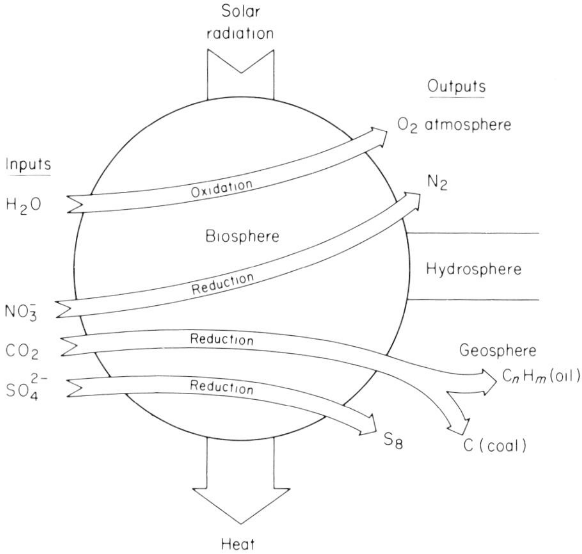

Fig. 5.2 Diagram showing major inputs and outputs of the biosphere in which solar energy is used

Decomposers include the fungi, actinomycetes and most bacteria, as well as a few aberrant plants which lack chlorophyll. Many animals also function as decomposers to some extent, and it is a nice point how much decomposition is effected by the animals themselves and how much by their symbiotic bacteria. In general decomposers obtain energy by the oxidation of organic matter to carbon dioxide and water, but other reactions may occur in anoxic situations. They are responsible for converting 90 % of the annual net production back to non-living matter. The biomass of decomposers is probably in the range 10^{11}–10^{13} kg DM. Alexander (1977) indicates that the masses of actinomycetes and bacteria in 1 m^2 soil are in the respective ranges 15–100 and 2·5–100 g DM, but gives no estimate for fungi. Ferguson and Rublee (1976) found $6{\cdot}6 \times 10^5$ bacterial cells per cm^3 sea water, corresponding to about 2×10^{13} kg DM in the entire ocean; this seems rather large and it is likely that the average concentration of bacteria in the open ocean is much lower.

A few organisms such as photosynthetic bacteria, iron and sulphur bacteria and nitrogen-fixing microbes are not covered by this simple classification, but will be mentioned where appropriate below.

It should now be clear that the biosphere is able to take the element carbon through biological cycles, analogous to but quite separate from the hydrological and geological cycles described in chapters 2 and 3. Indeed the biosphere carries all the elements through cycles, but these are thought to be of minor quantitative importance except for carbon, hydrogen, nitrogen, oxygen and sulphur, together with calcium, phosphorus and silicon. The first five elements named undergo *organic cycles* since the element concerned makes and breaks bonds with carbon during the cycle, while the last three elements do not. Other elements known to undergo organic cycles include As, Br, Cl, Hg, I, Pb, Se, Sn and Te, and this list will doubtless be extended in the future. A noteworthy feature of organic cycles is that they involve extensive oxidation and reduction, catalysed by enzymes which usually contain iron, copper or some other metal. The primary foods of plants, carbon dioxide, nitrate and sulphate are all reduced following uptake. On the other hand predators and decomposers obtain their energy by oxidising the reduced forms of carbon in organic matter. This reflects the fundamental photosynthetic reaction, which involves splitting water into dioxygen (an oxidant) and using the hydrogen as a reductant. Living cells are able to maintain oxidising and reducing conditions in existence simultaneously in separate compartments, and to use the energy provided by their interaction with suitable substrates in a controlled manner.

C, H, N, O and S in living organisms

The content of these elements in organisms is needed in order to calculate data relevant to their elemental cycles. Analytical data are somewhat meagre, and are given in Table 5.2. Note that the oxygen contents have almost all been calculated by difference, since until recently there have been no means of determining this important element directly. Analyses of marine plankton are technically difficult because many diatoms etc. have tests made of calcium carbonate or silica which give them a high ash content. The values quoted for arthropods, coelenterates, echinoderms and molluscs in Table 5.2 refer to the soft parts only.

It appears that most organisms contain similar amounts of carbon and hydrogen, but that animals contain more nitrogen and sulphur, but less oxygen, than do plants. Animals are, of course, richer in protein than are plants. Bacteria have a high nitrogen content and are also rich in phosphorus. Further data are needed for the sulphur contents of plankton and fungi. The figures quoted in Table 5.2 do not agree with estimates for the biosphere given by Deevey (1970) unless his figures for carbon and oxygen are transposed.

TABLE 5.2 Mean C, H, N, O and S content of organisms in g/kg DM

Group	C	H	N	O	S	Ash	Ref.
Seed plants	45·4	5·5	3·3	41·0	0·44	4·4	Rankama and Sahama (1950)
Brown algae	34·5	4·1	1·5	37·7	1·2	22·0	Vinogradov (1953)
Marine plankton	22·5	4·6	3·8	44·0	0·6	24·5	Vinogradov (1953)
Mean producer	45	5	3	40	0·5	6·5	
Bacteria	48·5	7·4	10·7	24·8	0·61	8	Spector (1956a)
Fungi	49·4	7	5·1	33·1	0·4	5	
Mean decomposer	49	7	8	29	0·5	6·5	
Coelenterata	43·6	4·5	6·3	27·1	1·9	16·6	Vinogradov (1953)
Annelida	40·2	5·9	9·9	34	1·4	8·6	Vinogradov (1953)
Nematoda	40	6	11	35·4	1·3	6·3	Vinogradov (1953)
Mollusca	40	6	8·5	39·1	1·5	4·9	Vinogradov (1953)
Echinodermata	40	4·5	4·4	45·6	0·5	5·0	Vinogradov (1953)
Arthropoda	44·6	7·3	12·3	32·3	0·44	3·1	King (1957)
Pisces	47·5	6·8	11·4	29·0	1·0	4·3	Vinogradov (1953)
Mammalia	48·4	6·6	12·8	18·6	1·6	12·0	Rankama and Sahama (1950)
Mean predator	45	6·5	10	30	1·3	7·2	

Representation of elemental cycles

The representation of elemental cycles by diagrams is not a trivial task, as a great deal of data (much of it imprecise or even contradictory) has to be summarised on a small chart. The following conventions are adopted here:

1. The cycle is considered as a series of reservoirs, represented by boxes, interconnected by input and output fluxes, represented by arrows joining the boxes.

2. Within each box there is a reference letter, the name of the reservoir, the chemical species, the phase (g, gas; l, liquid; s, solid; (susp.), suspended matter) and the mass of the element X in units of 10^{12} kg (N.B. M, $\times 10^6$).

3. To avoid confusion, the fluxes are not labelled on the charts, but are given in separate tables. For any pair of reservoirs M, N, (M, N) is the flux from M to N and (N, M) its inverse.

4. The following geochemical data have been assumed, see Table 3.1 etc.

Mass of atmosphere = $5{\cdot}3 \times 10^{18}$ kg
Volume of atmosphere reduced to STP = 4.1×10^{18} m^3
Mass of water in the ocean = $1{\cdot}39 \times 10^{21}$ kg
Area of ocean = $3{\cdot}61 \times 10^{14}$ m^2
Flux of river water to the ocean = $3{\cdot}7 \times 10^{16}$ kg/yr
Mass of Earth's crust = $2{\cdot}4 \times 10^{22}$ kg
Area of Earth = 510×10^{12} m^2
Density of Earth's crust = $2{\cdot}8$ g/cm^3
Mass of sedimentary rock = $3{\cdot}2 \times 10^{21}$ kg

The carbon cycle

The carbon cycle is represented by Fig. 5.3. Values for the masses given in it were obtained as follows:

A, B and C: calculated from data given in chapter 1.

D, E, F and G: calculated from data in Table 5.1, assuming that biological material contains 0·45 g C/g DM. Values for F and G are estimates.

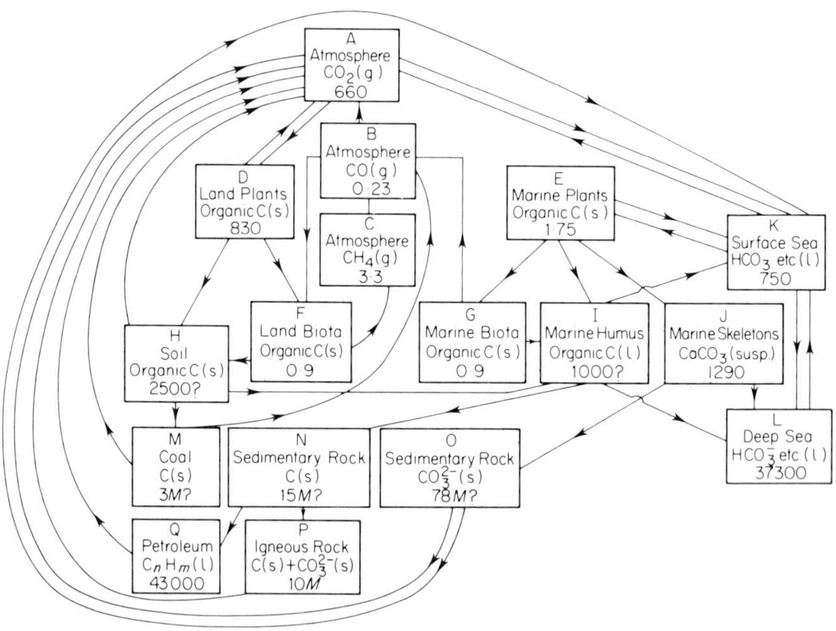

FIG. 5.3 Diagrammatic carbon cycle. Figures are masses in 10^{12} kg C. Biota, animals and microbes; M, $\times 10^6$; (susp.), suspended matter

H: taken from estimates by Whittaker (1975). Bolin (1970) gave estimates lower by a factor of about 4. For more exact work this reservoir should be divided into five compartments as described in chapter 4.

I: taken from estimates by Skirrow (1975) and Williams (1975). Whittaker (1975) gives a value greater by a factor of 10, Wedepohl (1969) gives a range from 280–3800, and Sharp (1973) estimates 1400.

J: taken from an estimate by Pytkowicz (1973).

K and L: taken from Skirrow (1975). The surface ocean is that part above the thermocline, which lies between 50 and 250 m deep. Broecker (1974) gives a valuable discussion of the relative proportions of CO_2, HCO_3^- and CO_3^{2-} in different parts of the ocean.

M, N, O and P: calculated from data in chapter 3; see also Wedepohl (1969). Whittaker (1975) gives 5000 for M and 10^7 for N, but estimates of coal reserves are notably variable, as recoverable coal is so much less than total coal. Pytkowicz (1973) gives $6{\cdot}8 \times 10^6$ for organic C in unconsolidated sediments, and 18×10^6 for reservoir O.

Q: includes oil and natural gas (90 % CH_4). Hobson and Tiratsoo (1975) estimate oil reserves of 5×10^{16} kg, of which $2{\cdot}7 \times 10^{14}$ kg is recoverable, and 4×10^{13} kg natural gas.

A few small reservoirs have been omitted from Fig. 5.3, for example elemental carbon from forest fires (Griffin and Goldberg, 1975). Estimates of the fluxes between reservoirs are given in Table 5.3. Many of these figures are controversial, and some are not based on experimental measurements but have been chosen to obtain consistency between inputs and outputs.

The following features of the carbon cycle are of interest:

1. The ocean acts as a gigantic buffer solution and prevents rapid changes in the carbon dioxide concentration in the atmosphere. Photosynthesis and sorption by the ocean are major processes in removing carbon dioxide from the atmosphere, where it has a residence time of about three years.

2. The carbon monoxide and methane subcycles are quantitatively much less important than the cycle of carbon dioxide. Bacteria such as *Methanobacterium* and *Methanosarcina* spp. can reduce CO_2 or CO to CH_4, while other species such as *Carboxydomonas oligocarbophila* can oxidise CO to CO_2 (Singer, 1970; Nriagu, 1976).

3. Although the predators and decomposers have a combined biomass of less than 1 % of the producers, they are able to reoxidise at least 99 % of the net production by green plants back to CO_2 or bicarbonate each year.

4. Since less than 10 % of the total budget of crustal carbon occurs in igneous rocks, the remaining 90 % must have been processed by living

TABLE 5.3 Fluxes in the carbon cycle (see Fig. 5.3)

Flux	Nature	10^{12} kgC/yr	Refs.
A, D	Gross land production	142	Table 5.1; Whittaker (1975)
A, K	Absorption of CO_2 by the sea	65	Pytkowicz (1973); Bolin (1970) gives 100
B, A	Tropospheric oxidation of CO	1·5	Weinstock and Chang (1976)
B, F	Sorption of CO by soil	0·19	
C, B	Oxidation of CH_4 to CO	1·3	
D, A	Land plant respiration	90	Whittaker (1975)
D, F	Land biota production	0·9	Estimate
D, H	Land humus formation	51	
E, G	Marine biota production	0·6	
E, I	Marine humus formation	23·6	Estimate; see Williams (1975)
E, J	Marine production of $CaCO_3$	1·1	Broecker (1974)
E, K	Marine plant respiration	12	Whittaker (1975)
F, C	CH_4 production by land biota	1·3	Weinstock and Chang (1976); Hidy (1973)
F, H	Land humus formation by biota	0·97	Estimate
G, B	CO production by marine biota	0·04	Weinstock and Chang (1976)
G, I	Marine humus formation by biota	0·27	Estimate
H, A	Oxidation of land humus to CO_2	51	Whittaker (1975)
H, I	Organic matter carried to sea	0·4	Williams (1975)
H, M	Conversion of humus to coal	0·0075?	Bowen (1966); very uncertain
I, K	Oxidation of marine humus to HCO_3^-	21	Whittaker (1975); Pytkowicz (1973)
I, L	Oxidation of humus in deep sea	2·8	Estimate
I, N	Kerogen formation from marine humus	0·5	
J, L	Dissolution of CO_2 in deep sea	0·73	Broecker (1974)
J, O	Net oceanic sedimentation of $CaCO_3$	0·35	
K, A	Release of CO_2 from sea surface	65	Pytkowicz (1973)
K, E	Gross marine production	37	Whittaker (1975); Williams (1975)
K, L	HCO_3^- transfer to deep sea	40	Bolin (1970)
L, K	HCO_3^- transfer to surface sea	43·5	Estimate
M, A	CO_2 from burning fossil coal	2·24	U.N. (1977)
M, B	CO from burning fossil fuels	0·375	Weinstock and Chang (1976)
N, P	Conversion of sediments to igneous rock	0·006?	Garrels and Mackenzie (1971)
N, Q	Rate of formation of petroleum	0·0002?	Estimate; see Hobson and Tiratsoo (1975)
O, A	CO_2 from lime burning	0·25	Estimate
O, K	Weathering of limestone	0·35	see also Pytkowicz (1973)
P, A	CO_2 from volcanoes	0·0082	Pytkowicz (1973)
Q, A	Rate of burning oil and gas	3·7	U.N. (1977) (if 80% crude oil is burnt)

organisms, which themselves contain only about 0·0005% of the total crustal carbon.

5. The net production of both coal and oil are very uncertain because they are estimated from the difference between two large, uncertain fluxes. Their rates of production are exceeded by current rates of human exploitation (chapter 12). Prior to large scale human intervention, coal and oil, and probably also limestone, must have been accumulating as a by-product of the carbon cycle. The burning of fossil fuels has resulted in a measurable increase in the amount of CO_2 in the atmosphere (chapter 12).

Keeling (1973) has described a 5 box model of the carbon dioxide cycle based on the following reservoirs: atmosphere, surface ocean, deep ocean, short-lived organic C and long-lived organic C. This can be used to test predictions of the effects of increasing inputs of carbon dioxide to the atmosphere on other parts of the cycle (Williams, 1978).

The hydrogen cycle

The hydrogen cycle, represented in Fig. 5.4, is dominated by the water cycle, which was discussed in chapter 2. Land plants play a surprisingly large part in the water cycle, since 10% of the solar energy they receive is used for

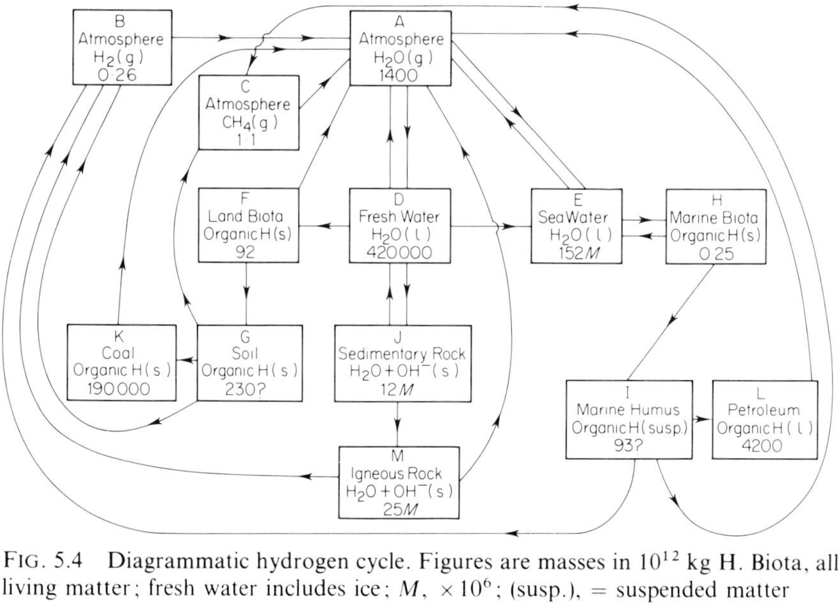

FIG. 5.4 Diagrammatic hydrogen cycle. Figures are masses in 10^{12} kg H. Biota, all living matter; fresh water includes ice; M, $\times 10^6$; (susp.), = suspended matter

transpiration, as compared to the 1·6% used in photosynthesis. The amounts of hydrogen converted to organic matter by plants are small compared with the mass of water transpired, and almost all the annual production is reoxidised to water by predators and decomposers. Small amounts of hydrogen are stored as humus, coal or oil, whose rates of formation are uncertain. Both dihydrogen and methane are given off by bacteria in anoxic environments, such as lake sediments and a few marine basins, estuaries and waterlogged soils (Nriagu, 1976). They may also be given off during the diagenetic conversion of peat to coal, or kerogen to

TABLE 5.4 Fluxes in the hydrogen cycle (see Fig. 5.4)

Flux	Nature	Flux/10^{12} kg H yr^{-1}	Refs and remarks
A, D	Rainfall on land	11 000	ch. 2
A, E	Rainfall on sea	39 000	ch. 2
B, A	Photooxidation of H_2 to water	0·052?	ch. 1
B, S	Loss of H_2 to outer space		
C, A	Photooxidation of CH_4 to water	0·28?	ch. 1
D, A	Evaporation from land	4000	
D, E	River discharge to sea	4100	ch. 2
D, F	Uptake of water by land plants	3000	
D, J	Water transfer to sediments	0·056	Garrels and Mackenzie (1971)
E, A	Evaporation from sea	42 500	ch. 2
E, H	Sea water turned to organic H	4·1	ch. 5
F, A	Transpiration from land plants	3000	
F, G	Organic H forming soil humus	4·7	ch. 5
G, B	H_2 formed by soil		
G, C	CH_4 formed by soil	0·32	ch. 5
G, K	Formation of coal	0·00047?	ch. 5
H, E	Water respired by sea plants	1·3	ch. 5
H, I	Organic H forming kerogen	2·2	ch. 5
I, B	H_2 formed by marine bacteria		
I, L	Formation of petroleum	0·00002?	see Hobson and Tiratsoo (1975)
J, D	Water set free by rock weathering	0·056?	Estimate, ass. 2% H_2O in rock*
J, M	Sediments forming igneous rocks	0·0023?	Garrels and Mackenzie (1971)
K, A	H_2O formed by burning coal	0·14	U.N. (1977)
L, A	H_2O formed by burning oil and gas	0·53	U.N. (1977), ass. 80% oil is burnt
M, A	Steam emitted by volcanoes	0·01?	Bowen (1966)
M, B	H_2 emitted by volcanoes	0·00011?	Bowen (1966)

* The amount of hydrogen in rocks can be calculated from the following data:

Mean for 546 granites: 0·84% H_2O (Daly, 1968)
Mean for 198 basalts: 1·62% H_2O (Daly, 1968)
Mean for sedimentary rocks: 3·4% H_2O (Garrels and Mackenzie, 1971)

petroleum. Methane is formed by bacteria living in the rumen of grazing mammals.

Dihydrogen is one of the few molecules which can escape from the Earth's gravitational field, and so be lost to space. Very little is so lost, as the gas is photo-oxidised to water in the atmosphere, where its residence time is about five years. Methane is also thought to be destroyed by photo-oxidation in the troposphere, forming water and carbon monoxide, and has a residence time in the range 2–5 years. It is surprising that so few organisms have been found to metabolise methane, in view of its likely occurrence in the primal atmosphere of the Earth, but a few bacteria are known which can oxidise hydrocarbons to water in the presence of oxygen.

The estimates of fluxes in the hydrogen cycle given in Table 5.4 are subject to the same uncertainties as those of carbon. Several of them have not yet been measured or even estimated.

The oxygen cycle

The oxygen cycle is the most complicated of all the organic cycles because of the large number of inorganic chemical species containing the element. In order of their abundances on the Earth's surface, these include inorganic oxygen (mostly silicates), water, carbonates, phosphates, sulphates, dioxygen, organic O, carbon dioxide and nitrates, with very small quantities of nitrous oxide, carbon monoxide, ozone and sulphur dioxide. Estimates of the masses of oxygen in these reservoirs are given in Table 5.5.

TABLE 5.5 Masses of oxygen in various reservoirs in units of 10^{12} kg O

Chemical species	Lithosphere	Hydrosphere	Atmosphere	Biosphere	Total
Inorganic O	$1{\cdot}08 \times 10^{10}$	7000	0·0007	0·4	$1{\cdot}08 \times 10^{10}$
$H_2O + OH^-$	$3{\cdot}0 \times 10^8$	$1{\cdot}26 \times 10^9$	11 000	4900	$1{\cdot}56 \times 10^9$
CO_3^{2-}	$2{\cdot}3 \times 10^8$	$1{\cdot}56 \times 10^5$			$2{\cdot}3 \times 10^8$
PO_4^{3-}	$3{\cdot}6 \times 10^7$	170		9	$3{\cdot}6 \times 10^7$
SO_4^{2-}	$1{\cdot}9 \times 10^7$	$2{\cdot}5 \times 10^6$	0·1		$2{\cdot}1 \times 10^7$
O_2		$4{\cdot}0 \times 10^3$	$1{\cdot}22 \times 10^6$		$1{\cdot}22 \times 10^6$
Organic O	9×10^4*?	740?		740?	9×10^4?
CO_2		890	1760		2650
NO_3^-	0·87	1900	?		1900
N_2O		0·25	0·73		1·1
CO		0·02–0·08	0·3		0·35
O_3			0·12		0·12
SO_2			0·05		0·05

* Uncertain figure based on mean oxygen content of coal

Despite the minute mass of the biosphere, it plays a major role in cycling the oxygen of the hydrosphere and atmosphere, both by the physical process of transpiration and by the photobiochemical oxidation of water to dioxygen. The dioxygen produced by photosynthesis almost exactly balances that consumed by respiration, but there must have been a small excess production to account for the composition of the atmosphere, assuming the primal atmosphere to have been anoxic.

Most organisms require dioxygen for respiration, but when this is absent, anoxic bacteria are able to use the oxygen atoms from nitrates to oxidise organic matter, giving dinitrogen as a by-product. When the nitrates are used up, organic matter decays by fermentation with the production of carbon dioxide and hydrogen. Some of the hydrogen may appear as ammonia, or it may be used by the bacterium *Desulphovibrio* to reduce sulphates to hydrogen sulphide. When the sulphates have been used up, *Methanobacterium* is able to obtain oxygen from carbon dioxide, forming methane as a by-product. Phosphates and 'inorganic oxygen' in the form of silicates are not regarded as a source of available oxygen for the biosphere, but the production of phosphine may occur in anoxic marshes.

Some of the fluxes in the oxygen cycle are given in Table 5.6, and others can be estimated from the fluxes given for the carbon and hydrogen cycles after multiplying by the appropriate factors. The first four entries in Table 5.6 concern the cycle of water, which is clearly in balance. The next five are concerned with the dioxygen budget. Although the inputs and outputs of the biosphere are not known with great precision, it seems that the current (1976) rates of burning fossil fuels have perturbed the dioxygen flux by 30%, but the effects of this perturbation are uncertain. Since the reservoir of atmospheric dioxygen is $1{\cdot}22 \times 10^{18}$ kg, it could be exhausted in a period of

TABLE 5.6 Some fluxes in the oxygen cycle

Nature of flux	Flux/10^{12} kg O yr^{-1}	Ref. and remarks
OH_2, rainfall on land	88 000	ch. 2
OH_2, river discharge to sea	33 000	
OH_2, evaporation from land	32 000	
OH_2, transpiration from land plants	23 000	
O_2 production by land plants	53	Whittaker, 1975
O_2 production by marine plants	14	
O_2 consumption by respiration	67	
O_2 consumption by burning fossil fuels	23	Calculated from U.N. (1977)
O_2, difference between production and respiration	0·005?	Singer (1970)
O_3 production in stratosphere	0·48	ch. 1
O_3 destruction in stratosphere	0·48?	

the order of 50 000 years; long before that time, all recoverable supplies of fossil fuels will have been exhausted. The last two entries in Table 5.6 illustrate our ignorance of the rates of formation and destruction of ozone in the stratosphere. There has been concern lest these rates be altered by the introduction of man-made contaminants into this part of the atmosphere.

The nitrogen cycle

While qualitative aspects of this organic cycle have been a feature of numerous textbooks for many years, its quantitative study has been held up by technical difficulties. Following the discovery that organisms which fix dinitrogen are also able to convert ethyne to ethene, it is now possible to

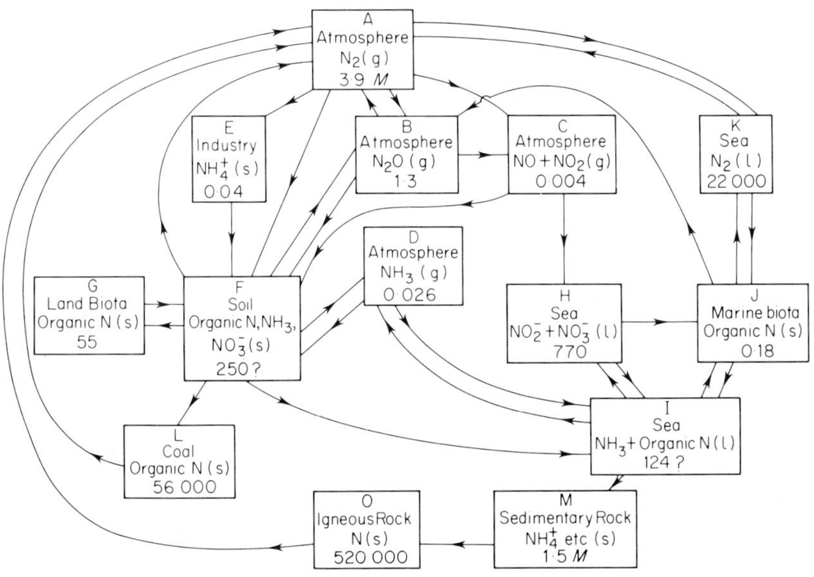

FIG. 5.5 Diagrammatic nitrogen cycle. Figures are masses in 10^{12} kg N. M, $\times 10^6$.

Masses were obtained as follows: A, B, C, D see chapter 1. E assumes that industrial stocks equal 1 year's production. F assumes a C/N ratio = 10 in soil; Burns and Hardy (1975) give 650×10^{12} kg N. G and J, see text. H and I, see Riley and Chester (1971); similar estimates are given by Baur and Wlotzka (1972). In H, marine nitrite-N is about 1% of marine nitrate-N. In I, soluble organic N is 100, ammonia-N 21 and particulate organic N 3 (all $\times 10^{12}$ kg) according to Williams (1975). K, see Kester (1975). L, M, O see chapter 3. M includes 680×10^{12} kg N in petroleum and $0{\cdot}1 \times 10^{12}$ kg nitrate-N (Baur and Wlotzka, 1972); marine sediments have an organic C/total N ratio between 1·5 and 3·3 (Müller, 1977). In O, it is uncertain how much nitrogen is N^{3-} and how much is trapped N_2; Burns and Hardy (1975) give a larger figure for this reservoir

measure the extent of nitrogen fixation in the field, but actual measurements are still few in number.

Figure 5.5 represents the nitrogen cycle, somewhat simplified by omitting a number of minor reservoirs and combining others. Note that there is considerable uncertainty in the mass of nitrogen in various important reservoirs such as soils, rocks and the sea. In comparison with all other

TABLE 5.7 Fluxes in the nitrogen cycle (see Fig. 5.5)

Flux	Nature	Flux/ 10^{12} kg N yr^{-1}	Refs and remarks
A, B	Atmospheric oxidation of N_2 to N_2O	0·015	Burns and Hardy (1975) (A, B and A, C)
A, C	Formation of NO by lightning	0·01	
A, E	Industrial production of NH_3	0·044	F.A.O. statistics (1975)
A, F	Fixation of N_2 by soil microbes	0·135	Burns and Hardy (1975)
A, K	Exchange of N_2 between air and sea		must = K, A
B, A	Photochemical reduction of N_2O	0·03	Burns and Hardy (1975) (B, A and B, C)
B, C	Photochemical oxidation of N_2O to NO	0·005	
B, F	Uptake of N_2O by soil		
C, F	Input of NO_3^- to soils in rain	0·013	ch. 2; Burns and Hardy (1975) (C, F to D, I)
C, H	Input of NO_3^- to sea in rain	0·047	
D, F	Uptake of NH_3 by soil	0·031	
D, I	Uptake of NH_3 by sea	0·109	
E, F	Agricultural use of fixed N	0·044	= A, E
F, A	Microbial denitrification of soil	0·19?	Burns and Hardy (1975) (F, A to F, D+I, D)
F, B+J, B	Microbial production of N_2O	0·02	
F, D+I, D	Microbial production of NH_3(g)	0·17	
F, G	Plant uptake of N from soil	3·5	From net productivity, ch. 5
F, I	River transport of N to sea	0·002	ch. 2; Burns and Hardy give 0·015
F, L	Conversion of soil humus to coal	0·00014?	ch. 5
G, F	Release of N to soil by land biota	3·5	= F, G
H, I	Conversion of NO_3^- to NH_3 in sea		
H, J	Uptake of NO_3^- by marine plants	1·7–(I, J)	ch. 5
I, H	Oxidation of NH_3 to NO_3^- in sea		
I, J	Uptake of NH_3 by marine plants	1·7–(H, J)	ch. 5
I, M	Net conversion of marine N to sediments	0·01?	Burns and Hardy (1975)
J, I	Release of NH_3 + organic N to sea by biota	1·7?	
J, K	Release of N_2 to sea by biota		
K, A			see (A, K)
K, J	Fixation of N_2 by sea biota	0·04	Burns and Hardy (1975)
L, A	Release of N_2 by burning coal	0·042	calculated from U.N. (1977)
M, O	Conversion of sediments to igneous rock	$<0·0006$	ch. 5
O, A	Volcanic N_2 released to atmosphere	0·005?	Burns and Hardy (1975); (Baur and Wlotzka, 1972, give $<0·0003$)

elements, the very large amount of nitrogen in the atmospheric reservoir is striking.

Stevenson (1965), Delwiche (1970), Robinson and Robbins (1970a), Burns and Hardy (1975) and Paul (1976) are among the many workers who have attempted to estimate fluxes in the nitrogen cycle, but many fluxes are still unknown or imprecise guesses. The numerical values given in Table 5.7 will doubtless be revised and improved by future work. It is still uncertain how much atmospheric dinitrogen arose from the degassing of the lithosphere, and how much from the activities of the biosphere, but it is believed that this reservoir has increased during geological time (Junge, 1972).

Dinitrogen fixation from the atmosphere is known to be carried out by the following organisms, many of which are symbiotic:

1. Blue-green algae such as *Nostoc*, in soils and in the sea, sometimes as symbionts.
2. Free-living bacteria, which may be oxygenic like *Azotobacter*, or anoxic like *Clostridium*, and a few actinomycetes like *Mycobacterium* and yeasts like *Rhodotorula*, which occur in soils and in the sea.
3. Symbiotic bacteria such as *Rhizobium* in the root nodules of leguminous plants, and probable actinomycetes in the root nodules of at least 14 genera of other plants, which appear to be the most important quantitatively.

A detailed description of all known nitrogen-fixing organisms is given by Hardy (1977).

The nitrogen in dead organic matter is converted into ammonia by bacteria and fungi. Although some of the ammonia may be fixed by clay minerals in the soil (chapter 4), most of it is rapidly oxidised, first to nitrite by *Nitrosomonas*, and then to nitrate by *Nitrobacter*. A small amount of soil nitrogen is tightly bound by soil humus, where it has a residence time in the range 30–75 years (Jansson, 1963). Nitrates can be converted to dinitrogen by a wide range of bacteria, mostly in anoxic conditions e.g. *Achromobacter*, *Bacillus*, *Micrococcus* and *Pseudomonas* spp. Nitrous oxide is a common by-product formed in this reaction.

The sulphur cycle

This is the last cycle we shall discuss in this chapter, and is represented in Fig. 5.6. The main reservoirs of sulphur are in the lithosphere, which cycles very slowly, and the ocean, where cycling is more rapid. Atmospheric sulphur compounds have residence times of the order of hours or days.

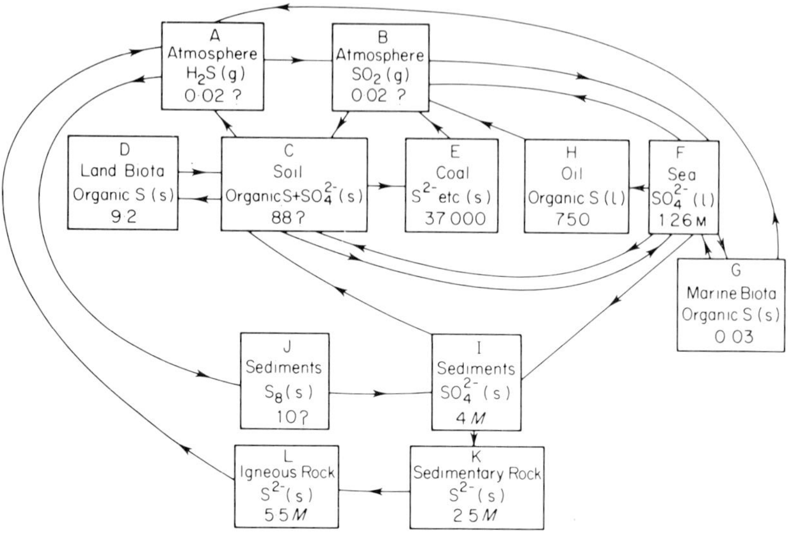

FIG. 5.6 Diagrammatic sulphur cycle. Figures are masses in 10^{12} kgS; M, $\times 10^6$. A and B are uncertain, based on 5 μg H_2S and 10 μg SO_2 per m^3 air; Friend (1973) gives lower values. C is equally uncertain, assuming C/S is 28·5 in soil. D and G assume C/S is 90 in biota. E, H, L, see chapter 3. F, see chapter 2. I and K, see Friend (1973). J, see Bowen (1977b). H includes an unknown mass of H_2S trapped with natural gas

General aspects of the cycle are considered in more detail by Eriksson (1960), Robinson and Robbins (1970b), Junge (1972) and Friend (1973).

The chemistry of sulphur in the atmosphere is more complex than is indicated in Fig. 5.6 and Table 5.8, which gives estimates of fluxes. In addition to H_2S and SO_2, particulate sulphates are usually present, which arise partly from sea spray and partly from the oxidation of SO_2 on the surface of particles of dust. Ammonium sulphate particles are often major constituents of aerosols near large cities. The atmospheric interconversions of sulphur are described by Junge (1972) and Spedding (1974). There is considerable uncertainty about the nature of the volatile sulphur compounds which must be emitted by the surface of the Earth in order to balance the rate of deposition of the element. Hydrogen sulphide is probably the main gas involved, though this awaits confirmation by detailed analyses. Other volatile compounds including CS_2, CH_3SH and $(CH_3)_2S$ have been identified in the atmosphere (Lovelock *et al.*, 1972; Banwart and Bremner, 1975). Very small amounts of sulphur dioxide are emitted by volcanoes, and the large scale release of this gas from burnt fossil fuels has markedly perturbed the atmospheric part of the sulphur cycle (chapter 12).

Sulphates in the soil solution or sea water are taken up by plants, and are then reduced to organic sulphur compounds such as the amino acid cysteine, an essential component of many proteins. Dead organic matter retains some sulphur tenaciously in humus, but most of the element is thought to be converted to hydrogen sulphide by decomposers. Hydrogen sulphide only accumulates in a few anoxic environments such as the deeper parts of the Black Sea and certain fiords and oil wells. In the presence of dioxygen it is rapidly oxidised to sulphates by bacteria of the genus *Thiobacillus*. Even in the absence of dioxygen, several families of bacteria, such as the Achromataceae, Beggiatoaceae, Chlorobacteriaceae and

TABLE 5.8 Fluxes in the sulphur cycle (see Fig. 5.6)

Flux	Nature	Flux/ 10^9 kg S yr^{-1}	Refs and remarks
A, B	Atmospheric oxidation of H_2S to SO_2	120	= (C, A)+(G, A)
A, J	Bacterial conversion of H_2S to S_8	0·00002	Estimate (Bowen, 1977b)
B, C	Transfer of $SO_2 + SO_4^{2-}$ from air to soil	116	Singer, 1970; Kellogg *et al.* (1972)
B, F	Transfer of $SO_2 + SO_4^{2-}$ from air to sea	96	Singer, 1970
C, A	H_2S formed by soil microbes	65	Estimate; Friend (1973), estimates 58
C, D	Uptake of soil SO_4^{2-} by land plants	5800	ch. 5
C, E	S fixed in coal	0·14?	ch. 3
C, F	River transport of S to sea	137	ch. 2; Friend (1973)
D, C	S returned to soil by land biota	5800	= C, D
E, B	SO_2 formed by burning coal	42	Calculated from U.N. (1977)
F, B	SO_4^{2-} in spray from sea to air	44	Eriksson (1960)
F, C	SO_4^{2-} spray transferred to land	4	Singer (1970)
F, G	Uptake of SO_4^{2-} by marine plants	280	ch. 5
F, H	S fixed as petroleum in sediments	0·0035	ch. 3
F, I	S fixed as sulphates in sediments	55/95	Garrels and Mackenzie (1971); Singer (1970)
G, A	Release of H_2S by sea biota	55	Estimate; Friend (1973), estimates 44
G, F	Release of organic S to sea by biota	23?	= (F, G – (G,)
H, B	SO_2 formed by burning oil	4·1	Calculated from U.N. (1977)
I, C	S released to soil by rock weathering	14/42	Singer (1970); Friend (1973)
	+ S applied as fertiliser to soil	11/26	Singer (1970); Friend (1973)
I, K	Reduction of SO_4^{2-} to S^{2-} in sediments	0–80	= (F, I)+(J, I)–(I, C)
J, I	Mining of S_8 to form sulphates	10	Bowen (1977b)
K, L	Conversion of sediments to igneous rock	7·8?	Garrels and Mackenzie (1971); estimate
L, A	Volcanic emission of S(g)	2?	Friend (1973); Berner (1971) gives 8 and Kellogg *et al.* (1972) give 1·5

Thiobacteriaceae, are able to oxidise hydrogen sulphide to elemental sulphur, and are thought to be responsible for the vast deposits of this element in Texas and perhaps elsewhere. The remarkable bacterium *Sulpholobus* prefers temperatures between 60 and 94 °C and is able to oxidise elemental sulphur under acid conditions (Brock *et al.*, 1972). On the other hand *Desulphovibrio* is a genus requiring anoxic conditions which is able to reduce sulphates to sulphides or hydrogen sulphide.

Cycling of other elements

Information on the cycling of elements in particular communities is scattered over the literature. Useful compilations and sources include the following: Tropical communities, see Golley *et al.* (1969) and Nye and Tinker (1977); Temperate communities, see Rodin and Bazilevich (1967) and Denaeyer-De Smet (1973); Agricultural communities, see Cooke (1975) and Frissel (1978).

In constructing elemental cycles, it must be borne in mind that organisms such as bacteria are able to catalyse redox reactions which would not be expected on thermodynamic grounds, or for purely geochemical processes. For example, the bacterium *Micrococcus lactilyticus* is able to oxidise dihydrogen to water; in the process, the following systems can be quantitatively reduced, and many more elements undergo partial reduction of their higher valency states (Woolfolk and Whiteley, 1962):

$$
\begin{aligned}
\mathrm{As(V)} &\rightarrow \mathrm{As(III)}\\
\mathrm{Bi(V)} &\rightarrow \mathrm{Bi(III)}\\
\mathrm{Fe(III)} &\rightarrow \mathrm{Fe(II)}\\
\mathrm{Mn(IV)} &\rightarrow \mathrm{Mn(II)}\\
\mathrm{Mo(VI)} &\rightarrow \mathrm{Mo(V)}\\
\mathrm{Pb(IV)} &\rightarrow \mathrm{Pb(II)}\\
\mathrm{Se(IV)} &\rightarrow \mathrm{Se(O)} \rightarrow \mathrm{SeH_2}\\
\mathrm{Te(VI)} &\rightarrow \mathrm{Te(O)}\\
\mathrm{Te(IV)} &\rightarrow \mathrm{Te(O)}\\
\mathrm{U(VI)} &\rightarrow \mathrm{U(IV)}\\
\mathrm{V(V)} &\rightarrow \mathrm{V(IV)}
\end{aligned}
$$

Summers and Sugarman (1974) have found strains of *Escherichia coli* which can reduce Ag(I), Au(III) and Hg(II) to the colloidal metals.

6 *The Elemental Composition of Living Matter*

In a short book of this kind it is impossible to list all the data available on this immense topic, but an attempt will be made to summarise 'best values' and ranges of composition for various groups of organisms and tissues. Before doing this, it is worth noting the practical problems in collecting and cleaning biological material, analytical problems and sources of error, and the methods of expressing the results.

Collecting and cleaning biological material

Most samples collected under natural conditions are more or less contaminated with the medium in which they live. Thus land plants retain particles of soil, marine organisms are contaminated with salt and the stomach contents of animals have very different composition from the living tissues. Cleaning is best carried out by dissection and/or washing. Dissection can remove the most obvious contamination, while washing is best carried out in the growth medium, at least for aquatic organisms. For land organisms washing usually involves some leaching of ions from the tissues, and in some tissues, such as erythrocytes, washing with distilled water can rupture the cell membrane and cause major changes in composition. Useful tests for soil contamination include abnormal ratios of ash weight to dry weight, and a high content of titanium (together with aluminium, scandium and zirconium). These elements are much more abundant in soils than in organisms; for example the ratio $[Ti]_{soil}/[Ti]_{plant}$ is usually of the order of a thousand.

Commensal organisms are often difficult to remove by cleaning, e.g. siliceous diatoms coating water plants, or small invertebrates on marine algae. It is impossible to separate some symbiotic organisms such as natural corals or lichens by dissection, or by any other simple technique.

Cutting instruments, and especially grinding devices, inevitably cause contamination, and methods for reducing this have been reviewed (Jones, 1972). As far as possible the tissues should only contact pure silica or hard, unpigmented plastic. Freeze-drying, or drying in a stream of warm, filtered air do not usually lead to serious contamination, but the use of preservatives, as in museum specimens, should be avoided. Stainless steel is a frequent cause of contamination.

Analytical errors

All analytical techniques are subject to errors, and extensive tests with a reference material (kale) have shown that about 30 % of recent analytical determinations of elements in this material are in error by more than $\pm 10\,\%$ (Bowen, 1975a). Earlier analyses are probably somewhat less reliable.

Losses of volatile elements such as Br, Cl, Hg, I and Se can occur during wet ashing, unless the solution is refluxed, and are even more likely during dry ashing. Thus mercury is quantitatively lost above 450 °C. Contamination is to be expected whenever biological materials or their ashes are dissolved in acids, even those of the highest purity (Kuehner *et al.*, 1972). In techniques involving the volatilisation of elements from graphite, contamination of the graphite can be a problem. More often the fact that different elements volatilise at very different temperatures and rates leads to problems in interpreting the spectra obtained.

Biological material frequently contains much lower concentrations of elements than does geological material, so that the amount present may be below the limit of detection or the reagent blank. The occurrence of spurious signals or 'matrix effects' may account for a number of cases where several parts per million have been reported to occur in tissues which are now known to contain far less of a particular element. Such spurious results have been noted especially for the following rare components of biological matter: Ag, Au, Be, Bi, Cd, Co, Ga, Nb, Se, Te, V and Zr.

Systematic errors, which are not surprising in view of the technical difficulties of the analyses involved, can be detected by analysing known reference materials, such as kale powder, or standard reference materials, such as N.B.S. orchard leaves or bovine liver. Data on the elemental composition of these materials is steadily accumulating, and there is no question of their value in revealing unsuspected sources of error. Improvements in limits of detection have been a notable feature of analytical chemistry over the last two decades, but they have not always been accompanied by improvements in accuracy, which might be obtained by regular use of reference materials in testing techniques.

Expression of the results

Different workers use different ways to express their results, but failure to give water or ash contents may mean that intercomparison with other results is not possible. This is particularly true of those who choose to quote results in units such as fat-free dry weight or sulphated ash. In this work concentrations are mostly expressed as mg/kg DM, and where possible other data has been recalculated from the literature into these units. Where no conversion factors are given, the mass of dry matter (DM) has been assumed to be $0{\cdot}25 \times$ mass of fresh matter (FM), which is a good approximation for soft tissues and organisms, but not for hard tissues such as bone. Useful compilations of DM/FM ratios are given by McCance and Widdowson (1960) for foods, and by Hamilton *et al.* (1973) and Snyder (1975) for human tissues. Ash contents are much more variable, and mean ash contents as a fraction of DM are given below, with the number of samples used to compute the mean in brackets.

Bacteria	0·0697 (21; Spector, 1956a)	Red algae	0·203 (25: Vinogradov, 1953)
Brown algae	0·288 (16: Vinogradov, 1953)	Seed plants	0·064 (7: Baumeister, 1958)
Bryophytes	0·0328 (8: Baumeister, 1958)	Annelida	0·024 (46: Vinogradov, 1953)
Equisetum sp.	0·20 (Cannon *et al.*, 1968)	Coelenterata	0·488 (16: Vinogradov, 1953)
Ferns	0·0771 (6: Baumeister, 1958)	Crustacea	0·217 (25: Vinogradov, 1953)
Fungi	0·0632 (11: Baumeister, 1958)	Echinodermata	0·608 (9: Vinogradov, 1953)
Kale	0·177 (20: Bowen, 1975a)	Mollusca	0·11 (70: Vinogradov, 1953)
Green algae	0·267 (5: Vinogradov, 1953)	Pisces (fish)	0·052 (630: Vinogradov, 1953)
Gymnosperms	0·024 (3: Baumeister, 1958)	Zooplankton	0·405 (Nicholls *et al.*, 1959)
Lichens	0·0156 (34: Solberg, 1967)		
Marine plankton	0·585 (8: Vinogradov, 1953)		

Ash contents of human tissues are given by Hamilton *et al.* (1973); see also Table 6.6

Although it is commonly assumed that data are normally distributed about the mean, this is probably more often than not a false assumption, but the degree of skewness is usually small. For most extensive series of measurements of elemental contents of organisms, the lognormal distribution fits the data rather better than does the normal distribution, so that median values may be better measures of the central tendency of the data than are mean values. In the great majority of published reports, too little data has been collected to test how the values are distributed, nor have the necessary statistical tests been applied. In this book, literature values have been summarised either as means or as ranges, and aberrant values have been rejected on subjective grounds. Uncertain values are distinguished by a (?), either because of analytical problems or sometimes because all the data come from a single report, and therefore need confirmation.

General remarks

In the tables which follow, some data in the literature have been deliberately rejected if they were collected for (*a*) diseased or deficient organisms; (*b*) contaminated organisms, or those growing in abnormal media (e.g. Ernst, 1974); (*c*) *accumulator organisms*, which concentrate particular elements in a striking manner, when compared with other species in the same environment. However, where a whole Phylum or other group accumulate an element, as appears to be the case for brown algae and iodine, the tables should reflect this fact.

Bacteria and non-vascular plants

The elemental composition of these groups is summarised in Table 6.1. Note that 'marine phytoplankton' are a heterogeneous group containing organisms from several Phyla, and that green algae are all marine species.

BACTERIA

Data are very sparse and mostly due to Porter (1946) or Schelhorn *et al.* (1976). If these are representative, bacteria appear to be rich in K, N and P but to have low contents of B and Mn compared with lower plants. Accumulator species are known for Fe, Mn, S and Si.

MARINE PHYTOPLANKTON

Samples of very different origin have been analysed by Riley and Roth (1971) and Martin and Knauer (1973); earlier analyses are reviewed by Vinogradov (1953). Compared to multicellular marine algae, plankton appear to be rich in Al, Ba, N, P, Ra, Si, Sn and Ti but poor in K. Many diatoms have a silica test which accounts for the large and variable Si content. Accumulator species have not been studied but several species accumulate iodine, while *Chaetoceros curvisetus* and *Rhizosolenia calcar-avis* are rich in barium (Wedepohl, 1969–1974). Moore and Boström (1978) found that the composition of plankton is more constant if normalised to $[Al]+[Fe]+[Zn] = 1$.

MARINE GREEN, RED AND BROWN ALGAE

Data for these very distinct Phyla has been collected from reports by Black and Mitchell, 1952; Vinogradov, 1953; Young and Langille, 1958; Fukai and Meinke, 1962; Lunde, 1970; Preston *et al.*, 1972; Bryan and Hummerstone, 1973; Fuge and James, 1974; Horowitz *et al.*, 1974; and others. Red and brown algae mostly accumulate bromine and iodine: brown algae are unique in accumulating strontium, so that the Sr/Ca ratio in their tissues is about 0·14, as against 0·02 in sea water. Some of the older work suggests that these algae accumulate arsenic.

LICHENS

Most of the data comes from surveys by Lounamaa (1956), who gives no conversion factors from ash weight to DM, LeRoy and Koksoy (1962), Solberg (1967) and Tuominen and Jaakola (1973). Other work has been reviewed by James (1973), and data for lichens growing in metal-rich habitats by Ernst (1974). Lichens appear to be rich in Pb and poor in Ga and B; Lounamaa's suggestion that they are rich in Cd and Sn has not been confirmed. Accumulator species that have been reported are *Acarospora smaragdula* for Fe, *Stereocaulon pileatum* for Pb, and *S. nanodes* and *Diploschistes scruposus* for Zn.

FUNGI

The data for this vast group are notably scanty, but suggest that the fruiting bodies of Basidiomycetes are rich in Cd, Cl, Cu, Hg, P and Se but poor in Ca and Mn (Sawada, 1965; Horowitz *et al.*, 1974; Hinneri, 1975; Byrne *et al.*, 1976). *Amanita* spp. accumulate Br, *Agaricus* spp. and *Hygrophorus puniceus* may accumulate Cd, *Boletus edulis* accumulates Se, *Amanita muscaria* accumulates V, and *Otidia onotica* accumulates Zn. Species of Tremellales are rich in Mn (Schmitt *et al.*, 1977).

BRYOPHYTES

While some mosses can be collected relatively clean, it is particularly difficult to separate soil particles from many leafy liverworts; this is reflected in abnormally high ash weights. The largest collection of data for bryophytes is based on specimens with a mean ash content of 15% DM, which suggests some contamination by soil (Shacklette, 1965). The relatively high contents of Al, Be, Fe, Si, Ti and Zr given by this author may plausibly be ascribed to such contamination. Mosses, like lichens, appear to be rich in Pb and poor in B, but the lead may come from atmospheric pollution. Accumulator species include *Mielichhoferia* spp. for Cu and perhaps also for V, and *Merceya* spp. for Cu (Persson, 1948; Shacklette, 1967).

TABLE 6.1 Elemental composition of cryptogamous plants in mg X/kg DM

X	Bacteria	Marine phytoplankton	Marine green algae	Marine red algae
Ag		0·4–3·3 Mk, Rr		0·11–0·28 H, Pr
Al	210 P	38–440 Mk		
As	0·1 Sa	12–36 Zd	1·2–6 Fm, Y	6 Y
Au			0·028 Fm	
B	5·5 P	33 Yy	160 V	100 V
Ba		20–290 Mk, Rr	1·5 Bo	2·8–50 Bo, Pu
Be				
Br			270 Gm	650–3300 Gm, Nf
C	540 k	225 k	338 k	283 k
Ca	5·1 k P	6·1 k Mk	6 k–24k Bo, E, Y, Ys	4–10 k Bo, E, Ys
Cd	0·31 Sa	0·4–2·3 Mk, Va	2·5 By	0·3 Pr
Ce				
Cl	2300 P			
Co	7·9 Sa	0·38 Vf	0·15–3·7 Gr, Sk, Vf, Y	0·16–4 H, Sk, Vf, Y
Cr	4 Sa	2·2–7·5 Mk, Rr, Va, Vf	1·8–12 Gr, Sk, Vf	2·1–13 Gr, H, Sk, Vf
Cs		0·11 Vf	0·044 Vf	0·05–0·4 H, Sm, Vf
Cu	150 Sa	8·5–36 Mk, Rr	4–58 By, Gr, Sk, Y	3–24 Gr, Pr, Sk, Y
Dy				
Er				
Eu				
F			6·9 Y	11 Y
Fe	170 Sa	220–1500 Mk, Vf	160–750 Gr, I, Sk, Vf	200–1800 Gr, H, I, Pr, Sk, Vf
Ga		1·5 Va		
Gd				
Ge				
H	74 k	46 k	52 k	72 k
Hf				
Hg		0·2 Kn, Mk		
Ho				
I		270 V	90 Y	360 Nf, Sl, Y
K	115 k P	13 k Mk	32 k Y	50 k Y
La				
Li				
Lu				
Mg	7 k P	14 k Mk	20 k Ys	6·4 k Ys
Mn	260 Sa	4–120 Mk, Rr, Va, Vf	16–300 Gr, Sk, Vf	16–500 Gr, Pr, Sk, Vf
Mo	9·8 Sa	1 Va	0·6–1·2 Sk, Y, Ya	0·33–2·1 Fm, Sk, Y, Ya
N	96 k	63–110 k Cd, Mk, V	18 k V	25 k V
Na	4·6 k P	6 k Va	41 k Y	26 k Y
Nb				

Table 6.1—*continued*

Marine brown algae	Lichens	Fungi	Bryophytes
0·18–0·28 Bu, H, Pr	0·11–4 H, La, Lr	0·18–7 H, Zi	0·01–0·88 H, Zk
62 V	20–6000 Lw, So, Tj	30–350 Bk, Lw	500–6000 Lw, Rt, Sh
8–30 L, Y		1·2–2·5 Bk, K	0·2–7 G
			0·0022 G
120 V	4·1 Sg	16 Kr	10–27 Rt, Sh
11–31 Bo, Pu	8–40 Tj		34–70 Rt, Sh
0·0028 Ms	0·04 Tj		0·9 ? Sh
220–900 Gm, Nf		2–36 Bk, Nf	
345 k	494 k		
9–23 k Bo, E, Ys	0·26–18 k Sg, So, Tj	0·2–1·7 k Hi, Ml, P, Sw	1·5–100 k Rt
0·4–1·3 Pr, Sn	0·1–0·8 So, Sq	0·1–75 Bk, Sn, Sv, Zg	0·1–1·2 Sq, Zk
	0·6–5·6 Tj, Yl	<0·1 Z	2·6 Yl
4700 V		10 000 Mc	1200 Sh
0·02–9 Fj, Gr, H, L, Sk, Vf	0·06–2 H, Tj	0·3–1 H, Ml	1·3–4·8 H, Sh
0·52–6 Gr, H, Sk, Vf	0·6–7·3 H, Lr, So, Tj	2·6H	6–12 H, Sh
0·04–0·13 H, Sm, Vf	0·007 Ya	0·7 H	0·8 H
2–68 Bh, Fj, Gr, L, Pr, Sk	9–24 Lr, Lw, So, Tj	7–160 Bk, Hi, Sw, Zg	6–40 Lw, Rt, Sh, Zk
	0·04–0·36 Tj, Yl		0·31 Yl
	0·01–0·19 Tj, Yl		0·15 Yl
	0·02–0·17 Tj, Yl		0·05 Yl
4·5			
13–1900 Bh, Fj, Gr, H, I, L, Pr, Sk, Vf	40–5200 H, Lw, So, Tj	90–2 k H, Hi, K, Lw, Sw, W	370–1400 H, Lw, Rt, Sh
0·5	2·2 La		2·7 Sh
	0·06–0·56 Tj, Yl	0·3 Z	0·31 Yl
		2·8 Sd	
41 k		55 k	
	0·02–0·4 Ey, Tj	<0·1 Z	0·16 Ey
0·03 Sx	0·34 Q, So	0·7–70 Bk, Q, Sv	
	0·004–0·07 Tj, Yl		0·05 Yl
430–2500 Nf, Sl, V	1·8 Sl	3·1–6·7 Bk, Sl	3·3–5 Sh, Sl
52 k	0·6–8·5 k Sg, Tj, Ya	22 k Lo	6·7 k Rt, Sh
10	0·4–3 Tj, Yl		1·3 Yl
5.4	0·02–0·3 Tj		
	0·001–0·02 Tj, Yl		0·05 Yl
9·4 k Ys	0·17–1 k Sg, So, Tj	1 k Hi, Sw	3 k Rt, Sh
1–400 Bh, Fj, Gr, Pr, Vf	3–190 Lw, Sg, So, Tj	10–400 Bk, Hi, K, Lw, Sw, Zh	50–360 Lw, Rt, Sh
0·2–0·75 Fj, L, Sk, Ya	0·03–3 Ey, Lr, Lw, Tj	1·4 Lw, Kr	0·33–2·6 Ey, Rt
18 k V	8·5–23 k Mi, Sg	34 k Ml	9–15 k Rt
33 k	40–1 k Tj	1·1 k Mc	0·8 k Rt
	0·02–0·4 Ey, Tj		0·45 Ey, Ty

Table 6.1—*continued*

X	Bacteria	Marine phytoplankton	Marine green algae	Marine red algae
Nd				
Ni		3·7 Mk	1–2 Sk, Y	1·2–2 Pr, Sk, Y
O	230 k	440 k	470 k	490 k
P	30 k P	4 k–18 k Cd, V, Va	4 k V	3 k V
Pb		5 k–8 M, Va	40 By	2·9 Pr
Pr				
Ra		0·4 μ–2 μ Cy, K	0·05 μ E	0·035 μ E
Rb			32 V	8 H, Sm, V
Re			0·014 Fm	
S	5·3 k P	3 k–6 k Gk, V	9·8 k Gk	14 k–30 k Gk, V
Sb			0·64 Vf	0·02–0·9 L, Vf
Sc				0·45 H
Se	0·016 Sa	3·5 Zd		
Si	180 P	6–210 k V		
Sm				
Sn		3·5 St		
Sr		70–700 Mk, Rr, Va	60–220 Bo, Ys	95–450 Bo, Ys
Tb				
Th		0·42 Cy	0·12–0·25 E, Sp	0·2–1·5 E, Sp
Ti		27–940 Mk, Rr, Va	10–41 Gr, Sk, Yf	5–37 Gr, Sk, Yf
Tm				
U		0·7 Cy	0·9 E	0·4 E
V		3–5 Rr, Va	1–11 Fm, Gr, Sk, Yf	3–16 Fm, Gr, Sk, Yf
W			0·035 Fm	
Y				
Yb				
Zn	83 Sa	20–280 Mk, Rr, Va, Vf	20–140 By, I, Sk, Vf, Y	21–150 H, I, Pr, Sk, Vf, Y
Zr	20 Va			

k, ×1000; μ, $\times 10^{-6}$; *References*, see p. 101; where no reference is given, see Bowen (1966)

Vascular plants

Plants in this group absorb most of their nutrients from the soil, but can also take up ions through their leaves. Some data on elemental composition is set out in Table 6.2.

HORSETAILS AND CLUBMOSSES

The elemental composition of horsetails (*Equisetum* spp.) has been reviewed by Cannon *et al.* (1968). They are all silicon accumulators and are also rich

Table 6.1—*continued*

Marine brown algae	Lichens	Fungi	Bryophytes
	0·24–3 Tj, Yl		1·7 Yl
0·4–5·3 Sk, Pr	1·5–3·8 La, Lr, Tj	1·8 K	12 ? Sh
470 k		340 k	
2·8 k V	740 Sg	5 k Kr, Ml	1·4–2 k Rt, Sh
2–38 Bh, Bu, Pr	1–78 Lr, So, Tj, Yl	0·2–40 K, Kr, Sv, W	8–25 Yl, Zk (−280 Sh)
	0·08–0·62 Tj, Yl		0·31 Yl
0·1 μ E	0·013 μ		
7–47 H, Sm	2·4–25 H, Tj	50 H	21H
8·3 k Gk	0·7 k–1·5 k Sg, So	4 k Mc	1 k–1·9 k Rt, Sh
0·05–0·5 L, Vf		0·05 Z	0·05–0·9 Sy
0·07 H	0·25 H	<0·002–0·3 H, Z	0·7 H
0·04–0·1 L, Sa		0·8–8 Bk, Sv	
1500			3500 Rt, Sh
	0·06–0·8 Tj, Yl		0·31 Yl
0·29 St	0·16–1·4 Lr, Tj	7–31 ? W	35 ? Rt
800–1600 Bo, Ys	0·8–250 Lr, Tj		68 Sh
	0·006–0·07 Tj, Yl		0·05 Yl
0·06–0·17 E, Sp	0·1–0·23 Ey, H, Yl	0·43 H	0·3–1·7 Ey, H, Yl
3·5–34 Gr, Sk, Yf		<8 Z	300 ? Sh
	0·001–0·026 Tj, Yl		0·04 Yl
0·8 E	0·1–0·25 Ey, Lr, Yl	<0·03 Z	0·15–0·45 Ey, Yl
1·2–10 Gr, Sk, Yf	0·3–4·8 Lr, Tj	0·4 Bk	11 Sh
	0·02–0·8 Ey, Tj		0·13 Ey
	0·2–2 Tj, Yl		1·3–7·5 Sh, Yl
	0·01–0·2 Tj, Yl		0·15–0·9 Sh, Yl
6–260 Bk, Fj, H, I, L, Pr, Sk, Vf	20–60 H, Lr, Lw, Tj	14–340 Bk, H, Hi, K, Kr, Lw, So, Sw, W, Zh	12–250 H, Lw, Rt, Sh, Zk
	10 La		19 ? Sh

in Ti. However they do not accumulate gold, as was suggested by early work.

Clubmosses are mostly aluminium accumulators, apart from the epiphytic species, and therefore grow on very acid soil (Sorenson *et al.*, 1974).

FERNS

Data is mostly taken from Lounamaa (1956) and earlier workers, reviewed by Bowen (1966). Their composition is close to that of seed plants, but they

TABLE 6.2 Elemental composition of vascular plants in mg X/kg DM

X	Horsetails (*Equisetum*)	Ferns	Woody gymnosperms	Woody angiosperms	Herbaceous vegetables	Kale (ref. B)
Ag	0·05 H	0·81 H	0·02–2·3 H, Sf	0·01–0·7 Ch, H, M	0·02–0·6 Ch	<0·01
Al		230 Lw	250–530 C, Hg	90–200 C, Hg, Lw	0·1–200 C, S, Sr	38
As	0·2 Cs	1·3 G, Sb	0·2–1·2 G, Sb, Za	2 G	0·01–1·5 Ch, Ha, Hm	0·12
Au	0·04 Cs	0·0017 G	0·001–0·003 G, Hj, Sf	0·0014 G, Hj	<0·00001–0·0016 O	0·0022
B	110 Cs	77 La	11–52 C, Hg	12–140 C, Hg	8–200 C, Ch, Jo	49
Ba	86 Cs	8 Bd	110–150 C, Hj	140 C, Hj	1–150 C, Ch, O	4·55
Be				0·037 Ms	0·0006 Ms	
Bi			<0·02 Hg	<0·02 Hg	0·06 Sh	
Br			7 Zb	15 Nf	<0·05–26 Ha, Ss	25·6
C		450 000	450 000	450 000	450 000	428 000
Ca	10 k Cs	3·7 k	4 k–13 k C, Hg, Hw	3 k–14 k C, Go, Hg, Hw	400–50 k C, Ch, S	40 800
Cd		0·13 Sn	0·1–0·9 C, H, Hg, Sq	0·02–2·4 C, Hg, Sn, Sq	0·05–0·9 C, Sq	0·85
Ce	2·9 Er	1–16 Eh	0·36 Yl	0·25 Yl	0·01–0·05 O	0·30
Cl	11 k Cs	6 k	550 Zb	700–27 k Ch	200–46 k Ch, S	3420
Co	0·1 H	0·66 H	0·02–0·7 C, H, Hg, Hj, Sq	0·005–1 C, Ch, H, Hg	0·01–4·6 C, Ch, S, T	0·585
Cr	0·45 H	1·9 H	0·1–0·5 C, H, Hg, Hj, Sf	0·03–10 C, Ch, H, Hg, Hj	0·016–14 C, Ch, Ha, T	0·33
Cs	0·03–0·1 H, Ya	0·1–0·44 H, Ya	0·03–0·1 H, Ya	0·067 H, Y	<0·001–0·08 Ha, O	0·074
Cu	12 Cs	15 Lw	5–13 C, Hg, Lw	6–14 C, Go, Hg, Lw	4–20 C, Ch, Jo	4·9
Dy	0·43 Er	0·05–0·6 Eh	0·11 Yl	0·14 Yl		
Er	0·38 Er		0·08 Yl	0·1 Yl		
Eu	0·13 Er		0·028 Yl	0·034 Yl	<0·005–0·07 O, Za	0·005
F			0·02–4 C, Zb	0·04–24 C, Ch	3–19 Ab, Ch	5·5
Fe	90–240 Cs, H	320–700 H, Lw	80–230 C, H, Hg, Lw	70–180 C, Go, H, Hg, Lw	2–250 C, Ch, Ha, Jo	120
Ga		0·23 La	0·03 La	0·01–0·1 C, Hg	0·03–1·8 Ch	0·046
Gd	0·48 Er	0·06–0·5 Eh	0·14 Yl	0·002–0·17 Yl, Ze		
Ge		1·3 Sd	1 Sd		2·4 Sd	
H		55 k	55 k	55 k	55 k	56 800
Hg			0·005–0·15 Sf, Za		0·013–0·17 Ha	0·166
Ho	0·11 Er		0·028 Yl	0·034 Yl		
I	2·7 Sl	5·4 Sl	0·15–4 C, Sl, Zb	2·7–4 C, Sl	0·07–10 C, S, Sl	0·126
In					0·0008 Za	0·00062
K	34 k Cs	18 k	5 k–20 k C, Hg, Hw	4 k–15 k C, Go, Hg, Hw	1 k–68 k C, Ch, Ha	24 700
La	1·7 Er	1·5–15 Eh	0·003–0·22 Sf, Yl	0·25 Yl	0·05–2 O, Za	0·0875
Li				0·5–3·4 C, Ch	0·8–1·3 Ch	1·6
Lu				0·03 Yl	0·0006–0·06 Za	0·02

X	Horsetails (*Equisetum*)	Ferns	Woody gymnosperms	Woody angiosperms	Herbaceous vegetables	Kale (ref. B)
Mg	3 k Cs	1·8 k	1·9 k–5 k C, H, Hg, W	1·3 k–9 k C, Go, Ha, Hw	700–5·6 k C, Ch	1590
Mn	50 Cs	700 Lw	300–650 C, Hg	17–600 C, Go, Hg, Lw	0·3–1000 C, Ch, Ha	14·9
Mo	<1 Cs	0·8–2·5 La, Lw	0·2–2·8 C, Hg, Lw	0·06–3 C, Ch, Hg, Lw	0·03–5 C, Ch, O	2·29
N		20 k	32 k	12 k–38 k Ch	30 k–75 k Ch	43 000
Na	1500 Cs	1400	60–110 C, Hg, Hw	34–950 C, Go, Hg, Hw	2–1400 C, Ha, S	2490
Nb			0·28 Yl	0·4 Ty		
Nd		0·3–7 Eh				
Ni	1·5 Cs, Er	1·5 La	2·1–5·3 C, Hg	1·4 C, Hg	0·02–4 C, Ch, T	0·95
O		430 k	440 k	410 k	410 k	397 000
P	1·7 k Cs	2 k	1 k–3 k C, Hg	120–2 k C, Go, Hg	130–10 k C, Ch, Ha	4490
Pb	2·4 Cs	2·3 La	0·9–13 C, Hg, La	1–8 C, Hg, M	0·2–20 Ch, Za	2·54
Pd						0·026
Pr	0·33 Er		0·056 Yl	0·075 Yl		
Ra				0·19 μ Vm	0·03–1·6 μ Si	
Rb	25–50 H, Ya	17–50 H, Ya	2–12 H, Ya, Za	3·4–10 C, H, Y	0·5–20 Ha, O	52
Ru						0·0045
S	2500 Cs	1000	1100	2000–8700 Ch	600–6000 Ch	16 000
Sb			0·002–0·034 Hj, Sf	0·024 Hj	0·0001–0·2 O, Za	0·069
Sc	0·01 H	0·13 H	0·02–0·035 H, Hj	0·012–0·021 H, Hj	0·00001–0·2 O, Za	0·0082
Se			0·026 C	0·029 C	0·001–0·5 C, Ha, O, Za	0·13
Si	62 000 Cs	5500		420 C	200–8000 S, Za, Zf	250
Sm	0·35 Er	0·1–0·8 Eh	0·14 Yl	0·15 Yl	0·01–0·1 Za	0·015
Sn		2·3 La	1 Hg	0·2 Hg, La, M	0·2–6·8 Bb, Ch	0·23
Sr	60 Cs	13 Bd	7–83 C, Hg	12–88 C, Bd, Go, Hg	3–400 C, Ch	96·5
Tb	0·12 Er		0·02 Yl	0·034 Yl	0·001 Za	
Th		0·42 H	0·033 H	0·14–1·3 Vm, Yl	<0·008–0·2 O, Za	0·0092
Ti	56 Cs	5·3	5–23 C, Hg	1·5–6·4 C, Hg	<0·02–2·6 C, O	0·3–3·3 ?
Tl			0·05 Za		0·03–0·3 Za	0·15
Tm	0·07 Er			0·004–0·027 Yl, Ze	0·004 Za	
U	0·03 Wh		0·015 Yl	0·005–0·038 Ca, Ko, Vm	0·01–0·06 Za	0·0108
V	<3	0·13	0·5–2 ? C, Hg	0·44–0·69 ? C, Hg	0·001–0·5 Bz	0·35
W					0·0005–0·15 Ba	0·058
Y		0·77 La	0·22 Yl	0·15 Yl	<0·35 C	
Yb	0·03 Er	0·04–0·6 Eh	0·065 C, Yl	0·1 Yl	0·02 Za	
Zn	120 H	60–400 H, Lw	20–90 C, H, Hg, Lw	34–68 C, Go, H, Hg, Lw	1·4–160 C, Ch, Ha, Jo	32·7
Zr	<10 Cs	2·3 La	0·3–1·6 C, La	0·48 La	<1·4 C, O	

k, $\times 1000$; μ, $\times 10^{-6}$; *References*, see p. 101; where no reference is given, see Bowen (1966), or Bowen (1974) for kale. Oakes *et al.* (1977) give lower values for Cd, Pb and V, with upper limits for Hf, Ta and Te, in herbaceous vegetables

are poor in Ca and S. No accumulator species have been confidently reported, but some species of *Asplenium* and *Notholaena* are indicator plants for rare minerals, and *Asplenium septentrionale* concentrates yttrium to some extent.

GYMNOSPERMS

Relatively few woody species have been analysed, but data is given by Lounamaa, 1956; Hanna and Grant, 1962; Horowitz *et al.*, 1974; Connor and Shacklette, 1975; and others. The composition is close to that of woody seed plants, but they may be richer in Al and poorer in Cu, Na, S and Zn. No accumulator species are confirmed, though *Pinus sibirica* (a variety of *P. cembra*) is said to accumulate tungsten (Peterson, 1971).

SEED PLANTS

Woody seed plants make much the largest contribution to the total biomass of any of the plant groups. Their composition has been sampled unsystematically, and tropical forest species have scarcely been examined. Much of the temperate data is for fruit trees grown in orchards. Wild forest species have been analysed by Lounamaa, 1956; Fleming, 1963; Golley *et al.*, 1969; and Connor and Shacklette, 1975. Data for wild species grown in botanic gardens are given by Hanna and Grant, 1962, and Horowitz *et al.*, 1974, and earlier work is summarised by Bowen, 1966. An interesting comparison of the amounts of Ca, K, Mg and Na in woody vegetation across a 0–3000 m altitudinal transect has been made by Hanawalt and Whittaker (1977).

Characteristics of the elemental composition of woody plants can be summarised very briefly. About 96% of the dry matter consists of the elements with organic cycles, C, H, N, O and S. Ca, Cl, K, Mg, P and sometimes Si each make up 0·1–2% DM. Roughly half of the elements, including Ag, Au, Be, Bi, Cd, Co, Cs, Ga, Ge, Hf, Hg, In, Ir, La–Lu, Mo, Nb, Os, Pd, Pt, Re, Rh, Ru, Sb, Sc, Se, Ta, Te, Tl, U, V, W, Y and Zr are normally present at concentrations of less than 1 mg/kg DM, unless accumulator species or metal-rich soils are involved.

Data for freshwater aquatic plants can be obtained from papers by Riemer and Toth (1968) and Cowgill (1973). The latter author gives concentrations for many rare elements such as Bi, Br, Ce, Cs, Dy, Er, F, Gd, La, Nb, Nd, Pr, Sc, Si, Sm, Ti, V, Y, Yb and Zr which seem abnormally high, possibly as a result of contamination by siliceous diatoms.

Cultivated vegetables and fruits have been analysed more than any other type of plant material, partly for the purpose of diagnosing agricultural problems and partly to obtain nutritional information. Extensive data are given by Beeson, 1941; Goodall and Gregory, 1947; McCance and Widdowson, 1960; Long, 1961; Haller *et al.*, 1968; and Connor and Shacklette, 1975, while useful critical reviews include those of Chapman (1966), Scharrer and Linser (1969) and Jones (1972). These should be consulted for information on the composition of individual vegetables or tissues.

A number of accumulator species have been reported from the seed plants, and these have been reviewed by Peterson (1971), Brooks (1972) and Ernst (1974). A list is given below, but it is by no means complete. Accumulator species for Au and Li have been reported, but require confirmation.

Generalisations are unreliable, but it seems that cereal grasses are all poor

Element(s)	Accumulator plants or families
Ag	*Eriogonum ovalifolium*?
Al	Diapensiaceae, Ericaceae, Melastomaceae, Symplocaceae, Theaceae; *Orites excelsa*
B	Many Cruciferae and Leguminosae are rich in B
Ba, Ra	*Bertholletia excelsa*
Cl, Na	Some Chenopodiaceae, Frankeniaceae and Plumbaginaceae; mangroves such as *Avicennia*, *Bruguiera* and *Rhizophora*
Co	*Clethra barbinervis*; *Crotalaria cobalticola*; *Nyssa sylvatica* (Brooks *et al.*, 1977b)
Cr	*Leptospermum scoparium; Pimelia suteri*
Cu	*Aeolanthus biformifolius*; *Becium homblei*; *Cryptosepalum maraviense*; *Elsholtzia haichowensis*; *Gypsophila patrinii*; *Lychnis alpina*; *Polycarpaea spirostylis*; *Silene dioica*; *S. vulgaris*; *Triumfetta welwitschi*; *Uapaca* spp.; *Vernonia glaberrima* (Malaisse *et al.*, 1978)
F	*Acacia georginae*; *Dichapetalum* spp.; *Gastrolobium grandiflorum* (Hall, 1977)
Hg	*Minuartia setacea*
I	*Feijoa sellowiana*
La–Lu	*Carya* spp.
Mn	Ericaceae; *Camellia* spp.
Mo	*Grindelia fastigiata* (Myers and Hamilton, 1961)?
Ni	*Alyssum bertolonii* and *A. murale*; *Dicoma* spp.; *Geissois* spp.; *Homalium* spp.; *Hybanthus floribundus*; *Pimelea suteri*; *Planchonella* spp.; *Psychotria* spp.; *Rinorea bengalensis*; *Sebertia* spp.; *Strychnos ignatii* seeds; *Trichospermum* spp. (Cole, 1973; Brooks *et al.*, 1977a; Brooks and Wither, 1977; Wither and Brooks, 1977)
Pb	*Amorpha canescens*; *Minuartia verna*
Re	*Atriplex confertifolia* (Myers and Hamilton, 1961)?
S	Cruciferae; *Allium* spp.
Se	*Astragalus racemosus* and 24 other species; *Lecythis ollaria* seeds; *Morinda reticulata*; *Neptunea* spp.; *Stanleya pinnata*; *Xylorhiza* spp. (Shrift, 1973)
Si	Cyperaceae, Gramineae, Juncaceae; *Moquilea* spp.
Sn	*Silene vulgaris*?
U	*Coprosma arborea*; *Uncinia leptostachya*
V	*Astragalus confertiflorus*?; *Petroselinum sativum* (Byrne and Kosta, 1978)
Zn	*Armeria maritima* subsp. *halleri*; *Minuartia verna*; *Silene vulgaris*; *Thlaspi alpestre*; *Viola tricolor* var. *calaminaria*

TABLE 6.3 Elemental composition of soft parts of marine animals in mg X/kg DM

X	Zooplankton	Annelida	Coelenterata	Crustacea	Echinodermata	Mollusca	Pisces (fish)
Ag	0·26 Mk	0·4 Bu	0·05 Rs	0·24 Bg	0·6 Rs	0·1–3 **Bg, Bl, Bp, Bu,** Sj, Wa	0·04–0·1 Br, Pa
Al	15 Mk	340	85 Rs	55Br	90–500 Rs	55–360 Bl, Bp, Sj	20 Br
As		6	20 Br	30 Br	5 Br	0·005–20 Br, F, Pe, Ho	0·2–10 Br, F, Ho
Au			0·007	0·0005 F		0·008 F	0·00012 F
B	32–130 Nc	2	63 Nc	15	46	20	20
Ba	25 Mk, Sz			0·2		3	0·017 Ps
Be				0·004–0·008 Ms	0·007 Ms	0·002 Ms	0·002–0·2 Ms
Bi			0·3		0·15	3·7 ? Wa	0·04
Br		100	1000	400	60	1000	400
C	416 k Bs	402 k	436 k	400 k	400 k	400 k	475 k
Ca	20 k–40 k Ma, Mk, Sz	11 k	1 k–110 k Rs	2 k–38 k Hl	14 k–170 k Rs	1 k–11 k Bp, Hl, Sj	76–20 k Ps
Cd	2·3–13 Fl, Ma, Mk		0·4–4 Fl, Mu, Rs	0·15–4 Fl, Hl, Mu, Pe	1–9 Fl, Mu, Rs	0·8–120 Bj, Bl, Bp, By, Fl, Hl, Mu, Pe, Sj	0·1–3 Br, Fl, Hl, Mu, Pe
Ce							0·006–0·03 Pa
Cl			90 k	6 k	7 k	5 k	6 k
Co	0·08–13 Ma, Nc, Vf		0·06–2 Io, Nc, Rs	0·02–7 Bg, Hl, Io, Vf	0·2–1·5 Io, Rs	0·1–8 Bl, Bp, Hl, Io, Sj, Vf, Wa	0·006–0·05 Io, Pa, Vf
Cr	<0·6–2 Nc, Vf		0·37 Rs	0·32 Vf	0·46 Rs	0·1–2 Bg, Bl, Bp, Sj, Vf, Wa	0·03–2 Hl, Pa, Vf
Cs	0·032 Vf		0·017 Io	0·02–0·04 Io, Vf	0·022 Io	0·02–0·32 Io, Vf	0·04–0·2 Io, Pa, Ps, Vf
Cu	2–240 Ma, Mk, Nc		10–57 Rs	20–110 Hl	4–130 Bj, Rs	2–450 Bj, Bl, Bp, By, Hl, Pe, Sj, Wa	0·7–15 Br, Hl
Dy						0·00035 Ka	0·0014 Ka
Eu						0·06 Zc	0·0002–0·0009 Pa
F				2 Mt		2 Mt	1400
Fe	100–1800 Vf	630	250–730 Rs	10–160 Bg, Br, Vf	16–1800 Rs	90–780 Bg, Bl, Bp, By, Sj, Vf	9–88 Pa, Vf
Ga			0·5		0·6		0·15
Ge			1·5		0·3		0·3
H	50 k	59 k	45 k	60 k	45 k	60 k	68 k
Hg	0·11 Kn, Mk		0·1 Br	0·4–1·3 Br, By	0·3 Br	0·04–1 Bg, Br, Ha, Pe	0·4 Br
I			15	1	6	4	1
K	10 k Mk	16 k	9·3 k Rs	13 k	6·3 k Rs	12 k Sj	15 k Ps

X	Zooplankton	Annelida	Coelenterata	Crustacea	Echinodermata	Mollusca	Pisces (fish)
La						0·0011 Ka	0·014 Ka
Li						1	
Mg	9 k–15 k Ma, Mk	6 k	10 k Rs	2 k	15 k Rs	4 k Bp, Sj	1·2 k
Mn	4–32 Ma, Mk, Vf		4–9 Rs	1–1700 Br, Hl, Vf	6–82 Rs	2–150 Bl, Bp, By, Hl, Sj, Vf, Wa	0·3–5 Hl, Pa, Vf
Mo	1·9–5·1 Nc		2 Br	0·6	2·5	1·5 Br	1 Br
N	88 k Bs, Cd	99 k	63 k	84 k	44 k	35 k–82 k Rs	80 k–96 k Bs, Cd
Na	80 k Mk		35 k–85 k Sj	4 k	13 k–23 k Rs	16 k	8 k
Ni	0·6–42 Ma, Mk, Nc	11	3–17 Rs	0·4–20 Hl	0·3–4 Br, Bs	1–170 Bl, Bp, By, Hl, Sj, Wa	0·1–4·2 Br, Hl
O	285 k	340 k	271 k	400 k	456 k	390 k	290 k
P	7·5 k Bs, Cd	8 k	3·2 k Rs	9 k	3 k Rs	6·6 k Sj	18 k
Pb	2–130 Ma, Mk, Nc	5–45 Bu	3–24 Rs	0·3–49 Hl	0·5–7 Rs	0·5–42 Bj, Bl, Bp, By, Hl, Pe, Sj, Wa	0·001–15 Hl, Ps
Pr							0·0032 Ka
Ra	0·08 μ–0·13 μ Kt, Sz			0·007 μ	0·04 μ	0·15 μ Ta	0·015 μ Ta
Rb				4·8 Bg		5	0·8–5·6 Pa, Ps
Re				0·0005 F		0·006 F	0·0008 F
S		9 k Gk	19 k	6 k Gk	5 k Gk	16 k Gk	4·7 k Gk
Sb	2–24 Vf		0·022	0·016–0·48 Bg, Vf	0·1–0·24	0·007–0·6 Bg, Vf	0·004–0·2 Pa, Vf
Sc						0·2 Bg	
Se				2–8 Bg	2–4·5 Bg	0·4–4 Br, Ho	0·17 Ho
Si		150		300		1000	70
Sn	<0·2–25 Nc		2·5 Nc	<0·2 Nc	0·2 Br	0·48 St	
Sr	340–720 Ma, Mk, Sz	20	10–220 Rs	500	260–550	12–80 Sj	0·2–20 Pa, Ps
Ta				0·027 Bm	0·021 Bm	0·04 Bm	
Th	0·32 Cy		0·03		0·003		0·32 Cy
Ti	2·5 Nc		1·9 Nc	1·8 Nc	5	3·2 Nc	0·2
Tl						0·34 Ho	0·08 Ho
U	0·38–0·56 Cy, Kt						0·04–0·08 Cy, Pa
V	3·2–55 ? Nc	1 ? Be	2 ? Be	0·4–2 ? Be, Br	2 ? Be	0·1–2 Be, Br, Bz	0·03 Bz
W				0·0005 F		0·05 F	0·0014 F
Zn	180–530 Ma, Mk, Vf		46–280 Io, Rs	52–230 Bg, Hl, Io, Pe, Vf	68–350 Bj, Io, Rs	40–8500 Bg, Bj, Bl, Bp, By, Hl, Io, Pe, Sj, Vf, Wa	9–82 Hl, Io, Pa, Pe, Vf

k, ×1000; μ, $\times 10^{-6}$; *References*, see p. 101, and Vinogradov (1953) and Bowen (1966)

in inorganic nutrients, while cabbages (*Brassica*), tomato (*Lycopersicon*) and tobacco (*Nicotiana*) are nutrient-rich. In general leaves are richest in Ca, while seeds are rich in nitrogen and phosphorus and poor in Na. The reference material kale, consisting of leaf lamina from *Brassica oleracea*, has been analysed by many workers and its elemental composition is fairly well known (Bowen, 1974). It is not representative of all vegetables, as its ash content is unusually high, as is its content of Br, Ca, Na, Rb, S and perhaps also Au, Cd, Hg and Pb.

Marine animals

The elemental composition of marine animals, based largely on the compilations of Vinogradov (1953), is given in Table 6.3. Some of the earlier analyses, especially for Ag, Bi, Co, Ni, Pb, Sn and Zn may have suffered from systematic errors. Filter-feeding molluscs, and other organisms of muddy habitats, can concentrate heavy metals from contaminated coastal waters and sediments. Such organisms may have abnormally high contents of some of the following metals: Ag, Cd, Cu, Hg, Ni, Pb and Zn. Patterson and Settle (1977) consider that all determinations of lead represent contaminated material.

The zooplankton analysed were heterogeneous and calcareous or siliceous tests were included, when present. Miscellaneous accumulator species have been reported by Bowen and Sutton (1951), Vinogradov (1953), Bryan (1976) and others, as follows:

Element	Accumulator Phylum and species
Ba	The protozoan *Xenophyophora* has a test of $BaSO_4$
Br	Some Porifera; many coelenterates (Corals and Alcyonarians); a few molluscs (*Aplysia*, and Muricidae)
Cd + Zn	Molluscs of the genus *Ostrea* (Oysters), and the liver of *Pecten*
Cu	The Poriferan *Chondrilla nucula*; the blood and liver of Crustacea and Mollusca
F	The Poriferan *Dysidea crawshayi*; possibly the mollusc *Archidoris britannica*
Fe	The blood of some Annelida, and echinoderms
I	Some Porifera; many coelenterates; and their predators
Mn	Poriferan *Terpios zeteki*; Annelid *Hermione*; kidney of mollusc *Pinna* (Chipman and Thommeret, 1970; Ghiretti *et al.*, 1972); Ascidians *Didemnum candidum* and *Halocynthia* (Carlisle, 1968)
Nb	The ascidian *Molgula manhattensis* (Carlisle, 1968)
Ni	Commensals of the Poriferan *Dysidea crawshayi*
Si	Protozoa (Foraminifera); many Porifera; Oncidiidae (Mollusca)
Sn	The Poriferan *Terpios zeteki*
Sr	Tests of the Protozoan *Acanthometra* are $SrSO_4$
Ta	The ascidian *Styela plicata* (Kokubu and Hidaka, 1965)
V	Ascidians (Orders Phlebobranchia and some Aplousobranchia) (Biggs and Swinehart, 1976). Reports for other organisms were errors
Y	The Poriferan *Melithoea* spp.
Zn	Kidney of mollusc *Pecten*; Barnacle *Balanus*

Hard tissues of marine animals

The elemental composition of various hard tissues from marine organisms is given in Table 6.4. Organisms secreting silica (Foraminifera and some Porifera) are mostly found in temperate parts of the ocean, where the solubility of silica is least. Organisms secreting calcium carbonate are mostly found in warm and/or shallow parts of the ocean, which are often supersaturated with $CaCO_3$. The occurrence of these organisms has been comprehensively reviewed by Lowenstam (1974). They can be subdivided into four groups according to their content of Mg and Sr; hard tissues of aragonite are usually poor in Mg and rich in Sr, while calcite is usually poor in Sr, but there are exceptions.

Rich in Mr and Sr: Coelenterate corals, groups Alcyonaria and Octocorallia.

Rich in Mg, poor in Sr: Calcareous red algae, family Corallinaceae; calcareous Foraminifera; Porifera; Echinodermata.

Poor in Mg, rich in Sr: Chysophyta (yellow-green algae–Cocolithophorids); calcareous green algae, e.g. *Halimeda;* Coelenterate corals, groups Hydrozoa and Madreporaria.

Poor in Mg and Sr: Brachiopoda; Arthropoda, subgroup Cirripedia (barnacles); Mollusca.

Tissues of apatite are formed by marine vertebrates, such as fish, and by two families of brachiopods (Craniidae and Lingulidae). Brachiopod shells may have a fluorine content of 2%, while fish bones have less fluorine but may accumulate Cu, La–Lu, Pb, Sn, Th, Y and Zn (Arrhenius *et al.*, 1957; Haskin *et al.*, 1968; Patterson and Settle, 1977). Pteropod molluscs accumulate Ce, Co, Cr, Eu, La, Sc and Th in their shells (Turekian *et al.*, 1973), while corals accumulate U (Cherry and Shannon, 1974).

TABLE 6.4 Elemental composition of hard parts of marine biota in mg X/kg DM

X	Foraminifera SiO_2	Porifera SiO_2	Red algae $CaCO_3$	Foraminifera $CaCO_3$	Porifera $CaCO_3$	Corals $CaCO_3$	Echinoderms $CaCO_3$	Molluscs $CaCO_3$
Ag	3	1				0·04 Ve	1·5	0·01–0·15 Bg, Sj
Al		9000	1100	2900 Bf				70–470 Sj
As		3·5					8	0·23 Ho
Au		0·01					0·007	
B			90	500		75 Li	30	3
Ba	700		50	20 Cj		9–85 Bo, Li	35	7
Be								0·01–0·04 Ms
Bi		0·6					0·03	
Br		2400			9600	5000		50
C	5 k	9 k	100 k	116 k	115 k	116 k	116 k	120 k
Ca	17 k	17 k	311 k Ys	355 k	340 k	350 k Bo, Li	350 k	395 k

Table 6.4—*continued*

X	Foraminifera SiO_2	Porifera SiO_2	Red algae $CaCO_3$	Foraminifera $CaCO_3$	Porifera $CaCO_3$	Corals $CaCO_3$	Echinoderms $CaCO_3$	Molluscs $CaCO_3$
Cd		1·5	0·7	2			0·03	0·03–2 Sj
Ce						2·7 Hf		0·8
Cl		52 k	5 k			1·7 k		35
Co		0·05				0·005–2 Li, Ve	2	0·03–1·2 Bg, Sj
Cr		0·2				2 Li	0·02	0·12–0·37 Bg, Sj
Cu	25	15		23–30 Bf, Cj		1–20 Li	18	0·1–42 Sj
Er						0·019 Hf		0·036 Hf
Eu						0·058 Hf		0·012 Hf
F			5	75–500 Cr		900 Cr		75
Fe	300	2500	3400	1 k–4·5 k Bf, Cj		5–6500 Li	38 Su	5–1600 Bg, Pi, Sj
Ga		0·2					0·7	
Gd						0·072 Hf		0·035 Hf
Ge		0·3					0·7	
Hg								0·3 Bg
Ho						0·005 Hf		0·01 Hf
I		2100	55		4800	5000		
K		5800	4000					18–260 Sj
La			0·05			0·08 Hf		0·13
Li						1 Li		
Lu								0·003 Hf
Mg	1000	7600	11 k Ys	4 k *or* 24 k Bf	24 k	1·7 k *or* 28 k	28 k	210–3500 Pi, Sj
Mn	8	75	55	130–330 Bf, Cj		2–2100 Li		1–26 Pi, Sj
Mo		0·7					0·1	
N			10 k					2 k
Na		25 k						2–5·5 k Pi, Sj
Nd						0·71 Hf		0·26 Hf
Ni	9	22		21 Cj		2 Li	2–50 Su	0·12–2·4 Sj
O	540 k	470 k	480 k	480 k	490 k	470 k	490 k	480 k
P		3300	1500				2600	10–630 Sj
Pb	10	5·5		20–140 Bf, Cj		2 Li	5	0·4–2 Sj
Pr								0·053 Hf
Ra			0·02 μ					0·05 μ Cy
S		7 k	2·7 k			2·5 k	3·3 k	1·7 k
Sb		0·08					0·18	0·025 Bg
Sc						0·003–0·3 Li		0·0045 Bg
Se							0·2 Ho	0·053 Bg
Si	410 k	437 k	1 k	12 k Bk		1 k Li	4 k	800
Sm						0·079 Hf		0·038 Hf
Sn		1·7					1·6	
Sr	400		10 k Ys	1100 Cj	1200	9 k Bo, Li	1600	70–1200 Pi, Sj
Tb						0·006 Hf		0·01 Hf
Th							0·001	0·005
Ti	150		25	400 Bf		50 Li	5	
Tl							0·11 Ho	
Tm						0·0032 Hf		0·0005 Hf
U						2–4 Cy, Li, Ve		0·09 Cy
V		1·7?		15–80? Bf, Cj			5?	
Y						0·12 Hf		0·22 Hf
Yb						0·018 Hf		0·019 Hf
Zn		100				1–70 Li	65	1–11 Bg, Sj

k, ×1000; μ, $\times 10^{-6}$; *References*, see below, or Vinogradov (1953), Bowen (1966)

References in Tables 6.1 6.4: Ab, Allcroft *et al.* (1965); B, Bowen (1974) and unpublished; Ba, Bowen (1960); Bb, Bowen (1972); Bd, Bowen and Dymond (1955); Be, Bertrand (1950); Bf, Belyayeva (1973); Bg, Bertine and Goldberg (1972); Bh, Bryan and Hummerstone (1973); Bj, Boyle and Lynch (1968); Bk, Byrne *et al.* (1976); Bl, Bryan (1973); Bm, Burton and Massie (1971); Bo, Bowen (1956a); Bp, Bryan *et al.* (1977); Br, Bryan (1976); Bs, Beers (1966); Bu, Bryan and Hummerstone (1977); By, Boyden (1975); Bz, Byrne and Kosta (1978); C, Connor and Shacklette (1975); Ca, Cannon (1960); Cd, Corner and Davies (1971); Ch, Chapman (1966); Cj, Chester and Stoner (1973); Cr, Carpenter (1969); Cs, Cannon *et al.* (1968); Cy, Cherry and Shannon (1974); E, Edgington *et al.* (1970); Eh, Erämetsä and Haukka (1970); Er, Erämetsä *et al.* (1973); Ey, Erämetsä and Yliruokanen (1971); F, Fukai and Meinke (1959); Fj, Fuge and James (1974); Fl, Fleischer *et al.* (1974); Fm, Fukai and Meinke (1962); G, Girling (1978); Gk, Goldhaber and Kaplan (1974); Gm, Gabrielli and Marletta (1969); Go, Golley *et al.* (1969); Gr, Gryzhankova *et al.* (1973); H, Horowitz *et al.* (1974); Ha, Haller *et al.* (1968); Hf, Haskin *et al.* (1966); Hg, Hanna and Grant (1962); Hi, Hinneri (1975); Hj, Hattula and Johanson (1978); Hl, Halcrow *et al.* (1973); Hm, Hamilton and Minski (1973); Ho, Hamaguchi *et al.* (1960); Hw, Hanawalt and Whittaker (1977); I, Ishibashi *et al.* (1965); Io, Ichikawa and Ohno (1974); Jo, Jones (1972); K, Karvanek and Janicek (1972); Ka, Kameda (1962); Kn, Knauer and Martin (1972); Ko, Kovalsky *et al.* (1967); Kr, Korbonskaya and Amanova (1975); Kt, Kharkar *et al.* (1976); L, Lunde (1970); La, Lounamaa (1956); Li, Livingston and Thompson (1971); Lo, Long (1961); Lr, LeRoy and Koksoy (1962); Lw, Lawrey (1977); M, Millman (1957); Ma, Martin (1970); Mc, McCance and Widdowson (1960); Mi, Millbank and Kershaw, (1973); Mk, Martin and Knauer (1973); Ml, Mlodecki *et al.* (1973); Ms, Meehan and Smythe (1967); Mt, Matsumura *et al.* (1955); Mu, Mullin and Riley (1956); Nc, Nicholls *et al.* (1959); Nf, Neufeld (1936); O, Oakes *et al.* (1977); P, Porter (1946); Pa, Pentreath (1977); Pe, Peden (1973); Pi, Pilkey and Harriss (1966); Pr, Preston *et al.* (1972); Ps, Patterson and Settle (1977); Pu, Puchelt (1967); Q, Quinche (1976); Rr, Riley and Roth (1971); Rs, Riley and Segar (1970); Rt, Rastorfer (1974); S, Scharrer and Linser (1969); Sa, Schelhorn *et al.* (1976); Sb, Schroeder and Balassa (1966a); Sc, Scadden (1969); Sd, Schroeder and Balassa (1967); Sf, Sheppard and Funk (1975); Sg, Solberg (1967); Sh, Shacklette (1965); Si, Smith and Mercer (1970); Sj, Segar *et al.* (1971); Sk, Saenko *et al.* (1976); Sl, Shacklette and Cuthbert (1967); Sm, Smales and Salmon (1955); Sn, Schroeder *et al.* (1967); So, Saeki *et al.* (1977); Sp, Strohal and Pinter (1973); Sq, Shacklette (1972); Sr, Sorenson *et al.* (1974); Ss, Stärk *et al.* (1971); St, Smith and Burton (1972); Su, Stevenson and Ufret (1966); Sv, Stijve and Besson (1976); Sw, Sawada (1965); Sx, Stock and Cucuel (1934); Sy, Steinnes (1977b); Sz, Szabo (1967); T, Thomas *et al.* (1974); Ta, Turner *et al.* (1958); Tj, Tuominen and Jaakola (1973); Ty, Tyutina *et al.* (1959); V, Vinogradov (1953); Va, Vinogradova and Kovalsky (1962); Ve, Veeh and Turekian (1968); Vf, Van As *et al.* (1975); Vm, Verkhovskaya *et al.* (1967); W, Wozniak (1975); Wa, Watling and Watling (1976); Wh, Wahlgren *et al.* (1976); Y, Yamagata *et al.* (1959); Ya, Yamamoto *et al.* (1968); Yf, Yamamoto *et al.* (1970); Yl, Yliruokanen (1975a); Ys, Yamamoto *et al.* (1969); Yy, Yamamoto *et al.* (1973); Z, Bergerioux and Zikovsky (1978); Za, Cali (1976); Zb, Stella *et al.* (1977); Zc, Hadzistelios and Papadopolou 1977); Zd, Kosta *et al.* (1978); Ze, Nadkarni (1977); Zf, Abedinzadeh *et al.* (1978); Zg, Meisch *et al.* (1977); Zh, Schmitt *et al.* (1977); Zi, Schmitt *et al.* (1978); Zj, Braman and Tompkins (1979); Zk, Ward *et al.* (1977)

Insects

Data on the elemental composition of this vast group is very scanty (King, 1957; Spector, 1956a; Bowen, 1966; Ponomarenko *et al.*, 1974). Their composition does not appear to differ much from other animals, though they may be rich in K, P, Si and Zn, and relatively poor in Ca and Mg.

Invertebrate blood

Table 6.5 gives some data, mostly rather old, on the elemental composition of invertebrate blood; further details may be found in Vinogradov, 1953; Spector, 1956a; Nicol, 1960; Florkin and Jeuniaux, 1964; Oglesby, 1969; and Jeuniaux, 1971. The most obvious feature is the much lower total concentration of elements in the blood of terrestrial and freshwater organisms than in marine organisms. The blood of many marine invertebrates is rather similar to sea water, with a high concentration of Cl and Na. Note that there is more Na than K in the blood of most invertebrates except for a few 'advanced' insects, which include all Lepidoptera, most Hymenoptera and a few Coleoptera.

TABLE 6.5 Elemental composition of invertebrate blood in mg X/l

X	Marine Annelida	Marine Coelenterata	Marine Crustacea	Marine Echinodermata	Marine Mollusca	Land Insecta	Land Mollusca	Land Nematoda
Ca	500	390	640	640	530	360–510	210	240
Cl	22 000	20 000	20 000	21 000	21 000	890–3900	2400	1900
Cu			76		180	27	130	
Fe			3·5			58		7·3
K	560	550	690	530	510	700–1600	180	980
Mg	1400	1200	680	1300	1400	270–830	34	120
Mn			2·6				46	
N						12 500		7400
Na	10 000	10 000	12 000	11 000	12 000	200–2600	1500	3000
P						1400		360
Zn			6·5			11		9

TABLE 6.6 Fresh weights, dry weights and ash weights of some tissues in 'Reference Man'

Tissue	FM/kg	DM/kg	Ash/g
Whole man	70	28	3700
Soft tissues	60	21	900
Adipose tissue	15	12·7	30
Blood	5·5	4·4	55
Brain	1·4	0·3	21
G.I. Tract	2·2	0·35	20
Heart	0·33	0·09	5
Kidney	0·31	0·07	3·4
Liver	1·8	0·5	23
Lung	1	0·22	11
Muscle	28	6	340
Skin	2·6	1	18
Hair	0·02	0·018	0·1
Nails	0·003	0·003	0·016
Skeleton	10	7	2800
Bone	5	4·15	2700
Marrow etc.	5	2·85	100
Teeth	0·046	0·042	34

Table condensed from fuller details given by Snyder (1975); G.I., gastro-intestinal

Mammalian tissues

There is an enormous literature on this topic. Monographs by Comar and Bronner (1962) and Underwood (1977) should be consulted for work on domestic mammals. For humans, useful summaries of data are given by Anspaugh *et al.*, 1971; Hamilton *et al.*, 1973; Snyder, 1975; and Iyengar *et al.*, 1978. 'Reference Man', as defined by Snyder (1975), has a fresh weight of 70 kg and consists of 60 kg of soft tissues plus 10 kg of bone. Further information on FM, DM and ash content of human tissues is given in Table 6.6.

Table 6.7 gives selected data on the elemental composition of Reference Man and six human tissues. There are still too few reliable analyses for many elements, and different workers disagree by an order of magnitude or more for the following elements: Ag, Al, As, Ba, Cd, F, Nb, Te, U and Zr. Accumulator tissues have been observed as follows:

Al	hair, lung and skin (probably contamination)
As	hair and nails
Au	hair (possibly contamination)
Ba	the choroid of the eye; bone
Ca	bone and tooth
Cd	kidney (increasing with age up to about 40 years)
Cs	muscle
F	tooth and bone
Fe	bone marrow and erythrocytes; liver
Hg	kidney; hair
I	thyroid; hair
La etc.	bone(?)
P	bone and tooth
Pb	bone
Po	hair
S	hair
Sb	hair
Si	lung and skin (probably contamination)
Sr	bone and tooth
Ti	lung and skin (contamination)
Zn	prostate gland; eye

Apart from these anomalies, 'Reference Man' can be approximately represented as the sum of two tissues, muscle and skeleton, since soft tissues all have much the same elemental composition as muscle.

TABLE 6.7 Elemental composition of six human tissues in mg X/kg DM, and of 'Reference Man' in mg X/70 kg FM

X	Kidney	Liver	Muscle	Bone	Hair	Nail	'Ref. Man'
Ag	0·005–0·26 I	0·005–0·25 I	0·009–0·28 I	0·01–0·44 I	0·005–3·8 I	0·003–1·4 I	
Al	1·5–27 I	3–23 I	0·7–28 I	4–27 I	4–29 I	130–930 I	61 S
As	0·007–1·5 I	0·023–1·6 I	0·009–0·65 I	0·08–1·6 I	0·06–3·7 IV	0·2–3 I	18 S
Au	0·014 I	0·0004 I		0·016 I	0·0017–1·8 I	0·03–0·8 I	
B	0·9–2·6 I	0·4–2·3 I	0·33–1 I	1·1–3·3 I	5 IV	33? I	
Ba	0·04–1 I	0·04–1·2 I	0·09 I	3–70 IY	0·55–4 IV	7·5 I	22 S
Be	0·0002 Ms	0·0016 Ms	0·00075 Ms	0·003 Ms	0·006–0·02 I	<0·01 I	0·036 S
Bi	<0·1–2 I	0·015–0·33 I	0·032 I	<0·2 I	2? I	1·3 I	
Br	16–30 I	0·2–7 I	7·7 R	6·7 R	0·65–53 IV	9 I	260 R
C	670 k S	670 k S	670 k S	360 k S	540 k S	540 k S	16 000 k S

Table 6.7—*continued*

X	Kidney	Liver	Muscle	Bone	Hair	Nail	'Ref. Man'
Ca	360–820 I	100–360 I	140–700 I	170 k I	150–3200 IV	370–3400 I	1000 k S
Cd	18–310 I	2–22 I	0·14–3·2 I	1·8 I	0·24–2·7 IV	0·08–3·4 I	50 S
Ce	0·013 I	0·29 I		2·7 Y	0·25 Za	0·6 I	
Cl	12 k I	3 k–7·2 k I	2 k–5·2 k I	900 I	1 k–4·8 k I	1 k–3·6 k I	95 k S
Co	0·035–0·31 I	0·06–1·1 I	0·028–0·65 I	0·01–0·04 I	0·2–1 I	<0·2 I	
Cr	0·05–4·7 I	0·02–3·3 I	0·024–0·84 I	0·1–33 I	0·13–3·6 IV	6? I	
Cs	0·028–0·049 I	0·04–0·054 I	0·07–1·6 I	0·013–0·052 I	0·37–1·1 I	7? I	
Cu	16 I	30 I	10 I	1–26 I	19 I	18 I	72 S
F	0·04–10 I	0·22–7 I	0·05 I	2 k–12 k I	<15 C		2600 S
Fe	170–710 I	250–1400 I	180 I	3–380 I	5–45 IV	27–350 I	4200 S
Ga	0·004 I	0·0025 I	0·0014 I		0·07 I		
Ge	40? I	0·15 I	0·14 I		2·3 I	1·2 I	
H	93 k S	93 k S	93 k S	52 k S	29 k S	29 k S	7000 k S
Hg	0·3–12 I	0·018–3·7 I	0·02–0·7 I	0·45 I	1·2–7·6 IV	0·07–7 I	
I	0·15 I	0·7 I	0·05–0·5 I	21? I	0·1–15 IV		
K	8 k I	8·5 k I	16 k I	2·1 k I	150–660 IV	360–2800 I	140 k S
La	0·013 I	0·3 I	0·0004 Z	<0·08–9	0·15–0·65 I	0·3 I	
Li	0·016–0·044 I	0·025 I	0·023 I				0·67 S
Mg	630 I	590 I	900 I	700–1800 I	19–160 IV	16–120 I	19 k S
Mn	3·3 I	3·6–9·6 I	0·2–2·3 I	0·2–100 IY	0·25–5·7 IV	0·04–2·1 I	12 S
Mo	0·9–3·1 I	1·3–5·8 I	0·018 Z	<0·7 S	0·06–0·2 IV	<0·15–16 I	
N	72 k S	72 k S	72 k S	43 k S	160 k S	150 k I	1800 k S
Na	10 k I	2 k–4 k I	2·6 k–7·8 k I	10 k I	18–1700 I	330–3 k I	100 k S
Nb	0·04–0·07 I	0·15 I	0·14 I	<0·07 I	2·2? I		
Ni	0·6–1·8 I	0·02–1·8 I	1–2 I	<0·7 I	0·2–6·5 IV	0·03–12 I	1 S
O	160 k S	160 k S	160 k S	285 k S	245 k S	255 k S	43 000 k S
P	7 k I	9·4 k I	3–8·5 k I	62–71 k I	83–165 IV	180–990 I	780 k S
Pb	1·2–6·8 I	3–12 I	0·23–3·3 I	3·6–30 I	3–70 IV	14–170 I	120 S
Ra	6n I	0·72n I	0·23n I	4n I	19n I		31n S
Rb	17–52 I	5–45 I	20–70 I	0·1–5 I	0·2–0·5 IV	3? I	680 S
S	9 k I	7 k–12 k I	5 k–11 k I	500–2400 IS	42 k–60 k BIV	37 k I	140 k S
Sb	0·026–0·22 I	0·011–0·42 I	0·042–0·19 I	0·01–0·6 I	0·09–3 I	<0·03–0·75 I	
Sc		0·0004–0·0014 I		0·001–6 I	<0·003 C		
Se	0·4–3·5 I	0·35–2·4 I	0·42–1·9 I	1–9 I	0·6–6 I	1–8 I	
Si	14–200 I	13–120 I	100–200 I	17 I	20–2 k IV	310–3500 I	
Sn	0·68–4·4 I	0·23–2·3 I	0·33–2·4 I	1·4 I	0·4–1 IV	12? I	
Sr	0·15–1·5 I	0·05–0·36 I	0·12–0·35 I	36–140 I	0·05–0·9 IV	0·02–0·65 I	320 S
Te	0·07? I	0·014? I	0·017? I		<2 C		
Th				0·002–0·012 I	<0·02 C		
Ti	1 I	1·2–4·7 I	0·9–2·2 I		0·08–14 IV	0·28 I	
Tl	0·013 I	0·004–0·033 I	0·07 I	0·002 I	0·016 I	0·0022 I	
U	0·009–0·024 I	0·003 I	0·0009 I	0·00016–0·07 I	0·00013 I		0·09 S
V	0·03 I	0·006 Bz	0·02 Bz	0·0035 Bz	0·0045–0·5 I	<0·15 I	0·11 Bz
W				0·00025 I	0·016 I		
Y	<0·0007 E	<0·01 E	0·02 I	0·07 I			
Zn	200 I	240 I	240 I	75–170 I	220 I	73–300 I	2300 S
Zr	0·066–0·09 I	0·11 I	0·08 I	<0·1 I	1·4 I	1·4 I	1 I

k, ×1000; n, $\times 10^{-9}$. Hamilton *et al.* (1973) and Ref. Y give additional data for human bone: Dy <0·36; Er <0·28; Eu 0·032; Gd <0·4; Hf 0·12; Ho <0·08; Lu <0·08; Nd <0·28; Sm 0·23; Ta 0·03; Tb 0·05; Tm <0·08; Yb <0·28

Chuang and Emery (1978) give for human hair: Dy <0·005; Er <1; Eu <0·005; Gd <2; Hf <0·02; In <0·03; Ir <0·02; Lu <1; Nd <2; Pd <2; Rh <0·1; Sm 0·005; Ta <0·01; Yb <0·03

References: B, Bacso *et al.* (1978); Bz, Bvrne and Kosta (1978); C, Chuang and Emery (1978); E, Erämetsä *et al.* (1968); I, compilation of Iyengar *et al.* (1978); Ms, Meehan and Smythe (1967); R, Cross and Smith (1978); S, compilation of Snyder (1975); V, compilation of Valkovic (1977); Y, Brätter *et al.* (1977); Z, D'Hondt *et al.* (1977); Za, Arunachalam *et al.* (1979)

Vertebrate blood

The commoner elements in the blood of various vertebrates have been tabulated by Long (1961). Vertebrate blood differs from invertebrate blood both in having a lower total salt content, and in having a higher ratio of univalent to bivalent cations, i.e. $[Na^{+}+K^{+}]/[Ca^{++}+Mg^{++}]$. It consists of a suspension of erythrocytes, which are red cells without nuclei, in a protein-rich solution called plasma. The distribution of Na^{+} and K^{+} between erythrocytes and plasma varies with species. The composition of blood is under homeostatic control, and most elements injected into the blood stream are rapidly removed from it during circulation.

The elemental composition of human blood, erythrocytes and plasma is given in Table 6.8. This assumes that blood contains 42·8% v/v erythrocytes and 57·2% v/v plasma. The number of atoms of an element X per erythrocyte can be calculated to be $4{\cdot}75\times10^{7}[X]/A$, where A is the atomic weight of X. The data of Table 6.8 show that only radium, and probably radioactive elements such as Ac, Pa, Po and Pu, are present in amounts of less than one atom per red cell. The other elements present are either known to be essential, or if not, they could have some undiscovered function or be contaminants.

TABLE 6.8 Elemental composition of human blood, red cells, plasma and serum in mg/l

Element	Whole blood	Red cells	Plasma	Serum
Ag	<0·003*	0·010	0·0036–0·044	0·003–0·2
Al	0·39	0·064?	0·32	0·23
As	0·0017–0·09	0·0027	0·0024	0·03?
Au	0·00001–0·00042	0·0008?	0·00006?	0·00008?
B	0·13		0·12	
Ba	0·068	0·007	0·058	0·06
Be	<0·00001			
Bi	0·016?		<0·0006	
Br	4·7	1·45	4·3	3·9
Ca	60·5	3·2	96	97
Cd	0·0052		0·0026	0·0028
Ce	<0·002			
Cl	2890	2040	3610	3650
Co	0·0002–0·04	0·00016–0·086	0·00007–0·012	0·0002–0·062
Cr	0·006–0·11	0·037	0·026–0·16	0·002–0·02
Cs	0·0038		0·0031	0·0016
Cu	1·01	0·98	1·12	1·19
F	0·5	0·45	0·2?	0·027?
Fe	447	1050	1·1	1·09
Ga	<0·08		0·00021	
Ge	0·44?	0·65?	<0·03	0·29?
Hg	0·0078	0·007	0·0065	0·012?

Table 6.8—*continued*

Element	Whole blood	Red cells	Plasma	Serum
I	0·057		0·069	0·066
K	1620	3550	157	191
Li	0·004	0·028?	0·03?	0·009?
Mg	37·8	57·2	21·2	21·7
Mn	0·0016–0·075	0·035	0·0006–0·068	0·00054–0·061
Mo	0·001?	0·017?	0·013?	0·006?
Na	1970	284	3160	3250
Nb	0·0047?		<0·01	<0·06
Ni	0·01–0·05	0·083	0·054?	0·003
P	345	706	121	142
Pb	0·21	0·46	0·13?	0·016–0·13
Ra	$6{\cdot}6 \times 10^{-9}$		$16{\cdot}6 \times 10^{-9}$	
Rb	2·49	4·7	1?	0·2
S	1800		330?	1200
Sb	0·0033	0·0046	0·0032?	0·0026
Sc	0·0076?			0·00015
Se	0·171	0·22	0·03–0·59	0·122
Si	3·9	4·1?	0·43?	6·9?
Sm	0·008?		<0·002	
Sn	0·29?	0·38?	<0·004–0·1	0·03
Sr	0·031	0·0072	0·036	0·037
Te	0·0055?		<0·03	
Th	0·0002–0·0024	0·00016	<0·00004	
Ti	0·054	0·11?	0·01–0·11	0·09?
Tl	0·00048		<0·0025	
U	0·0005	0·00024?	<0·00013	
V	<0·0002†	0·056?	0·056?	0·000066**
W	0·001			0·0004?
Y	0·0047		<0·01	
Zn	7·0	14·2	1·14	1·15
Zr	0·011	0·012?	0·0017?	0·1?

Data summarised from Iyengar *et al.* (1978). The large range of values in the literature for As, Au, Co, Cr, Mn and Th in whole blood makes it difficult to assign a best mean at present. Inconsistent or unconfirmed data are indicated by (?)

* Howlett and Taylor, 1978 † Byrne and Kosta, 1978 ** Cornelis *et al.*, 1979

Subcellular particles

Interest in the elemental content of subcellular particles has declined in the last decade, having shifted to the analysis of enzymes and enzyme systems from such particles (chapter 9). In practice, the preparation of pure samples of nuclei, mitochondria etc. is extremely difficult without contamination or loss. Thus Wacker and Vallee (1959) obtained a ribosome fraction from calf liver containing 0·1% of chromium, which presumably came from contamination by stainless steel. Up to 65% of the sodium and potassium in cell nuclei may be lost when the organelles are isolated in aqueous media (Itoh and Schwartz, 1956). Table 6.9 gives some analyses of nuclei, mitochondria and cytoplasm from mammalian tissues. Wester (1965) has measured the distribution of 17 elements in fractions isolated by centrifugation in sucrose solution from disrupted mammalian heart cells. Plant chloroplasts have also been analysed; for example, the following values were reported for chloroplasts from the green alga *Tolypella intricata*, in mg/g FM: calcium 0·84; chlorine 12; potassium 13; magnesium 0·34; sodium 0·83 (Larkum, 1968; see also Nakatani *et al.*, 1979). It appears that there is a large gradient in concentration for chlorine and sodium between the cytoplasm and the nucleus, and mitochondria appear to be rich in calcium with respect to the cytoplasm; potassium and magnesium are more uniformly distributed, and information for other elements is not available.

TABLE 6.9 Elemental composition of subcellular fractions from mammalian tissues in μg/g DM

Element	Isolating solution	Nuclei	Mitochondria	Cytoplasm	Refs
Al	Aqueous	170			DB
Ca	Aqueous	210	250–2800		DB, HR, Sp, TV
Ca	Non-Aq.	190		200	SL
Cl	Aqueous		500		HR
Cl	Non-Aq.	7200		120	SL
Co	Aqueous	0·46			SK
Cu	Aqueous	7	18–28		BK, DB, TV
Fe	Aqueous	140	230		DB, TV
K	Aqueous	8000	1000–7400		HR, IS, MS, Sp, TV
K	Non-Aq.	18 000		12 000	IS, SL
Mg	Aqueous	70–1100	330–5000		DB, HR, Sp, TV
Mg	Non-Aq.	750		700	SL
Mn	Aqueous	12	11		GR, TV
N	Aqueous	128 000	110 000		Sr
Na	Aqueous	830–2100	190–810		HR, IS, MS, Sp, TV
Na	Non-Aq.	3800		440	SL
P	Aqueous		3100		HR
P	Non-Aq.	25 000			WG
S	Aqueous	1500			Sr
Zn	Aqueous	3–140	39		DB, HR, TV

References: BK, Balevska and Kasabova (1971); DB, Dounce and Beyer (1948); GR, Gilbert and Radley (1964); HR, Honda and Robertson (1956); IS, Itoh and Schwartz (1956); MS, Macfarlane and Spencer (1953); SK, Siebert *et al.* (1951); SL, Siebert and Langendorf (1970); Sp, Spector (1953); Sr, Spector (1956a); TV, Thiers and Vallee (1957); WG, Williamson and Gulick (1944)

7 *Uptake and Excretion of Elements by Organisms*

Living cells are able to control their internal environment in several ways. All cells are surrounded by a membrane, which is more or less permeable to small molecules but much less so to charged ions. One current picture of the membrane is that it incorporates specific pumps, which can use metabolic energy to pump chemical species into or out of cells. However, there is still much to learn about how these pumps function, and also about how transport from one cell to another takes place.

Uptake and excretion of gases

Gas molecules enter cells by diffusion. Most of them are metabolised so rapidly that equilibrium across the cell membrane is never attained.

Dihydrogen is absorbed by certain bacteria, and by the leaves of seed plants, and oxidised to water or organic matter (Cole, 1976). Its main acceptor molecule is the enzyme hydrogenase (chapter 9). It is known to be excreted by many algae (Kessler, 1974).

Dinitrogen is absorbed by a few bacteria, most blue-green algae and some symbiotic actinomycetes, which can reduce it to ammonia. Its acceptor molecule is the enzyme system nitrogenase (chapter 9). It is excreted by some bacteria and probably by other organisms (Cole, 1976).

Dioxygen is absorbed and used for respiration by nearly all organisms, although it is thought likely that the most primitive cells were anoxic. Its acceptor molecule is the enzyme system cytochrome oxidase. It is excreted as a by-product of photosynthesis by all green plants.

Methane is oxidised by some oxygenic bacteria, such as *Methanomonas*, and is excreted by many anoxic bacteria (Mah *et al.*, 1977).

Carbon monoxide is absorbed by most cells and is more or less toxic, but it can be oxidised by the bacterium *Bacillus oligocarbophilus* (Cole, 1976). It is excreted by marine organisms (Weinstock and Chang, 1976).

Carbon dioxide is absorbed by most cells in the dark by one or more processes requiring metabolic energy, such as the carboxylation of phosphoenol pyruvate to oxalacetate. It is also absorbed, at a much faster rate, by all green plants in sunlight. Here the acceptor molecule is ribulose diphosphate and the first stable product is phosphoglycerate. All respiring cells excrete carbon dioxide, and it is a major product of fermentation.

Ammonia is absorbed by most, if not all, organisms. The main acceptor molecule is α-ketoglutarate, which forms glutamate in the presence of $NADH_2$. Ammonia is excreted by many bacteria and most animals.

Nitrous oxide is probably absorbed by soil bacteria; it is excreted by soil and marine micro-organisms as a by-product of the reduction of nitrates (Burns and Hardy, 1975).

Nitric oxide is readily absorbed by organisms and is toxic. It may be excreted by bacteria but this is uncertain.

Hydrogen sulphide is probably absorbed by most organisms and is toxic. It is an excretion product of a few anoxic bacteria (Kadota and Ishida, 1972).

Sulphur dioxide is readily absorbed by plant cells, where it is toxic, but it is not known to be excreted.

TABLE 7.1 Concentration factors for phytoplankton and brown algae from sea water

X	F_X (plankton)	F_X (brown algae)	X	F_X (plankton)	F_X (brown algae)
Ag	2500	1250	Mn	10–150 k	5–500 k
Al	5 k–50 k	7500	Mo	25	4–17
As		70–2000	N	40 k	8400
B		7	Na	4·14	0·7
Ba	400–4000	400	Ni	600–7000	440–3000
Be		125	P	17–70 k	12 k
Br		3	Pb	40–7300 k	17–320 k
Ca	4	6–15	Ra	1100	280
Cd	900–5500	900–2700	Pb	17–100	
Cl		0·06	Re		1250
Co	60 k	2·5–10 k	S	0·8–1·7	2·8
Cr	3300	800–5000	Sb	50[a]	83
Cs	1–5[a]	60	Sc		33 k
Cu	4 k–18 k	2·4 k	Se		50
F		1·2	Si	23 000	180
Fe	25 k–2000 k	15 k–250 k	Sn	90 k–900 k	7 k–30 k
Ga	13 k	4 k	Sr	2–20	25–50
Hg	1700	230	Th		75 k
I	1·2 k	1·8 k–10 k	Ti	7 k–230 k	3 k–8 k
K	8	32	U		62
La		740k	V	400	240
Li		7	Zn	1 k–14 k	2·5 k–13 k
Mg	0·6–2·7	1·4	Zr	170 k	0·35 k–1 k[a]

k, $\times 10^3$; [a], from tracer work by Mauchline and Templeton (1964)

Uptake of ions from solution

An important characteristic of living cells is their ability to take up ions from a solution against a concentration gradient. This is most clearly shown by free-living aquatic organisms, especially for autotrophic algae which obtain all their nutrients from solution. If we define a concentration factor F for each element X by

$$F_X = [X]_{\text{alga FM}}/[X]_{\text{water}}$$

and assume that algae contain 75% water, we can use the data in Tables 2.5 and 6.1 to calculate concentration factors from sea water. This has been done for marine phytoplankton and brown algae in Table 7.1. In interpreting this table, one must bear in mind the possibility of analytical errors, and the fact that the composition of sea water is not constant near the coast, where brown algae grow. Confirmation of the orders of magnitude of the concentration factors obtained has been given by Mauchline and Templeton (1964), who analysed organisms grown in radioactively labelled sea water: some of their data is also given in Table 7.1.

The ranges of the concentration factors found are as follows:

Range of F_X	X
<1	Cl, Na
1–10	B, Br, Ca, F, Li, Mg, S
10–100	Cs, K, Mo, Rb, Sb, Se, Sr, U
100–1 k	As, Ba, Be, Hg, Ra, Si (algae), V
1–10 k	Ag, Al, Cd, Co, Cr, Cu, Ga, I, Ni, Re, Zn
10–100 k	Mn, N, P, Sc, Si (plankton), Sn, Th, Ti, Zr
>100 k	Fe, La, Pb

Hence we can generalise as follows:

1. All elements are concentrated from sea water except for the two commonest ions in this medium, Cl^- and Na^+, but for F and Mg the concentration factors are of the order of unity.

2. Among cations, multivalent ions are more strongly concentrated than are univalent ions of Group IA, and divalent ions of transition metals are more strongly concentrated than divalent ions of Group IIA. In general the heavier elements in a Group are more strongly concentrated than related light elements, though Be is an exception. The order of concentration factors is not the same for brown algae and phytoplankton, and cannot be

correlated with the order of stability of complexes formed with specific organic ligands.

3. Among anions, nitrate, phosphate and silicate (for plankton) are concentrated by the largest factors. Multivalent ions are more strongly concentrated than univalent ions such as the halogens. The true concentration factor for nitrate is uncertain because of the many other chemical forms of nitrogen available for uptake from sea water.

4. There is no obvious relationship between concentration factors and essentiality. The essential elements Ca, Mg and S are scarcely concentrated, while the strongly concentrated La, Pb, Sc, Th, Ti and Zr are not thought to be essential (chapter 8).

Passive diffusion of ions

One way of trying to understand how living organisms can concentrate elements is to compare the uptake of ions by living and dead material. The latter can show preferences for certain ions. Thus cation-exchange resins containing carboxyl groups have affinities for the alkali metals in the order $Li^+ > Na^+ > K^+$, while similar resins containing sulphonate groups have affinities in the opposite order. Dowex-1 anion exchange resin has affinities for the halide ions in the order $I^- > Br^- > Cl^-$, which is the same as that found for plankton and brown algae. At present these resin phenomena cannot be predicted, as they are caused by small differences in free energy when the ions are held in the body of the ion-exchanger.

The cytoplasm within the cell is believed to be an efficient cation-exchanger and a less efficient anion-exchanger. This is so because of the preponderance of acidic groups in the structural polymers of cells, e.g. carboxyl groups in nucleic acids, proteins and pectins and sulphonate groups in mucopolysaccharides. There are relatively few substituted ammonium groups to function as anion exchange sites. It is also known that there is a potential difference of about 0·05–0·1 V between the inside and outside of algal and other cells, with the inner surface negative, confirming the prevalence of anionic groups within the cell. When a cell dies, the membrane surrounding it breaks and the potential difference disappears as ions leak out.

The equilibrium between two phases containing diffusible cations, one of which contains immobile anions, is an example of the Donnan equilibrium. For potassium chloride at equilibrium between a resin bead (*r*) and a solution (*s*), the ionic concentrations obey the relation:

$$[K^+]_r[Cl^-]_r = [K^+]_s[Cl^-]_s$$

The Donnan equilibrium is unaffected by the presence of oxygen and other gases, varies slightly with temperature, and is rather sensitive to pH, which may suppress the ionisation of weakly acid groups such as carboxyl. Donnan equilibria are indeed found for dead tissues, but are wholly inadequate to explain most of the facts about the uptake of ions, such as the selective uptake of K^+ and rejection of Na^+ by most cells.

Vacuoles

A feature of plant cells that distinguishes them from most animal cells is the common occurrence of vacuoles. These are cavities within the cytoplasm, bounded by membranes, containing a liquid sap. Most mature plant cells, and also the cells of many protozoa, contain a single large vacuole which may be formed by the coalescence of smaller vacuoles. Immature plant cells do not have vacuoles, but develop them as they mature. When studying the uptake of an ion X by a cell, it is desirable to measure the concentrations of X in the nutrient medium, cytoplasm and vacuolar sap, together with the rates of transfer of X across the outer cell membrane and the vacuolar membrane. Freshwater charophytes are useful for such studies, because their stems consist of cylindrical cells about 1 cm long and 1 mm in diameter.

Two possible functions of vacuoles are 1. as waste-bins for unwanted ions and molecules, and 2. to increase the surface/volume ratio in cells. Even if the first function is the true one, and vacuolar sap is an internal excretory product, it is rarely analysed separately and contributes to the overall elemental content of plant tissues.

Active transport

Active transport covers all modes of uptake of ions which require the expenditure of metabolic energy, and has the following characteristics:

1. Ions can be taken up against a concentration gradient. This involves passing through the cell membrane, which is about 7 nm thick and consists of a double layer of lipid molecules, with polar groups pointing outwards and hydrocarbon 'tails' pointing inwards (Fenton, 1977).
2. Active uptake rates are roughly doubled by raising the temperature from 10 °C to 20 °C, which would only increase passive diffusion by about 20 %.
3. Active uptake is relatively independent of the pH of the nutrient

medium over the range 4–9. It does not involve the appearance of an equivalent number of exchanged ions in the medium.

4. Active uptake requires oxygen, and is inhibited by many substances such as cyanide ion which are known to inhibit the respiration of cells.

The minimum energy needed for active uptake of 1 mole of ions can be calculated from the relation

$$E = 2{\cdot}303\ RT \log [X]_i/[X]_o$$

where *i* and *o* refer to inside and outside the cell respectively. For the active transport of potassium ion into the cytoplasm of *Nitella translucens*, the energy requirement is about 13 kJ/mol, which is a small fraction of the energy evolved in respiration (MacRobbie, 1974).

Radioactive tracers have been used to measure the actual rates of transport of a few ions across cell membranes in both directions. Some data for the charophyte *Nitella translucens* are shown in Fig. 7.1, and data for other species can be found in a review by MacRobbie (1974). Extensive research on the giant cells of charophytes has revealed the following general features of the uptake of ions by these tissues.

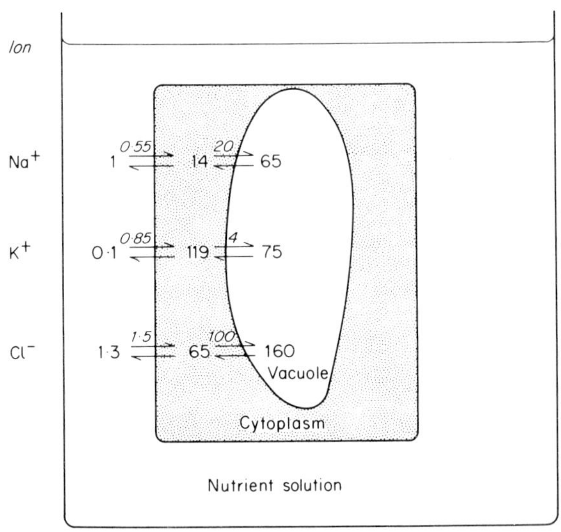

FIG. 7.1 Uptake of Na^+, K^+ and Cl^- by a vacuolated cell of the alga *Nitella translucens*. Numbers are concentrations in mmol/l; italicised numbers above arrows are rates of crossing membranes in pmol/cm^2 s. Data from MacRobbie (1962, 1966 and 1974) and Spanswick and Williams (1964)

1. Potassium may either be actively transported across the cell membrane, or be incorporated by passive diffusion. Chloride ions are actively absorbed, while hydrogen and sodium ions are actively excreted. Less detailed evidence suggests that Ca^{2+}, Mg^{2+}, HCO_3^-, SO_4^{2-} and PO_4^{3-} are all actively absorbed. The uptake of $Si(OH)_4$ by the diatom *Nitzschia* is an active process (Johnson and Volcani, 1978).

2. Both sodium and chloride (but few other ions) are actively excreted from the cytoplasm into the vacuole, so that their concentrations in vacuolar sap usually exceed the concentrations in cytoplasm or nutrient medium. In the marine red alga *Desmarestia*, sulphate ions are actively transported into the vacuolar sap, which contains 0·2 M sulphuric acid (O'Kelley, 1974).

Secretion of inorganic matter by algae

Closely related to the uptake of ions is the formation of hard tissues by algae. What little is known about this process has been summarised by Darley (1974a), who shows that the hard tissues are of three main types:

(*a*) Calcite, formed by some red algae (Cryptonemiales), a few charophytes and blue-green algae, and coccolithophorids. In the latter group, calcification seems to be initiated by the Golgi body within the cell.

(*b*) Aragonite, formed by other red algae (Nemaliales) and a few green and brown algae.

(*c*) Opal, or amorphous silica, formed by many diatoms and silicoflagellates, some Chrysophyceae and Xanthophyceae, and a few green and brown algae.

In addition to massive hard tissues, crystals of barium sulphate, calcium oxalate and calcium sulphate are found in some cells, probably as 'internal' excretion products (Borowitzka, 1977). Fungi are known to excrete calcium oxalate crystals to the soil (Graustein *et al.*, 1977).

Uptake of ions by bacteria and fungi

In a stimulating review, Harold (1977) has emphasised that microbial cells constantly excrete protons, making their surroundings more acid but leaving the cytoplasm both alkaline and negatively charged. An example is the active transport system found in the fungus *Neurospora crassa* by Slayman and Slayman (1968). This takes up K^+ at a rate of 150 nmol m^{-2} s^{-1}, and excretes Na^+ and H^+ at rates of 97 and

56 nmol m^{-2} s^{-1} respectively, and is an ATPase analogous, but not identical, to the ATPases found in the membrane of mammalian cells. The bacterium *Escherichia coli* accumulates both Mg^{2+} and Mn^{2+} by specific active transport systems, but the mechanism remains to be discovered (Silver *et al.*, 1970; Silver and Clark, 1971). An interesting protein has been isolated from *Salmonella* by Pardee (1966). This has a molar mass of 32 000 and binds sulphate ions in a 1:1 molar ratio, so it could be involved in the active transport of that ion.

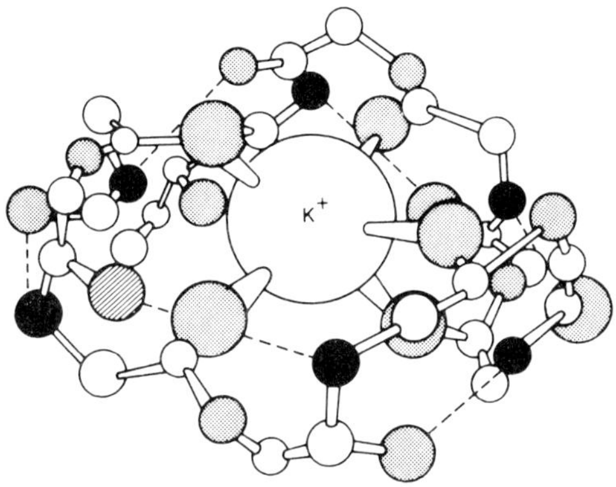

FIG. 7.2 A potassium ion enfolded by the cyclic peptide valinomycin. ○, C; ◍, O; ●, N; ---, hydrogen bond. H atoms are omitted. After Fenton (1977)

Substances of great current interest are the ionophores which have been discovered in numerous extracts from fungi and actinomycetes. Ionophores are neutral molecules which are able to enfold or chelate specific ions, rendering the latter soluble in organic solvents and so allowing them to cross lipid membranes (Prelog, 1971; Pressman, 1976; Fenton, 1977; Burgermeister and Winkler-Oswatitsch, 1977). They include macrocyclic peptides such as valinomycin (from *Streptomyces*), which has a 36-membered ring and is able to fold itself in such a way as to give a cavity which can entrap large ions of K^+ or Rb^+ but not the smaller Na^+ ion (Fig. 7.2.) Similar uncharged substances include nactins from *Actinomyces* and enniatins from *Fusarium*, which are less specific ionophores for K^+. A second class includes macrocycles with carboxylate groups, such as dianemycin, grisoryxin, monensin, nigericin, A23187 and X537A. The last of these forms lipid-soluble complexes with all known cations, while A23187 is very selective for divalent cations. Emery (1971) isolated a cyclic hexapeptide from *Ustilago sphaerogena* which he called ferrichrome, since it is an ionophore for iron(III), but it also transports Al and Ga. While none of these have yet been extracted from algae or higher plants, it is highly likely that these or similar compounds occur there. A peptide isolated from mitochondria, with a molar mass of 1600, acts as an ionophore for Ca^{2+} and K^+ (Blondin *et al.*, 1977).

Uptake of ions by multicellular plants

Multicellular plants can take up ions in two ways. Free-floating aquatic species take up ions directly from solution and differ little in this respect from unicellular algae. Timofeeva-Resovskaya *et al.* (1961) studied the uptake of 19 elements from radioactive solutions by 32 aquatic species including algae, bryophytes and seed plants, with the results summarised in Table 7.2. Mean concentration factors were found to lie in the range 150–7500. Most bryophytes have unicellular rhizoids and can absorb some nutrients from the soil, but they also take up ions from rainfall through their leaves, as can higher plants.

Most vascular plants absorb the bulk of their nutrients, other than carbon, through their roots. Roots take up both water and salts, but there is

TABLE 7.2 Mean concentration factors for 19 elements taken up from freshwater by 32 plant species (Timofeeva-Resovskaya *et al.*, 1961)

Element X	Ion	Mean F_X
Ca	Cation	265
Cd	Cation	1620
Ce	Cation	7100
Co	Cation	4425
Cr(III)	Cation	695
Cs	Cation	480
Fe	Cation	305
Ge	Anion	305
Hg	Cation	5915
I	Anion	370
Nb	Anion	7640
P	Anion	5480
Rb	Cation	1230
Ru	Cation	1700
S(VI)	Anion	165
Sr	Cation	475
Y	Cation	475
Zn	Cation	4600
Zr	Cation	6230

no clear connection between these two processes. The uptake of ions by roots requires dioxygen and is an active process; some plants such as mangroves have evolved special mechanisms for transporting dioxygen to their roots when these penetrate anoxic soils. Absorption of ions takes place mainly through unicellular root hairs which are constantly being replaced behind each root tip as it grows. The efficiency of absorbing potassium and phosphate ions is often increased when symbiotic fungi called mycorrhiza are present; these fungi certainly excrete acids, and perhaps also ionophores, into the soil.

TABLE 7.3 Concentration factors for 12 elements in kale (*Brassica oleracea*) compared with soil solution (Bowen, 1960)

X	mg X/kg FM in kale	mg X/kg soil solution	CF
Ca	10 200	750	14
Cl	850	29	29
Cu	1·2	0·04	30
Fe	30	0·25	120
K	6170	5.4	1140
Mg	400	22	18
Mn	3·7	0·072	51
N	10 800	215	50
Na	620	39	16
P	1120	0·70	1600
S	4000	205	20
Zn	8·2	0·2	41

When plants are grown artificially in nutrient solutions, it is possible to calculate concentration factors as described above. For example, Collander (1941) showed that plants could either concentrate or actively reject sodium ions from nutrient solutions according to the concentration supplied, but that halophytes always took up much more sodium than did other species. Recent work on uptake from nutrient solutions has been reviewed by Nissen (1974).

For plants growing naturally in soil, the concentration factor can be calculated in several different ways, and there is no agreement among agricultural scientists as to how much of a given nutrient in a soil is available to plants. If concentration factors are calculated with respect to the soil solution, which plant roots are assumed to contact, values like those in Table 7.3 can be obtained. These are notably smaller than those found for aquatic plants, implying the possible existence of a barrier to ions at the root surface or within it. It is thought that uptake of potassium ions from soils is an active process, since it is markedly reduced by cooling, which does not affect the rate of uptake of water (Bowling, 1968). Uptake of sulphate ions is also an active process (Smith, 1976). The evidence for active uptake of some essential micronutrients has been reviewed by Moore (1972). He concludes that the uptake of both iron and manganese is linked to respiration and is therefore active. The uptake of boron, copper and zinc can sometimes be linked to metabolism, but further work is needed, as in the case of zinc passive diffusion may also be important.

Translocation of ions in vascular plants

The main channel of translocation from root to shoot is generally thought to be the xylem cells in the vascular system (Crafts, 1961). Xylem sap is not often analysed, but some data is collected in Table 7.4, together with data for phloem sap. The chemical species present in plant saps have also been examined, with the following results (see also Ziegler, 1975):

K, Mn and Na are probably free cations (Tiffin, 1972; MacRobbie, 1974).

Ca and Mg may be complexed by organic acids such as citric acid (Molloy and Richards, 1971); 35% of the Mn in phloem sap of *Ricinus* is organically bound (Van Goor and Wiersma, 1976).

Cl exists as a free anion (MacRobbie, 1974).

TABLE 7.4 Elemental composition of xylem and phloem saps from various plants in mg/l

X	[X] xylem sap *Zea mays*	[X] xylem sap *Lupinus*	[X] phloem sap *Lupinus*	[X] phloem sap *Yucca*	[X] phloem sap many spp.
Ca	34	17	21	14	10–20
Cu			0·4	0·4	
Fe		1·8	9·8	4·1	
K	235	90	1540	1680	800–3300
Mg		27	85	51	56–560
Mn		0·6	1·4	0·5	
Mo				0·01	
Na		60	120	4·1	1·4–7
N	260	10			
P	69			310	90–300
S	51				
Si	85				
Zn		0·4	5·8	2·1	
Ref.	L	P	P	T	M

References: L, Lowry *et al.* (1936); M, MacRobbie (1971); P, Pate (1975); T. Tammes and Van Die (1974)

Fe is anionic; citrate complexes are known in *Glycine* and *Helianthus* (Tiffin, 1972).

Cu, Ni and Zn may exist as one or more anionic complexes (Tiffin, 1972; Van Goor and Wiersma, 1976).

Cr is known to occur as the trioxalato-chromium(III) anion (Lyon *et al.*, 1969).

B probably exists as complexes of *cis*-diols with borate.

Si probably exists as orthosilicate (Lewin and Reimann, 1969); the evidence for organo-silicon compounds is doubtful (Lovering and Engel, 1967).

Transport through the phloem, which redistributes carbohydrates and inorganic ions within the plant, is poorly understood, and the high rates of flow have not been explained. The limited data given in Table 7.4 illustrate the high K/Na ratio in phloem sap, which resembles cytoplasm in this respect. It is not understood why Cl, K, Mg, Na, P and S are mobile in phloem, while B, Ca and Pb are immobile.

Excretion of salts by plants

A number of halophytic seed plants, notably mangroves such as *Aegialitis*, *Avicennia* and *Sonneratia*, but also many Plumbaginaceae, Frankeniaceae, Tamaricaceae and Chenopodiaceae, together with salt marsh grasses such as *Aeluropus*, *Chloris* and *Spartina*, have evolved special salt glands which are able to excrete sodium chloride from the leaf surface. This adaptation is obviously useful to plants which grow in saline soils, and the mechanism presumably involves an extension of the active transport system described above. It appears that the salt glands actively excrete chloride ions, and that sodium ions follow to preserve electrical neutrality (Hill and Hill, 1973; Lüttge, 1975). Other excretion products of plants include crystals of calcium oxalate, known as raphides (Arnott, 1973).

Uptake of ions by invertebrate animals

The most primitive animals, the unicellular protozoa, take up ions from solution in the same way as algae. The majority of the species can form vacuoles, which can be opened at intervals as a means of excretion.

Multicellular invertebrates can be classified into two groups as far as uptake of ions is concerned, those with permeable skins and those with impermeable skins. The majority of marine invertebrates (annelids, coelenterates, echinoderms and molluscs) have permeable skins through which ions can diffuse. In such cases the blood is similar to sea water in its composition, and the cells maintain elevated concentrations of K^+ and $H_2PO_4^-$ by actively excreting Na^+ and Cl^-. Much experimental work has been carried out using giant cells from the nerve axons of squids, because normal sized cells offer many experimental problems (Baker, 1972, 1977). Several active transport systems or ion-pumps have been discovered in this tissue, e.g.

1. Active uptake of K^+, inhibited by tetra-ethylammonium ions, associated with an ATPase in the cell membrane.
2. Active excretion of Na^+, inhibited by ouabain and also by tetrodotoxin, associated with an Na–K ATPase in the cell membrane.
3. Active excretion of Ca^{2+}, inhibited by Co^{2+}, Mg^{2+}, Mn^{2+} or La^{3+}, associated with a different ATPase.

Table 7.5 compares the elemental composition of cytoplasm from nerve axons with that of squid blood, and the data suggest that active excretion of Mg^{2+} may also occur.

TABLE 7.5 Elemental composition of nerve axons (cytoplasm) and blood in the squid (Baker, 1977)

X	mg X/kg cytoplasm FM	mg X/kg blood FM
Ca	16	440
K	15 500	400
Mg	200	1300
Na	1100	9200

Another way of classifying Invertebrates is by their feeding habits. *Filter-feeders* (e.g. Coelenterates and some molluscs) consume plankton and mud which they filter from sea water: *carnivores* feed on other animals. Although it might be expected that filter-feeders would have a very different elemental composition from carnivores, the differences are small in unpolluted environments. In polluted estuaries, filter-feeders become contamined with heavy metals more rapidly than do carnivores (Bryan, 1976).

Simkiss (1976) has commented on the fact that many invertebrates have small granules containing calcium in their cytoplasm. These granules may either be amorphous or consist of crystalline calcium carbonate, and may represent a mode of excretion for calcium. Similar granules consisting of amorphous zinc phosphate and an unknown copper/sulphur compound have been found in the cells of the barnacle *Balanus* on polluted shores

(Walker *et al.*, 1975; Walker, 1977). They also appear to be a mode of excretion. It is possible that these granules are related to the process of shell formation (Schraer, 1970).

Invertebrates with impermeable skins, such as arachnids and insects, exclude ions by a waxy layer which has pores to admit air. These organisms take up ions by mouth, from their food. Univalent ions readily diffuse across the gut wall and enter the bloodstream, while multivalent ions diffuse less readily and are partly absorbed and stored in blind-ended Malpighian tubules which open off the gut wall. Ions are excreted in both urine and faeces. The concentration of ions in the urine is controlled by special glands in the hind gut, which are able to produce urine either more concentrated or more dilute than the blood. The curious difference in composition of blood from primitive and advanced insects has been illustrated in chapter 6. Advanced insects are unique in having high concentrations of potassium and low concentrations of sodium in their blood, but the mechanism by which this is achieved need not require any novel type of active transport system.

Uptake and excretion of ions by cells of vertebrates

Despite the experimental difficulties in studying animal cells, several tissues have been studied with regard to ion transport. For example, Lee and Armstrong (1972) showed that active transport of Na^+ takes place across the gut wall, from digested food to the bloodstream. They found that within the epithelial cells of bull frog intestine virtually all the potassium was present as K^+, but nearly half the sodium was present in a non-ionic form. The actual concentrations found were $[K] = [K^+] = 3{\cdot}35$ g/l; $[Na] = 0{\cdot}62$ g/l; $[Na^+] = 0{\cdot}33$ g/l.

The active uptake of potassium and excretion of calcium and sodium by erythrocytes has been reviewed by Dahl and Hokin (1974) and Schatzmann (1975). The overall reactions pumping ions from outside (o) to the inside (i) of cell membranes are:

$$\underset{o}{ATP^{4-}} + \underset{o}{2K^+} + \underset{i}{3Na^+} \xrightarrow[\text{Na/K ATPase}]{} \underset{o}{ADP^{3-}} + \underset{o}{P^-} + \underset{i}{2K^+} + \underset{o}{3Na^+}$$

and

$$\underset{o}{ATP^{4-}} + \underset{o}{2Ca^{2+}} + \underset{i}{Mg^{2+}} + \underset{i}{2K^+} \xrightarrow[\text{Ca ATPase}]{} \underset{o}{ADP^{3-}} + \underset{o}{P^-} + \underset{i}{2Ca^{2+}} + \underset{o}{Mg^{2+}} + \underset{o}{2K^+}$$

The sodium pump is an Na/K/Mg ATPase which is an intrinsic component of the cell membrane. It has a molar mass of about 300 000, but can be dissociated into two components, one of which is a glycoprotein with a molar mass of about 55 000. It is poisoned by ouabain. The calcium pump is a Ca/Mg ATPase which is also a component of the cell membrane. It is not poisoned by ouabain, but is inhibited by La^{3+}, and is able to pump Sr^{2+} as well as Ca^{2+} (Martonosi, 1972; MacLennan and Wong, 1971; Kretsinger, 1976).

The active excretion of chloride ions from the oxyntic cells lining the vertebrate stomach is well known. In man the fluid excreted contains about 0·17 mol HCl/l, which is used in digesting food. The classic explanation is that the oxyntic cells pump bicarbonate ions into the bloodstream and hydrogen ions into the stomach; the chloride ions follow the latter to maintain electrical neutrality.

Uptake and excretion by entire vertebrate animals

Vertebrates ingest most of their nutrient elements through the mouth, though there is some absorption through the lung. The Amphibia have permeable skins which take up some ions, and fish can absorb ions through their gills, but other vertebrates have impermeable skins. All vertebrates have blood which is more dilute than sea water; the kidneys of mammals and birds can produce urine which is more concentrated than their own blood, which is not the case for Amphibia, fish or reptiles. A few marine vertebrates have special salt-excreting organs which are situated in the gills of fish, above the eyes of reptiles and in the nasal glands of birds (Lockwood, 1963).

A diagrammatic outline of the passage of food through a vertebrate body is given in Fig. 7.3. After digestion by enzymes and/or hydrochloric acid in

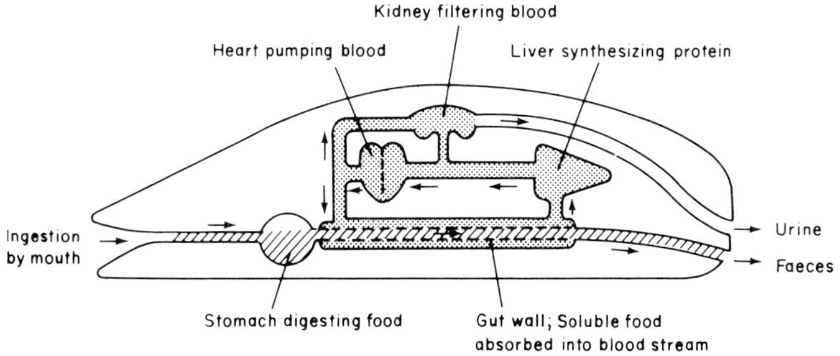

FIG. 7.3 Diagrammatic section of a mammal to show how foods are ingested, absorbed and excreted

the stomach, the food passes in aqueous suspension into the gut, which is permeable to ions. The ions which pass through the gut wall are absorbed into the bloodstream, from which a small amount may be transferred back into the gut again. Solid food which has been neither digested nor absorbed is excreted from the far end of the gut as faeces.

The bloodstream which carries away the elements absorbed through the gut wall is pumped round the body by the heart. Some ions are removed by the liver and other tissues such as bone, but the main organ which purifies the blood is the kidney, which filters human blood about once every 50 minutes. Waste products from the blood are excreted by the kidney as liquid urine. A rough guide to the percentage of a given ion in food which has been metabolised is given by the ratio of urinary excretion to oral ingestion. This is clearly imperfect, as we should allow for fluxes in both directions across the gut wall, as well as the fluxes to and from storage tissues, but such information is rarely available. The percentage of daily intake of 49 elements by 'Reference Man' which is excreted in the urine is given in Table 7.6 (Snyder, 1975). Some of the data may be unreliable and this is indicated by a query sign. Although the chemical species present in food, blood and urine are rarely known, the elements have been arbitrarily classified as anions or cations. Some of the multivalent elements may in fact be present as colloidal species, which would not behave in the way we might expect.

According to Table 7.6, nearly all anions readily diffuse across the gut wall: vanadates and perhaps arsenates and antimonates appear to be exceptions. Some anions, notably arsenite and chromate, but also antimonite, polonate, selenate and tellurate react with specific proteins in erythrocytes and remain in these cells until they die. Fluoride, phosphate and tungstate are rapidly incorporated into growing bone, while iodide is absorbed by the thyroid gland. Molybdate and vanadate are temporarily accumulated by the liver, but are not stored there. Anions such as bromide are much less quickly removed from the bloodstream by the kidney than are cationic or colloidal impurities.

Univalent cations also diffuse readily across the gut wall, become more or less uniformly distributed in the soft tissues and are then rapidly eliminated by the kidney. Sodium is the dominant cation in plasma, while potassium predominates within erythrocytes and other cells. Small amounts of salt (sodium chloride) are excreted as sweat by man and a few other mammals, but the concentration of salt in sweat is lower than it is in blood. Among the rare elements, thallium(I) is highly toxic and silver(I) has a unique pattern of metabolism. Ag is almost quantitatively absorbed by proteins in the plasma, is removed from circulation by the liver and is mostly excreted into the faeces (Durbin, 1960).

Bivalent cations are less readily absorbed across the gut wall, especially if

TABLE 7.6 Percentage of daily intake of Reference Man which is excreted in urine

X	Anions % (Reference)	X	Cations % (Reference)
As	5 (S); 25 (Sn)	Ag	0·8 (W); 13 (S)
B	77 (S)	Al	0·22 (S)
Br	93 (S)	Ba	6·7 (S)
Cl	84 (S)	Be	8·3 (S)
Cr(VI)	46 (S) ?	Bi	8 (S)
F	55 (S)	Ca	17 (S)
Ge	93 (S)	Cd	4·5 (T, W); 67 (S) ?
I	85 (S)	Co	0·7 (Cs); 22 (C); 67 (S)
Mo	50 (S)	Cr(III)	0·4 (Cs)
N	97 (S)	Cs	90 (S)
Nb	58 (S) ?	Cu	1·4 (S)
P	64 (S)	Fe	1·6 (S)
S	94 (S)	Hg	7 (T, W) ?; 33 (S) ?
Sb	3 (W); 80 (S)	K	85 (S)
Se	33 (S); 80 (C)	Li	40 (S)
Si	50 (S) ?	Mg	39 (S)
Te	88 (S) ?	Mn	0·006 (Cs); 0·81 (S)
V	0·75 (S) ?	Na	75 (S)
		Ni	2·7 (S)
		Pb	10 (−85 in infants) (S)
		Ra	3·5 (S)
		Rb	86 (S)
		Sc	5 (C)
		Sn	0·5 (S) ?
		Sr	18 (S)
		Th	3·3 (S); 0·7 (Z)
		Ti	39 (S)
		Tl	33 (S)
		U(VI)	16 (S)
		Zn	3·8 (S)
		Zr	3·6 (S)

%, 100 × (daily excretion in urine/daily intake in food)
References: C, Clemente (1976); Cs, Cornelis *et al.* (1975); S, Snyder (1975)—data mostly from Sn; Sn, Schroeder and Nason (1971); T, Task group (1973); W, Wester (1973); Z, ICRP (1972)

they have large atomic weights. Once in the bloodstream Ca^{2+}, Sr^{2+}, Ba^{2+}, Ra^{2+} and Pb^{2+}, together with some Mn^{2+} and Zn^{2+}, are absorbed by growing bone and are then excreted very slowly; lead is unusual in that infants absorb it far more efficiently across the gut wall than do adults. Mg^{2+} behaves rather like sodium, while Be^{2+} is very poorly absorbed and so behaves like aluminium. Cadmium, mercury and the uranyl cation (UO_2^{2+}) have little affinity for bone but react with proteins in the kidney, where the two former elements may be retained for long periods.

Multivalent cations are almost all poorly absorbed across the gut wall, possibly because they hydrolyse to form colloidal suspensions at neutral pH. They are therefore poorly metabolised and are mostly excreted in the faeces. However, the small fraction that is absorbed tends to be very tenaciously retained, usually by bone. This has been confirmed by tracer work, often involving intravenous injection, as for example the study of the metabolism of seventy elements by the rat (Durbin, 1960). A surprising finding is the selectivity of the body towards the lanthanides. All are strongly taken up by bone from the bloodstream, but the larger cations (La–Gd) are also retained by the liver, which is not so for the smaller cations (Y, Tb–Lu). The actinides Ac, Am, Cm and Pu are also held up by the liver, and later deposited in the bone (ICRP, 1972).

The extent of absorption across the gut wall must depend on the chemical species present in the digested food, and can be modified by other components of the diet. For example polygalacturonates, phytates, sodium alginate and vitamin D all affect the absorption of calcium, ascorbic acid affects the uptake of iron, aluminium ions affect the uptake of phosphate and barium ions that of sulphate (Comar and Bronner, 1962). The interactions of elements and organic matter form too large a topic for discussion here.

Current thinking about the formation of bone mineral have been discussed by Simkiss (1975). Both calcium and phosphate ion concentrations in the blood are homeostatically controlled, but if thermodynamic equilibrium were attained then calcium phosphate should precipitate out, which in practice only occurs on surfaces of growing bone. One theory is that bone growth is modified by an inhibitor, possibly the pyrophosphate ($P_2O_7^{4-}$) ion, which prevents further crystallisation unless it is destroyed by an enzyme such as alkaline phosphatase. Another theory holds that bone is normally covered by a layer of cells which prevent direct contact with the bloodstream. It also seems likely that collagen fibres, which contain calcium-binding proteins (chapter 9), play an important part in the formation of bone.

Excretion by the kidney is a complex process, but it is known that active excretion of ions such as chloride, sulphate and sodium can occur there. The kidney also secretes granules of calcium phosphate, which can take up substantial amounts of lead when the diet is rich in this metal (Carroll *et al.*, 1970). Other modes of excretion by vertebrates, such as sweating, or transfer to hair, feathers or nail are of minor quantitative significance. A few non-metals such as As, I, Po, S, Sb, Se (and perhaps also noble metals such as Au and Hg) are characteristically excreted into hair and nail by mammals.

Summary of active transport of ions

Our knowledge of active transport is still in a developing phase. While both ionophores and ATPase ion pumps are certainly involved in some aspects, the presence of ionophores has yet to be demonstrated in many tissues. Some workers have speculated that cell membranes may contain tunnels of cyclic peptides which can act as pores allowing ions to enter the cell, but this does not easily account for the directional nature of active transport. Much work on the biochemical mechanisms of ion-pumps is still in progress.

8 *Essentiality, Deficiencies and Toxicities of the Elements*

The criteria which are necessary and sufficient to establish the essentiality of an element X for an organism may be formulated thus (Arnon, 1950).

1. The organism can neither grow nor complete its life cycle without X.
2. X cannot be wholly replaced by any other element.
3. X has a direct influence on the organism, and is involved in its metabolism.

In practice, certain explanatory notes should be added. Since X can never be entirely removed, 'without X' must be understood to mean 'if X is supplied at a sufficiently low concentration'. Secondly there are a few cases where a pair of closely related elements may be interchangeable with regard to their essentiality. While none of these cases is completely proved, there is no theoretical reason to doubt that they could occur, in view of the fact that a number of enzymes can be activated by several metals. Examples are the interchangeability of potassium and rubidium in bacteria (Lester, 1958) and the green alga *Ulva* (Eppley, 1962); calcium and strontium in the green alga *Chlorella* (Walker, 1953) and the fungus *Allomyces* (Ingraham and Emerson, 1954); and possibly vanadium and molybdenum in some bacteria such as *Azotobacter* (Bortels, 1930; Nicholas *et al.*, 1960). Finally a few cases of conditional essentiality are known. For example, many micro-organisms find molybdenum essential if they have nitrate as a source of nitrogen, but not if they are fed with ammonium salts (Ichioka and Arnon, 1955). A different example is the essentiality of cobalamin (vitamin B_{12}) to mammals, which cannot be met by feeding cobalt.

Deficiency is observed as abnormal or stunted growth when an essential element is present in inadequate concentrations in the nutrient intake. Serious deficiencies can lead to growth or reproductive failure, or to death. Deficiency disorders can usually be corrected by adding the deficient element to the food supply.

Toxicity can result from too high a concentration of an element in the nutrient intake, and also gives rise to abnormal growth, disease or even death. All elements, including the essential ones, are toxic at high concentrations. For example, the essential nutrient copper is toxic at quite low concentrations, and is widely used as an algicide and fungicide. Idealised growth response curves for essential and inessential elements are given in Fig. 8.1 (Smith, 1962). For essential elements the optimum range of concentrations in the nutrient may be narrow, as it is for Cu and Se, or broad, as it is for Ca and K.

It is well known that toxicity depends on the chemical nature of the species containing the element concerned, and is readily modified by other components in the nutrient solution or diet.

Miettinen (1975) has emphasised the variability of individuals in biological populations towards toxic elements. In practice this is allowed for by measuring a dose-response curve for a number of individuals, when a curve resembling that shown in Fig. 8.2 should be obtained; this figure also shows how the dose needed to cause 50% of the population to respond, or RD_{50}, is estimated. If the response measured is death, the lethal dose

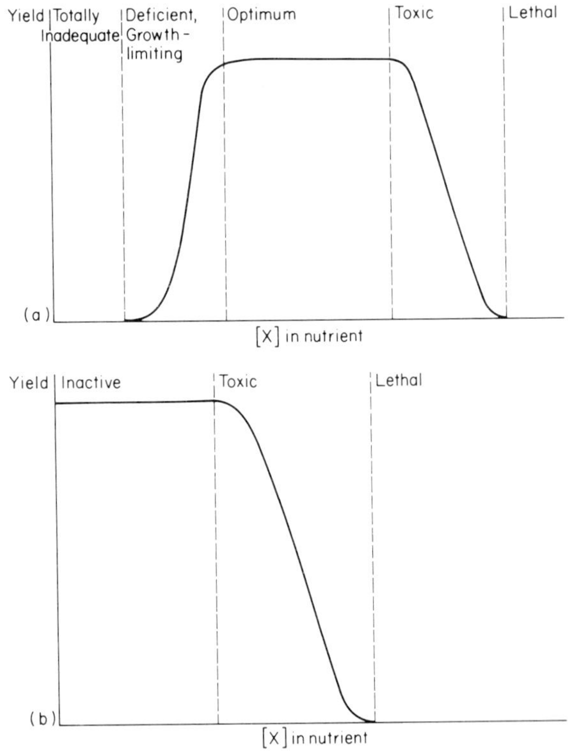

FIG. 8.1 Idealised diagrams showing the yield of an organism as a function of the concentration of an element X in the nutrient supplied: (a) X essential, (b) X inessential

quoted is usually the LD_{50}. Any RD_{50} or LD_{50} value should preferably be qualified by specifying:

(*a*) The chemical nature of the poisonous element X.
(*b*) How X was fed.
(*c*) The age or developmental stage of the organism.
(*d*) The time period between dosage and response.

It should be clear that the RD_{50} is by no means a perfect measure of response, but no better measure has been proposed.

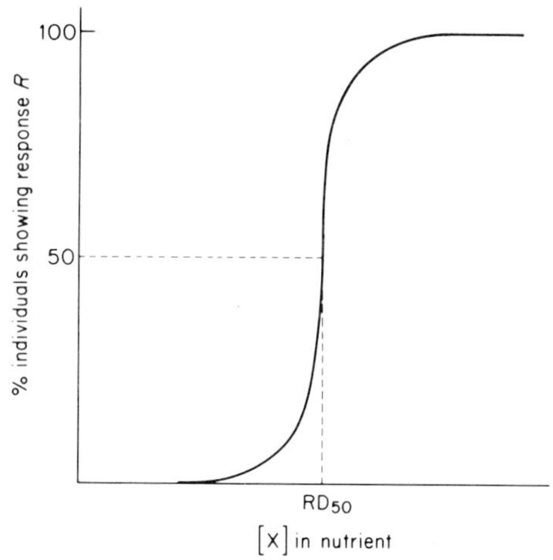

FIG. 8.2 Idealised dose-response curve to show how RD_{50}s may be obtained

In the remainder of this chapter, organisms will be considered under two main headings. Bacteria, fungi and green plants can be grown in sterile nutrient solutions of known composition, and many are autotrophic so that they require only inorganic nutrients. Animals, on the other hand, are all heterotrophic and cannot exist on a purely inorganic diet.

Plants, bacteria and fungi

Although it might seem to be a relatively simple matter to test the elemental requirements of organisms which can be grown in solutions of inorganic salts, there are many practical problems which become highly intractable for those elements required in extremely small amounts. A distinction is often made between macronutrients, such as Ca, K, Mg, N, P and S, which are usually supplied at concentrations of 10 mg/l or more, and micronutrients, which are usually needed at concentrations of less than 1 mg/l. Practical problems are rarely encountered in work with macronutrients, but the following types of problem can arise in micronutrient studies:

1. Contamination from the macronutrient salts or water used, necessitating careful purification and analysis.
2. Contamination from the containing vessel, especially for B, Na and Si.
3. Contamination from air and dust; filtered air was needed to prove the essentiality of chlorine for plants (Broyer *et al.*, 1954).
4. Contamination from propagules: for example a tomato seed weighing 4 mg may contain up to 10 ng Mo, which is enough to sustain a fair-sized plant (Bowen and Cawse, 1962).
5. Losses can arise owing to unexpected precipitation of substances such as magnesium or iron phosphate in the nutrient solution.

As an example of the problems of purification of nutrients, consider the element cobalt, which may be needed in organisms at concentrations of about 10 ng Co/g FM. Since the concentration factor for cobalt from the nutrient solution may be 10^4 or more (Table 7.1), the nutrient should contain less than 1 ng Co/l, and methods of analysis for picogramme quantities of cobalt will be needed. The difficulties involved in preparing and purifying nutrients for work with higher plants have been thoroughly discussed by Hewitt (1966). Microbiologists have paid less attention to micronutrient problems, perhaps because of the organic components of their nutrient media are rich sources of trace elements. Hutner (1972) has suggested that many recommended microbial nutrients may be deficient, and speculates that this could be one of the reasons why so few parasitic micro-organisms can be grown on non-living media.

ESSENTIAL ELEMENTS FOR PLANTS, BACTERIA AND FUNGI

Table 8.1 summarises what is known about the essentiality of elements for several groups of organisms. It must be borne in mind that only about 300 of the half million species of plants have been investigated with respect to their elemental requirements, and the evidence for essentiality sometimes rests on experiments with a single species. However, the elements C, H, N, O, P and S are universally regarded as essential to all organisms and are omitted from Table 8.1. The facts in this table can be summed up as follows:

Probably essential to all plants: C, Ca, Cl, Co, Cu, Fe, H, K, Mg, Mn, N, Na, O, P, S, Se and Zn.

Essential to some groups, not necessarily for all: B, I, Si.

Further work needed: As, Br, Cr, F, Ni, Sn, V.

TABLE 8.1 Elements essential for bacteria, fungi, algae and seed plants (other than C, H, N, O, P and S). (References are given in brackets)

Element	Bacteria	Fungi	Blue-green algae	Green algae	Seed plants
B	e ?(G)	e ?(He)	e (Ok)	e ?(Ok)	E (Warrington, 1923)
Ca	*I*[Sr] (N)	e ?(Hu)	E (Ok)	E (Ok)	E (Bo)
Cl	e (Y)		E (Ok)	E (Ok)	E (Johnson *et al.*, 1957)
Co	E (Hu)	E (Hu)	E (Ok)	e ?(Ok)	e (Hallsworth *et al.*, 1965)
Cu	E (Hu)	E (Hu)	E (Ok)	E (Ok)	E (Sommer, 1931)
Fe	E (Hu)	E (Hu)	E (Ok)	E (Ok)	E (Gris, 1844)
I					e ?(Lehr *et al.*, 1958)
K	*I*[Rb] (L)	E (Nc)	E (Ok)	*I*[Rb] (K)	E (Bo)
Mg	E (Nc)	E (Nc)	E (Ok)	E (Ok)	E (Bo)
Mn	E (Hu)	E (Hu)	E (Ok)	E (Ok)	E (Bertrand, 1905)
Mo	*I*[V] (B)	E (Hu)	E (Ok)	E (Ok)	E (Arnon and Stout, 1939)
Na	e (Y)	e (Hu)	e (OK)	e (Br)	e (Br)
Ni	e (Hu)				e (Dixon *et al.*, 1975)
Se	e (Hu)				e ?(Broyer *et al.*, 1966)
Si				e (M)	e (Raleigh, 1939)
V	*I*[Mo] (B)	e ?(Be)		e (A)	
Zn	E (Hu)	E (Hu)	E (Ok)	E (Ok)	E (Mazé, 1914)

E, essential for several, probably for all species; e, essential for at least one species, not necessarily for all; *I*[X], interchangeably essential with element X. The reference given is not necessarily the first proof of essentiality.

Other essential or beneficial elements: Al is beneficial to some trees (*Camellia* and *Eucalyptus*: Mullette, 1975; Matsumoto *et al.*, 1976). As is essential to the red alga *Asparagopsis* (Van Stosch, 1963). B is also essential to diatoms (Ok). Br is essential to some red algae, and can replace Cl in some unicellular algae (McClachan and Craigie, 1967; Fries, 1975). Co is also essential to diatoms, red algae and brown algae (Ok). I is essential to red algae and brown algae (Ok; Van Stosch, 1963; Pedersen, 1969). Si is essential to diatoms (Ok).

References: A, Arnon and Wessel (1953); B, Bortels (1930); Be, Bertrand (1941); Bo, Boussingault (1851); Br, Brownell and Wood (1957); G, Gerretsen and de Hoop (1954); He, Herzinger (1940); K, Kellner (1955); L, Lester (1958); M, Moore and Traquair (1976); N, Norris and Jensen (1957); Nc, Nicholas (1963b); Ok, O'Kelley (1974); Y, Young *et al.* (1944)

NUTRIENT SOLUTIONS FOR BACTERIA AND LOWER PLANTS

The scrappy and disconnected data on the optimum ranges of elemental composition of nutrient solutions is brought together in Table 8.2. The data are for organisms whose natural habitat is freshwater, since marine organisms prefer sea water, whose composition (Table 2.5) is more or less constant. There is insufficient information to specify the composition of deficient and toxic nutrients.

The effects of mineral deficiencies on lower plants have been reviewed by Nicholas (1963b). The requirements of micro-organisms for nutrients when grown on deficient media can be sufficiently reproducible to allow them to be used for bio-assays of elements, using *Streptococcus faecalis* for K and

TABLE 8.2 Ranges of concentrations in nutrient solutions, in mg X/l, giving healthy growth

X	Bacteria	Fungi	Blue-green algae	Green algae
B	0·1–0·5	0·01–0·1	0·1–1	<0·0005–4
Ca	0·1–40	0·2–40	20	10–120
Cl	50–2000	120	60	0·1–1000
Co	0·01–0·1	0·002–0·1	0·01	0·003–1
Cr	0·1	0·1		
Cu	0·005–0·4	0·01–0·4	0·02	0·0006–1
Fe	0·02–6	0·1–6	4	0·02–10
K	200	40–300	2000	20–4000
Mg	0·5–50	25–50	25	2·5–500
Mn	0·01–5	0·005–5	0·5	0·01–10
Mo	0·01–2	0·02–2	0·1–0·2	0·002–0·8
N	250	1000	300	0·35–350
Na	40–1200	40	16	2–250
Ni	0·1	0·1		
P	30–150	30–250	450	15–600
S	30	3–60	30	0·01–1000
Se				0·01
Si				<6
V	<0·2	0·01–0·2	0·01	0·01
Zn	0·1–5	0·2–5	0·005	0·01–10

Iverson *et al.* (1975) have isolated a bacterium (*Pseudomonas* spp.) strains of which can tolerate the following nutrient concentrations, in mg X/l: Al 1000; As(V) 10–50; Co 50; Cr(III) 10; Hg 50; Mg 10 000; Pb 100; Sn(II) 10; Sn(IV) 1000; Te(VI) 10

References: For bacteria and fungi, see Spector (1956a); Nicholas (1963a); Hutner (1972); for blue-green algae, see the above and Wiessner (1962); for green algae, see Eyster (1967); Droop (1969); Aaronson (1970); San Pietro (1971)

Mg, *Lactobacillus arabinosus* for Mn, or various strains of *Aspergillus niger* for Cu, K, Mg, Mn, Mo, P and Zn. The limit of detection for molybdenum is 100 pg, which is better than most chemical techniques.

The effects of metal toxicity have not been studied much for algae and bacteria, but are better known for the fungi (Scharrer, 1955; Horsfall, 1956). It is essential to standardise experimental conditions when attempting to measure toxicities, since these may arise from many causes in addition to the presence of metals, e.g. 1. Incorrect osmotic pressure in the nutrient solution, which should be in the range 0·4–1 bar for freshwater species, or about 23 bar for marine species. 2. Extremes of pH, which should be between 4 and 9, though many species are more exacting. Excessive growth of organisms can reduce the pH of the medium. 3. Deficiency of dioxygen, which may arise if excessive respiration is occurring.

All these problems can be overcome by frequently, or in some cases continuously, replenishing the nutrient solution. Experimental work on the toxicity of elements to lower plants is beset with two further problems. One is the innate variability of individual organisms, which can sometimes adapt themselves to normally poisonous media. For example, certain fungi or yeasts have been selectively adapted to grow in high concentrations of As (Cerbón, 1970), Co (Perlman and O'Brien, 1954), Cr (Roberts and Marzluf, 1971), Li (Laskowski, 1956), Cu, Mn, Ni or Zn (Horsfall, 1956). The second problem is that the toxicity of metal ions in solution is profoundly modified by the presence of ligands or chelating agents, some of which may be excreted by the organisms themselves. In general a complexed or chelated metal is much less toxic than the free ion, but there are notable exceptions. Methyl mercuric chloride is 65 times as toxic to fungi as is mercury(II) chloride, and higher alkylmercury chlorides are more toxic still (Webb, 1975). Methylarsenicals can be more poisonous than arsenites, and trialkyl tin derivatives are more toxic than tin(IV) compounds (Lyr, 1977).

Green algae are poisoned by tin(II) or copper(II) at concentrations of 0·002 and 0·01 mg/l respectively (Young, 1935; Wiessner, 1962), but can tolerate manganese(II) at concentrations up to 2 mg/l (O'Kelley, 1974). Probably the most effective algicides are dichlorine or dibromine at 0·2 mg/l.

Ag(I), Au(III) and Hg(II) are the most toxic metals towards the bacterium *Thiobacillus ferrooxidans*, while Cu(II), Tl(I), U(VI) and Zn(II) are somewhat less toxic (Murr *et al.*, 1978).

The most toxic elements towards fungi are Ag(I), Cu(I), Cu(II), Hg(I) and Hg(II). Of these only copper(II) is used on a large scale for spraying vineyards (Bordeaux mixture), while phenylmercury derivatives are used to treat seeds. All of them are believed to act by reacting with free –SH groups present in the cell membrane, causing a rapid loss of potassium from the cells. They are rather unspecific poisons at concentrations of about 1 mg/l in the applied solution. Somers (1959) has compared the relative toxicity of a number of metals towards the spores of *Botrytis fabae*. The order of toxicity found was:

$$\mathrm{Hg > Cu > Pb > Pd > Ni > Co > Be > Zn > Mn > Sr > Mg, Ca > Ba}$$

which correlates fairly well with the order of electronegativities, or the order of solubility products of the metal sulphides.

Elementary sulphur, S_8, is very toxic towards most bacteria and fungi, but not to green algae. It readily penetrates cell membranes, and is then reduced to H_2S by the respiratory enzymes within the cell, but is unable to penetrate the cell walls of higher plants (Lyr, 1977).

Since we do not know how a healthy cell functions, it is not surprising that the mechanism of toxicity of most elements is unknown or speculative. The following suggestions deserve further investigation.

1. *Interference with the permeability of the cell membrane.* This probably occurs with free ions of Ag, Au, Cd, Cu, Hg, Pb and UO_2^{2+}, and also with free halogen molecules (F_2, Cl_2, Br_2 and I_2). B-metals such as Ag, Hg and Pb are thought to react with free thiol groups, since their toxicity is alleviated by prior complexation with cysteine (Stirpe and Novello, 1971).

2. *Antimetabolites.* Antimetabolites are supposed to occupy sites in the cell which are normally filled by essential metabolites, e.g. carbon monoxide prevents haemoglobin from taking up dioxygen, and sulphur dioxide interferes with the absorption of carbon dioxide by plants. Beryllium may compete with magnesium; it is strongly bonded to some proteins and inhibits many enzymes (Reiner, 1971). Other probable antimetabolites include arsenates and antimonates (for phosphates), chlorates (for nitrates), cyanides (powerful enzyme inhibitor), selenates and tellurates (for sulphates) and tungstates (for molybdates). Fluoride toxicity to *Streptococcus* may arise from the inhibition of enolase activity (Kashket and Bunick, 1978).

3. *Reactions with essential metabolites.* Sulphide ions may precipitate ZnS and cause zinc deficiencies in plants (Kittrick, 1976). Barium ions may precipitate sulphates, and many ions may precipitate phosphates or react with ADP or ATP (e.g. Al, Be, La–Lu, Sc, Y, Zr).

4. *Partial replacement of essential ions.* Examples are Br and Cl, Li and Na, Cs or Rb and K, and Sr and Ca. All of these have very low toxicity, which only arises when a large proportion of the essential ion has been replaced. The toxicity of deuterium is of this type. Strontium rickets in mammals may arise because Sr fails to trigger part of the homeostatic control mechanism for Ca (Omdahl *et al.*, 1977).

NUTRIENT SOLUTIONS FOR VASCULAR PLANTS

Table 8.3 attempts to summarise what is known about deficient, optimum and toxic concentrations of elements in nutrient solutions, but it must be recalled that different plant species vary somewhat in sensitivity. For information on the requirements of individual species, the following reviews may be consulted; Scharrer, 1955; Stiles, 1958; Chapman, 1966; and Hewitt, 1966.

The most toxic element to seed plants appears to be As(III), but Al, B, Be, Cd, Co, Cr(VI), Cu, I, Mo, Ni, Se(IV) and Tl can all be toxic at low concentrations, and it is noteworthy that many of these are also essential. As(V), Bi, Br, F, Fe, Li, Mn, Pb, Te, V, W and Zn are moderately toxic, while Ba, Ca, Cl, K, Mg, N, Na, Rb, S and Sr are only toxic at very high concentrations. Plant tolerance of Al^{3+} and Mn^{2+} is particularly variable. The concentration of these cations in the nutrient solution depends critically on pH, and in Nature they are only abundant in acid lakes and bogs with a pH < 5. Some plant species, such as the whole group of clubmosses (Lycopodiaceae), have adapted to high concentrations of aluminium, while the heath family (Ericaceae) are tolerant of high concentrations of manganese as well. A few individual species or ecotypes have evolved adaptations to high concentrations of other elements, such as As, Co, Cu, Ni, Pb, Se or Zn, as described in chapter 12.

The data given in Table 8.3 show that information on the optimum concentrations of many elements in nutrient solutions is far from complete: the references quoted give information on further complications such as interactions between nutrients. The situation with regard to plants growing on soil is even less satisfactory. As a first approximation one might regard the soil solution as the nutrient medium in which plant roots grow, but in practice soil solutions have been too rarely analysed to test their predictive value concerning potential deficiency or toxicity. It is known that the total elemental composition of a soil is a poor guide to such predictions (Bear, 1964), and in practice agricultural chemists use two techniques to test soil nutrients:

1. Empirical correlations based on extracting the soil with specific reagents, and comparing the amount of each element extracted with plant growth (chapter 4). Unfortunately no universal reagent has been found which gives useful results with every nutrient.
2. Empirical correlations based on the analysis of plants growing in the soil, and comparing the concentrations of elements in healthy, deficient and excessively loaded tissues (Chapman, 1966; Gauch, 1972). This approach works but requires a large amount of background data for each element and for each plant species, and also needs careful attention to factors such as the age of the tissue to be analysed.

Animals: nutrient requirements

The difficulties of purifying diets for heterotrophic organisms, which require organic nutrients such as vitamins or growth factors, are much greater than those encountered for autotrophic plants. The problems involved have been discussed by House (1961) for insects, and by Underwood (1976) for mammals. It may be necessary to raise animals in plastic cages, using highly purified diets and filtered air, in order to demonstrate essentiality. The limiting feature is often the sensitivity of the method of analysis for the element concerned.

Table 8.4 gives a list of those elements now thought to be essential for

TABLE 8.3 Concentrations, in mg X/l, in nutrient solutions affecting the growth of seed plants

X	X deficient	Normal growth	X toxic	References
Al		<0·1	0·1–32	C, St; McLean and Gilbert (1927)
As(III)		<0·02	0·02–7·5	S, St
B	<0·1	0·02–1	1–5	C, H, St
Ba		<500	500	C
Be		<0·5	0·5	Romney *et al.* (1962)
Bi			27	S
Br		<15	15–600	C
Ca	0·015–4	0·12–200	>500	C, H; Wyn Jones and Lunt (1967); Loneregan *et al.* (1968)
Cd		<0·05	0·2–9	S; Page *et al.* (1972)
Cl	0·01	4–200	>600	C, H; Johnson *et al.* (1957)
Co	0·0006	0·001–0·01	>0·1–3	C, H, St; Hallsworth *et al.* (1965); Wilson and Nicholas (1967)
Cr(VI)		<0·5	0·5–10	C; Turner and Rust (1971); Mertz (1974)
Cu	<0·01	0·01–0·1	0·5–8	C, H, S, St
F		<5	>5	S
Fe	<0·5	0·5–5	10–200	C, H
Ga		0·01–0·02		S
I	<0·000001	0·1–0·25	1	C, St
K	0·03–0·12	0·1–320	>500	H; Asher and Ozanne (1967)
Li		<10	>30	C, S

X	X deficient	Normal growth	X toxic	References
Mg	<5	12–50		H
Mn(II)	0·0025–0·02	0·1–1	1–100	H, S, St; Schroeder *et al.* (1966)
Mo	<0·01	0·01–0·5	0·5–2	C, H
N	<1·4	70–210	>1000	H
Na	<0·06	0·06–60	>400	H, St
Ni		<0·5	0·5–2	S
P	<10	20–150	>500	H
Pb		<3	3–20*	S; Miller and Koeppe (1970); Höll and Hampp (1975)*
Rb		<90	>90	S
S	<0·7	3–270	>8000	C, H
Se(IV)	<0·02	<1	1–2	S; Broyer *et al.* (1966)
Si	<0·09			Woolley (1957)
Sn			>40	Schroeder *et al.* (1964)
Sr		<400	>400	S
Te(IV)		<6	>6	S
Tl		<1	1	St
V	<0.0004	<10	10–40	C; Schroeder *et al.* (1963); Walch and Huffman (1973)
W		<1	>10	S
Zn	<0·0006	0·002–0·2	60–400	H, S; Carroll and Loneregan (1968)

C, Chapman (1966); H, Hewitt (1966); S, Scharrer (1955); St, Stiles (1958)

* 0·2 mg Pb/l reduces CO_2 fixation by isolated chloroplasts

insects, birds and mammals, but there is no reason to believe that it is complete. Essential elements include most of the light elements (except Al, B, Be and Li) and first long period elements (except Sc, Ti, Ga, Ge and Br), but only I, Mo and Sn among the heavier elements.

TABLE 8.4 Essential elements for insects, birds and mammals (excluding C, H, N, O, S and P)

X	Insects	Birds	Mammals
As			e (Nielsen *et al.*, 1975)
Ca	e (Sang, 1956)	E	E (Eagle, 1956)
Cl	E?		E (Eagle, 1956)
Co	E	E	E (Underwood and Filmer, 1935)
Cr			E (Schwarz and Mertz, 1959)
Cu	E (Gordon, 1959)	E (Beck, 1956)	E (Hart *et al.* 1928)
F			E? (Schwarz and Milne, 1972a)
Fe	E (Gordon, 1959)	E (Planas, 1967)	E (Underwood, 1977)
I		E (Creek *et al.*, 1954)	E (Chatin, 1852)
K	E (Sang, 1956)	E	E (Eagle, 1956)
Mg	E (Sang, 1956)	E	E (Eagle, 1956)
Mn	E (Gordon, 1959)	E (Wilgus *et al.*, 1936)	E (Kemmerer *et al.*, 1931)
Mo	E?	E (Davies *et al.*, 1960)	E (Higgins *et al.*, 1956)
Na	E (Sang, 1956)	E	E (Eagle, 1956)
Ni		E (Nielsen and Sauberlich, 1970)	E (Nielsen and Sandstead, 1974)
Se		E (Thompson and Scott, 1969)	E (Schwarz and Foltz, 1957)
Si		E (Carlisle, 1972)	E (Schwarz and Milne, 1972b)
Sn			E? (Schwarz *et al.*, 1970)
V*		E (Hopkins and Mohr, 1973)	E (Schwarz and Milne, 1971)
Zn	E (Gordon, 1959)	E (O'Dell *et al.*, 1958)	E (Keilin and Mann, 1940)

E, essential for several, probably for all species; e, essential for at least one species, not necessarily for all; *V may be essential for some Ascidians (Biggs and Swinehart, 1976). The reference given is not necessarily the first proof of essentiality.

Other potentially essential elements: Ba and Sr may be essential for mammals (Rygh, 1949); Br may be essential for mammals (Huff *et al.*, 1956); Cd may be essential for mammals (Schwarz and Spallholz, 1979)

A limited amount of work has been carried out on the toxicity of elements to fish (Jones, 1964; Katz, 1975). The commonest toxicant in aquaria is probably ammonia, which affects fish at concentrations of 0·3 mg/l (Spotte, 1970). Factory effluents containing phosphorus, P_4, have caused local damage to fisheries (Zitko *et al.*, 1970). Jones (1939) has given a useful list of the toxic concentrations of metals towards the freshwater stickleback (*Gasterosteus aculeatus*), summarised in Table 8.5: note the high toxicities of silver, copper and mercury. As usual, the toxicity of metals is affected by many factors, including pH, temperature, calcium content of the water and the presence of natural chelating agents. Information on the toxicity of metals to invertebrates is given in reviews by Katz (1975) and Bryan (1976).

A great deal of experimental work has been devoted to determining the elemental composition of mammalian diets. Available information on deficient, normal and toxic concentrations in the diets of two species, the

TABLE 8.5 Toxicity of metals to the stickleback by chronic exposure, in water containing 1 mg Ca/l (Jones, 1939)

X	Lethal concentration/mg l^{-1}	X	Lethal concentration/mg l^{-1}
Ag	0·003	Hg	0·008
Al	0·07	K	50
Au	0·4	Mg	300
Ba	400	Mn	40
Ca	800	Na	500
Cd	0·2	Ni	0·8
Co	10	Pb	0·1
Cr(III)	1·2	Sr	1200
Cu	0·015	Zn	0·3

laboratory rat and Man is summarised in Tables 8.6 and 8.7. Different mammals have different sensitivities, but in comparing the response of different species the dose per unit body weight is a fairly constant parameter. Note that the amounts of most elements required for normal metabolism are often much lower than the dietary intake. Other points to notice about human diets are:

(*a*) The lack of information concerning deficiencies and toxicities in the diet.

(*b*) The observation that, of the essential elements, Cr, Fe and Zn may be more or less deficient in 'normal' diets.

(*c*) The low margin of safety between concentrations in 'normal' diets and those known to be toxic for Cd, F and Pb.

(*d*) The relatively large amounts of even the most poisonous elements (As, Hg, Tl) needed to cause death, compared with the amounts in the diet.

(*e*) The greater sensitivity of unweaned infants to elements such as Cd, Hg, Pb, Pu and Ra (Jugo, 1977).

In view of the uncertainty arising from poor and variable absorption across the gut wall, as described in chapter 7, toxicities have also been measured following intravenous injection. The amounts needed to cause death by intravenous or other forms of injection are smaller than those taken orally. Some values are given in Table 8.8 for small laboratory mammals, and since these are expressed in mg X/kg body weight they should be of the same order of magnitude for all mammals.

Fiabane and Williams (1977) have drawn up a table of the effects of deficiencies and excesses of elements in mammals, an amplified form of which is given in Table 8.9. Effects on humans are treated in much more detail by Browning (1969), Beliles (1975), Hoffmann (1977) and many other medical texts, while those on domestic animals have been reviewed by

TABLE 8.6 Elements in the diet of the rat (*Rattus norvegicus*) in mg X/day

X	Deficient	Normal range	Toxic	Lethal	Ref.
Al	<0·001		200 ?	60–220	V
As(III)	0·0003		0·6	1·3–5	V, Nielsen *et al.* (1975)
Au				10–130	V
B	<0·0006		0·15	130–270	
Ba				33–100	V
Be				3·3	V
Bi			1·5	160	
Br	<0·005			800	
Ca	1	45–60		150–400	V
Cd	<0·00006	0·0007–0·015	0·5	16–55	CEC (1978)
Ce				450	V
Cl	<0·4	5–30		500	Boyd (1975)
Co	<0·0003		0·7	27	U, V
Cr(VI)	0·001	0·02–0·05	5	44	U, V, Roginski and Mertz (1967)
Cs				370	V
Cu	0·01	0·05–0·2	1–5	20–130	U, Christensen (1972)
Dy				370	V
Eu				570	V
F	0·005		0·1	30	Schwarz and Milne (1972a)
Fe		0·1–0·5		100–200	V
Ga			10	620	V
Gd				580	V
Ge				>170	V
Hf			>500		V
Hg			>0·05	8–9	V, Christensen (1972)
I	<0·0007	0·001–0·002	5–20		U
In			30	200–700	V
K	0·3	50		480	Boyd (1975)
La				720–800	V
Li		0·0001 ?		40	V
Mg	0·1	2–5		240	V

X	Deficient	Normal range	Toxic	Lethal	Ref.
Mn(II)	0·0003–0·06	0·03–0·2	20	280	U, V, Schroeder *et al.* (1966)
Mo	0·0005	0·0005–0·001	5	40–120	V
Na	0·2	5–50		320	Boyd (1975)
Nb				220	V
Nd				300	V
Ni	<0·00015	0·03	50	80	U, V, Smith (1969)
P		35–45			
Pb				45–270	Christensen (1972)
Pr				400	V
Rb		0·5	10	280	V
Sb				14–90	V
Se(IV)	<0·0003	0·0003–0·004	0·004–0·06	1–2	V
Si	<0·05	>0·5			Schwarz and Milne (1972b)
Sm				330	V
Sn(II)	<0·01		2·5	150	U, V
Sr			8	900	
Ta				300	V
Te(VI)			0·25	1–70	V
Th				>150	V
Ti				>50	V
Tl				2–6	Zitko (1975)
U(VI)				36	
V	0·00035		0·25–0·5	1·5–23	V, Hudson (1964); Myron *et al.* (1975)
W				30–220	V
Yb				400	V
Zn	0·016	0·02–0·04	50	150–260	Christensen (1972)
Zr				280–540	V

A rat weighs about 300 g FM, and eats about 15 g FM food per day

U, Underwood (1977); V, Venugopal and Luckey (1978). Where no reference is given, data on diets come from Comar and Bronner (1962), and data on lethal doses from Spector (1956b)

TABLE 8.7 Elements in human diets in mg X/day

X	Deficient	Normal range	Toxic	Lethal	Ref.
Ag		0·0014–0·08	60	1300–6200	H, S, V
Al		2–45	5000		H, S, T, U, V, Wi
As(III)	0·07	0·04–1·4	5–50	50–340	H, L, S, U, V, Nakao (1960)
B		1–3	4000		H, S, U, Wa
Ba		0·6–1·7	200	3700	H, S, U, V, Wa
Be		0·01			H, S
Bi		0·005–0·02			H, S
Br		0·8–24	3000	35 k–350 k	H, L, S, U, Gillespie *et al.* (1970)
C		300 k			S
Ca	400	600–1400			H, S, T, Wi
Cd		0·007–0·3*	3–330	1500–9000	H, M, S, U, CEC (1978), Fox (1976), Linnmann *et al.* (1973), Piscator (1976), Rantu and Sporn (1970), Yamagata and Shigematsu (1970)
Cl	70	3000–6500			H, S, T, Wi
Co	0·0002	0·005–1·8*	500		C, M, S, T, U, Wa, Yamagata *et al.* (1963), Z
Cr	0·005	0·01–1·2*	200	3000–8000	C, H, M, S, U, V, Wi, Doisy *et al.* (1976), Z
Cs		0·004–0·03			C, H, S, Y, Z,
Cu	0·3	0·5–6		175–250	H, L, S, T, V, Wa, Wi, Z
F		0·3–5	20	2000	H, S, Wi, Jones *et al.* (1971)
Fe	6	6–40	200	7 k–35 k	C, H, S, T, V, Wa, Wi
Ge		0·4–1·5			H, S, U,
H		350 k			S
Hg		0·004–0·02	0·4	150–300	H, L, S, T, Wa, Abbott and Tatton (1970), D'Itri (1972)
I	0·015	0·1–0·2	2	35 k–350 k	H, S, U, Wi
K		1400–4700	6000 ?	14 k	H, L, S, T, V, Wi, Y
Li		0·1–2	90–200		H, S, U, V, Wa
Mg	12	250–380			H, S, T, Wi, Shils (1964)
Mn		0·4–10			H, L, M, S, T, U, Wa, Wi
Mo		0·05–0·35			H, S, U, Wi
N		16 k			S
Na	45	2 k–15 k		230 k	H, S, T, V, Wi
Nb		0·02–0·6*			H, S,

X	Deficient	Normal range	Toxic	Lethal	Ref.
Ni	0·0006	0·3–0·5			H, S, Wa
O		2600 k			S
P		900–1900			H, S, Wi
Pb		0·06–0·5	1	10 k	H, S, U, Wa, Imamura (1967), Jacobs (1967), Kehoe (1965), Lehnert *et al.* (1969), Patterson (1965)
Po		6×10^{-13}			S
Ra		2×10^{-9}			S
Rb		1·5–6			C, H, S, U, Wa, Y, Z
S		850–930			H, S
Sb		0·002–1·3*	100		C, H, M, S, U
Sc		0·00002–0·0005			C, Z
Se	0·006	0·006–0·2*	5		C, H, S, T, U, Wa, Wi, Burk (1976)
Si		18–1200			H, S
Sn		0·2–3·5 (−17*)	2000		H, S, U, Wa, Wi
Sr		0·8–5			H, S, U, Wa
Ta		0·001			H
Te		0·6*		2000	S
Th		0·00005–0·003*			H, S
Ti		0·8			H, S
Tl		0·0015		600	H, S
U		0·001–0·002			H, S
V		0·04 (−4*)	18		S, U, Wi, Byrne and Kosta (1978)
W		0·001–0·015			H, Z
Y		0·016			H
Zn	5	5–40	150–600	6000	C, H, M, S, T, U, V, Wa, Wi
Zr		0·05–4*			H, S, U

Reference Man weighs 70 kg FM and eats 750 g DM/day (Snyder, 1975); * Discrepancies between different workers may be analytical errors.

C, Clemente (1976); H, Hamilton and Minski (1973); L, Lee *et al.* (1976); M, Murthy *et al.* (1971); S, Snyder (1975); T, Tanner and Friedman (1977); U, Underwood (1977); V, Venugopal and Luckey (1978); Wa, Waslien (1976); Wi, Williams (1977); Y, Yamagata (1962); Z, Wester (1971, 1974) who gives additional data for normal diets: Au, 0·000004–0·0005; Ce, 0·016–0·24; La, 0·005–0·008; Sm, 0·02–0·07

TABLE 8.8 Acute lethal doses of elements injected into mammals in mg X/kg body weight (FM)

X	Chemical form	LD_{50}/mg X $(kg FM)^{-1}$	Animal	Mode of injection	Ref.
Ag	$AgNO_3$	5–60	Rabbit	Intravenous	S
As	As_2O_3	6	Rat	Subcutaneous	V
Au	$AuCl_3$	10	Rabbit	Intravenous	S
Ba	$BaCl_2$	13	Rat	Intravenous	S
Be	$BeCl_2$	4·4	Rat	Intraperitoneal	V
Cd	$CdCl_2$	1·3	Rabbit	Intravenous	V
Ce	$CeCl_3$	55	Rat	Intravenous	S
Co	$CoCl_2$	50	Rabbit	Subcutaneous	S
Cr	$CrCl_3$	90	Rabbit	Intravenous	S
Cr	$Na_2Cr_2O_7$	15	Rabbit	Intravenous	S
Cs	CsCl	1200	Rat	Intraperitoneal	S
Er	$Er(NO_3)_3$	30	Rat	Intravenous	S
Ga	$Ga(NO_3)_3$	20	Rat	Subcutaneous	S
Ge	GeO_2	500	Rat	Intraperitoneal	S
Hg	$HgCl_2$	1·5	Rabbit	Intravenous	V
I	I_2	180	Rabbit	Subcutaneous	S
I	NaI	1100	Rat	Intravenous	S
In	$In(NO_3)_3$	1·6	Rat	Intraperitoneal	V
Li	LiCl	90	Rabbit	Subcutaneous	S
Mg	$MgCl_2$	60	Rat	Intraperitoneal	S
Mn	$MnCl_2$	18	Rat	Intravenous	S
Mo	Na_2MoO_4	140	Rat	Intraperitoneal	S
Nd	$NdCl_3$	125	Rabbit	Intravenous	S
Ni	$NiSO_4$	110–220	Rabbit	Subcutaneous	S
^{237}Np	?	3	Rat	Intravenous	M
Pb	$Pb(C_2H_3O_2)_4$	70	Rat	Intraperitoneal	S
Pd	$PdCl_2$	11	Rat	Intravenous	V
Pr	$Pr(NO_3)_3$	4	Rat	Intravenous	S
Pt	$PtCl_2$	23	Rabbit	Intravenous	V
^{239}Pu	Pu(IV)citrate	1	Rat	Intravenous	M
Re	$NaReO_4$	1000	Rat	Intraperitoneal	S
Rh	$RhCl_3$	100	Rat	Intravenous	V
S	Na_2S	2·5	Rabbit	Intravenous	S
Sb	$K_2Sb\ C_4H_6O_8$	25	Rat	Intravenous	S
Se	Na_2SeO_3	1·3	Rat	Intravenous	S
Sn	SnI_4	35	Rat	Intravenous	S
Sr	$SrCl_2$	220	Rat	Intravenous	S
Te	Na_2TeO_4	25	Rat	Intraperitoneal	S
Te	Na_2TeO_3	2·3	Rat	Intraperitoneal	S
Th	$ThCl_4$	18	Rat	Intravenous	S
Tl	$TlNO_3$	15	Rat	Subcutaneous	V
U	$UO_2(NO_3)_2$	1	Rat	Intravenous	S

Np, Pu, Th and U are radioactive, but it is thought that their chemical toxicity is the cause of death. With shorter-lived radionuclides such as ^{241}Am and ^{244}Cm, ionising radiation is probably the cause of death, and the LD_{50} doses by intravenous injection are 34 and 0·034 μg X/kg FM (after 30 days; Moskalev, 1972).

M, Moskalev (1972); S, Spector (1956b); V, Venugopal and Luckey (1978)

TABLE 8.9 Some effects of elemental deficiencies and excesses on mammals

X	Effects of deficiency	Effects of excess
As	Enlarged spleen; poor hair growth	Stomach pain; convulsions; goitre*
Be	None	Lung cancer by inhalation†
Ca	Deformed bones; tetany	Atherosclerosis; cataract; gall stones*
Cd	Reduced growth?	Hypertension ?; nephritis†
Co	Anaemia	Heart failure; polycythemia
Cr	Corneal opacity; poor glucose metabolism*	Lung cancer by inhalation†
Cu	Anaemia; hair kinky or depigmented	Jaundice; Wilson's disease†
F	Poor bones and teeth*	Mottled teeth; bone sclerosis*
Fe	Anaemia*	Haemochromatosis; siderosis*
Hg	None	Encephalitis; neuritis†
I	Hypothyroidism; goitre*	Hyperthyroidism*
Mg	Convulsions	Anaesthesia
Mn	Bones deformed; gonads affected; hair reddens	Ataxia†
Mo	Reduced xanthine oxidase activity	Growth depression
Na	Addison's disease; cramps*	
Ni	Dermatitis; liver changes	Dermatitis; lung cancer by inhalation†
Pb	None	Anaemia; brain damage; neuritis; kidney cancer†
Sb	None	Heart disease
Se	Infertility; liver necrosis; muscular dystrophy; leucocytes inefficient	Cancer; deformed nails and hair*
Si	Poor bone and cartilage growth	Kidney stones; lung disease†
Sn	Reduced growth	
V	Serum cholesterol lowered	Reduced growth
Zn	Dwarfism; immature gonads;* dermatitis	Anaemia

* Found in some human populations
† Caused by industrial overexposure

Underwood (1977). Schroeder and Darrow (1973) have described the carcinogenic properties of the elements towards mammals, which is still a somewhat controversial topic. It appears that only Pb, Pd, Rh and Se are known to cause cancers by ingestion in food, while Be, Cr and Ni can cause lung cancers when inhaled. All the radioactive elements are potential carcinogens because of the ionising radiations that they emit.

Residence times of the elements in Man, and cumulative poisons

From data on human intake (Table 8.7), multiplied by the fraction absorbed across the gut wall (Table 7.6), and a knowledge of the elemental composition of Reference Man (Table 6.7), it is possible to estimate residence times in the human body. At present such estimates are relatively crude, but give the following orders of magnitude (Yamagata *et al.*, 1963; ICRP, 1972; Bowen, 1977a):

X	Residence time in the body/days
H_2O	14
Hg, K, Mo	50–100
Co, Cs, Mn, Rb	100–200
As, Sb, Se	200–400
Cd	400–800
Ba, Ca, Cu, Ni, Ra, Sn, Sr	800–1600
Ac, Am, Cm, Fe, Pb, Pu, Th, Zn	>1600

Better estimates of residence times can be made by analysing or counting the excreta from volunteers for a long period after a single dose of a stable or radioactive tracer. This has been done for Ba, Ca and Ra (Newton *et al.*, 1977, Fig. 8.3), Cd (Rahola *et al.*, 1972), Cs (Ekman, 1967), Hg (Rahola *et al.*, 1973), Mn (Cotzias *et al.*, 1968), Pb (Rabinowitz *et al.*, 1973) and Sr (Rundo, 1967). If [X] is the concentration of tracer X retained in the body at time t, a graph of log [X] against t is either a straight line or can be resolved into two or more straight lines, the slopes of which correspond to residence times. If there is more than one residence time, different compartments of the body are assumed to be involved. It can be shown by direct counting of a labelled subject that tracers have different residence times in different organs and tissues (Cotzias *et al.*, 1968). The very long residence times found for Cd in kidney, or for Pb in bone, mean that these elements can be classified as cumulative poisons (Task group on metal accumulation, 1973).

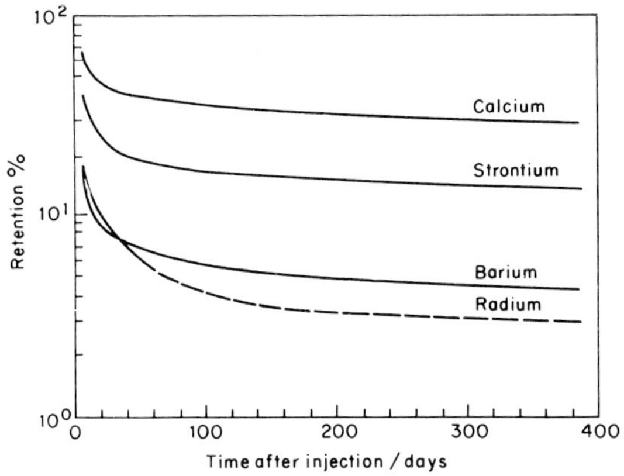

FIG. 8.3. Retention of Ca, Sr, Ba and Ra by a human subject, injected with radioactive tracer at day O. The dotted line is an extrapolation for Ra. From Newton *et al.* (1977)

Toxicities of gases to mammals

The absorption of inorganic gases through the lungs is approximately equivalent to an intravenous injection, since the gases are directly transferred to the bloodstream. Data on the toxicity of gases to various small mammals has been collected by Spector (1956b). The relative concentrations of various gases causing death are given below:

Lethal concentration range/mg l^{-1}	Gases
0·001–0·01	O_3
0·01 –0·1	AsH_3, B_2H_6, $COCl_2$, HF, H_2Se, PH_3
0·1 –1	$AsCl_3$, Br_2, Cl_2, HCN, H_2S
1 –10	CO, GeH_4, HCl, NH_3, NO, $Ni(CO)_4$, SO_2, SnH_4
10 –100	CCl_4, CS_2, I_2
100 –1000	CO_2

Absorption of particulate matter by the lung, such as asbestos, BeO or SiO_2, can also lead to chronic injury, described in medical texts (e.g. Browning, 1969; Bendz and Lindqvist, 1978).

9 *Chemical Forms and Functions of the Elements*

In this chapter we intend to review the forms and functions of the inorganic elements, other than C, H, N, O, P and S, in living matter. Any such review must be incomplete, disconnected and full of uncertainties. New bio-inorganic compounds, from small molecules to large proteins, are being discovered and isolated at such a rate that no individual can keep track of the latest developments. One should neither underestimate the technical difficulties of the isolation procedures, nor neglect the possibility of their forming artefacts, which have led to controversy on more than one occasion. In an attempt to bring some kind of order into a complex and heterogeneous subject, we shall classify current knowledge under the following headings:

1. Inorganic structural materials: Ca, F, P, Si and rarely Ba, Sr, Mg.
2. Electrochemical and messenger functions of ions: Ca, Cl, K, Mg, Na.
3. Small molecules, including antibiotics (Br, Cl, Cu, F, Fe, I), porphyrins (Co, Cu, Fe, Mg) and miscellaneous (As, B, Ca, Hg, Se, Si and V).
4. Large molecules, almost all proteins, some of which are enzymes with catalytic properties (Ca, Cl, Co, Cr?, Cu, Fe, K, Mg, Mn, Mo, Na, Ni, Se, Zn), while others have storage, transport or unknown functions (Ca, Cd, Co, Cu, Fe, Hg, I, Mn, Ni, Se, Zn); a few contain more than one metal.
5. Organelles or part organelles, e.g. the nitrogenase enzyme system, mitochondria and chloroplasts, in which several metals may function co-operatively.

Inorganic structural materials

The importance of exo- and endo-skeletons to invertebrates and vertebrates respectively needs no emphasis. Here we can only give a brief summary of the occurrence of inorganic matter in these materials.

Calcium carbonate (calcite or aragonite) is produced by a few blue-green, green, yellow-green and brown algae, and by an important group of red algae; among the invertebrates, it is formed by many Foraminifera, some Porifera, all Coelenterate corals, most Echinoderms and Molluscs, together with some Brachiopods and barnacles (Arthropods) (Darley, 1974a,b; Lowenstam, 1974; Borowitzka, 1977).

Calcium phosphate (apatite) occurs in all vertebrate bones and teeth, and in a few Brachiopods; it always contains fluorine as a structural component (Schraer, 1970; Zipkin, 1973; Lowenstam, 1974).

Calcium alginate occurs in the cell walls of brown algae (Siegel, 1962), and *calcium pectate* in the cell walls of seed plants (Wyn Jones and Lunt, 1967).

Silica (opal) occurs in many diatoms, a few green, yellow-green and brown algae, and in the cell walls of many vascular plants, such as all Equisetaceae and many Gramineae; it is also secreted by many Foraminifera and Porifera (Siegel, 1962; Darley, 1974).

Magnesium hydroxide (brucite) is found in the coralline alga *Goniolithon* (Schmalz, 1965).

Barium sulphate (baryte) occurs in the test of the Protozoan *Xenophyophora* (Vinogradov, 1953). Minute crystals are found in the vacuoles of rhizoids of the green alga *Chara*, where they may function in the perception of gravity (Schröter *et al.*, 1975).

Strontium sulphate (celestine) occurs in the test of the Protozoan *Acanthometra* (Bütschli, 1908).

Electrochemical and messenger functions of ions

Elements with an electrochemical function occur inside cells as free ions at concentrations different from those in the nutrient solution. Hence they allow the cell to simulate a storage battery, with a potential difference across its bounding membrane. This p.d. is usually about 0·1 V, and is given by the Nernst equation:

$$E = 2{\cdot}303\ \mathrm{RTF}^{-1}\ \log[\mathrm{X}^+]_i/[\mathrm{X}^+]_o$$

for a univalent ion X^+, where i = inside and o = outside the cell. One of the first results of cell death is the restoration of equilibrium and the disappearance of this potential difference.

The cations Ca^{2+}, K^+, Mg^{2+} and Na^+, together with the anions Cl^-, HCO_3^-, SO_4^{2-} and PO_4H^{2-}, probably all have electrochemical functions, as does the proton H^+. It is well known that the cytoplasm of cells is rich in K^+ and Mg^{2+}, but poor in Ca^{2+} and Na^+, compared with the nutrient medium; an example is given in Fig. 7.1. It is difficult to measure the actual concentrations of free ions inside the cytoplasm, but it is thought that Cl^-,

K^+ and Na^+ are largely free both inside and outside the cell, while Mg^{2+} and Ca^{2+} are mainly free outside and bound inside. The concentration of free Ca^{2+} in cytoplasm is probably about 2 ng/g (Baker, 1977).

Several suggestions have been made to specify more precisely the real nature of the term 'electrochemical function', and these are given below, but all are somewhat speculative.

1. As a source of free energy during cell stimulation. This is suggested by the profound changes in $[K^+]_i$ and $[Cl^-]_i$ found in the turgor movements of plant cells, as in the leaf movements of *Albizzia* (Schrempf *et al.*, 1976), or the closing of stomatal guard cells (Humble and Raschke, 1971).
2. To maintain the osmotic pressure of cells, as in the turgor movements above.
3. To maintain electrical neutrality within cells, which is unlikely to require specific ions.
4. To assist in the ordering of water molecules in the cytoplasm (Österberg, 1976). Na^+ and Mg^{2+} increase the regularity in the arrangement of water molecules, K^+ has little effect, and Ca^{2+} tends to destroy the order in water.
5. To stabilise emulsions within the cytoplasm. It is known that univalent ions stabilise oil-in-water emulsions, while divalent ions stabilise water-in-oil emulsions. Globulins and other cell proteins are insoluble in pure water, but dissolve in dilute salt solutions. Note that tri- and quadri-valent ions, which are very rare in cells, tend to coagulate proteins from emulsions.

'Messenger' functions have been ascribed to the calcium ion in many situations. A sudden increase in the free calcium ion concentration in the cytoplasm, perhaps released from the mitochondria (Borle, 1975), is thought to be associated with a variety of phenomena, e.g.

(*a*) The contraction of animal muscle (Briggs and Solaro, 1976).

(*b*) The release of acetylcholine at the neuromuscular junction, which triggers nerve impulses in animals (Baker, 1977).

(*c*) The luminescence of certain Coelenterates (Cormier *et al.*, 1974).

(*d*) The generation of a potential difference in the electric eel (Chang and Neumann, 1976).

(*e*) The formation and dissolution of the mitotic spindle (Petzelt and Auel, 1977).

(*f*) (possibly) The alteration in permeability of cell membranes, leading to plant movements (Marschner, 1975).

A gradient in calcium ion concentration is thought to stimulate the growth of pollen tubes down the style (Mascarenho and Machlis, 1964).

Calcium uptake from the nutrient medium precedes and is needed for 'capacitation' when a sperm fuses with an ovum (Singh *et al.*, 1978).

Small molecules: transport complexes and miscellaneous

ORGANIC MOLECULES CONTAINING HALOGENS

The functions of these compounds are uncertain, but it seems likely that many of them are natural antibiotics.

ORGANIC FLUORIDES

A few very toxic fluoro-acids have been isolated from seed plants, namely fluoroacetic acid from *Acacia georginae*, *Dichapetalum cymosum* and *Gastrolobium grandiflorum* (Oelrichs and McEwan, 1961; McEwan, 1964; Peters and Shorthouse, 1964), ω-fluoro-oleic acid and ω-fluoropalmitic acid from *Dichapetalum toxicarium* (Ward *et al.*, 1964), together with ω-fluorocapric acid and ω-fluoromyristic acid (Hall, 1977). A fluorinated carbohydrate derivative, nucleocidin, is made by the microbe *Streptomyces clavus* (Neidleman, 1975).

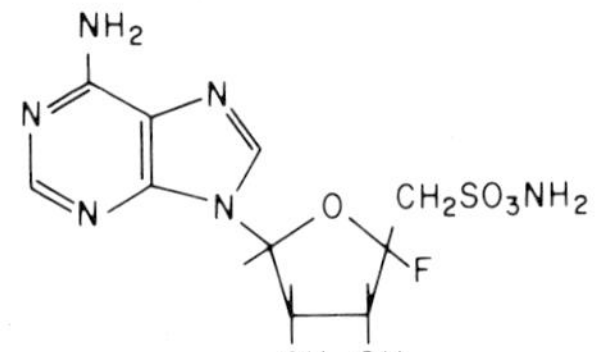

Nucleocidin

Other organic fluorides in plant tissues (Chang and Thompson, 1965; Peters and Shorthouse, 1967) and in serum (Taves, 1968) await characterisation.

ORGANIC CHLORIDES

Over a hundred organochlorine compounds have been reported from natural sources (Neidleman, 1975). Most of them are chlorophenols such as chlortetracycline or griseofulvin from fungi (*Aspergillus*, *Penicillium* etc.) or the related depsides from lichens such as *Anaptychia*, *Buellia* and *Lecanora*, which probably function as antibiotics. The following miscellaneous compounds deserve mention: hexachloroacetone and homologues from the red alga *Asparagopsis* (Burreson *et al.*, 1976); caldariomycin from the fungus *Caldariomyces fumago*, which possesses a chloroperoxidase enzyme capable of forming carbon–halogen bonds, and 2-amino 4,4′dichloro-butyrate from *Streptomyces* (Neidleman, 1975). Mollisin from the fungus *Mollisia caesia*, and drosophilin from the fruit fly *Drosophila* are probably antibiotics (Neidleman, 1975); islanditoxin, from the fungus *Penicillium islandicum*, is a very poisonous dichlorocyclopeptide isolated from 'yellow rice' (Marumo, 1959); 4-chloroindolyl 3-acetate is a plant growth hormone found in the seeds of *Pisum sativum* (Marumo *et al.*, 1971). Toxic constituents of higher plants containing chlorine include acutumine and acutumidine from *Menispermum* and chlorolactone from *Eupatorium*, together with the alkaloid jaconine from *Senecio jacobaea* (Addy and Parker, 1963; Hewitt and Smith, 1975). The functions of the chlorinated dodecosanes and tetracosanes from the flagellate *Ochromonas danica*, chlorotyrosines from sponges and a chlorobromoterpene from the mollusc *Aplysia* remain obscure (Faulkner *et al.*, 1973; Faulkner and Andersen, 1974; Neidleman, 1975).

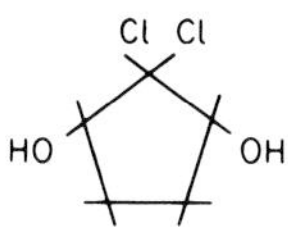

Caldariomycin

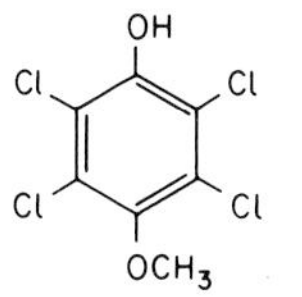

Drosophilin

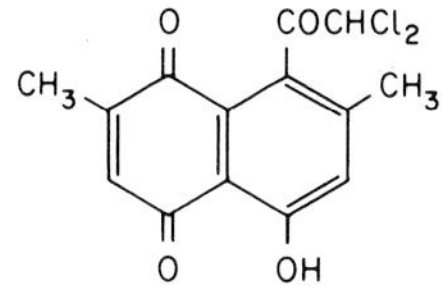

Mollisin

ORGANIC BROMIDES

Many organic bromides, mostly bromophenols, have been found in red algae, especially in the family Rhodomelaceae and the genera *Halopitys*, *Laurencia*, *Odonthalia*, *Plocamium*, *Polysiphonia* and *Rhodomela* (Augier and Mastagli, 1956; Faulkner and Andersen, 1974; Neidleman, 1975). A commonly found example is lanosol, which is toxic to some unicellular algae (Pedersen *et al.*, 1974). Over seventy aliphatic halogen derivatives, mostly bromides, have been isolated from the red alga *Asparagopsis*. They include $CHBr_3$ (the main volatile component), CBr_4, CH_2Br_2, CHBrClI, CH_3COCBr_3, $CHBr_2CONH_2$, $CHBr_2CHOHCHBr_2$, and many other derivatives of isopropanol or but-3-en-2-ol (Burreson *et al.*, 1976; Woolard *et al.*, 1976). Seven distinct bromophenols have been found in the calcareous green alga *Cymopolia barbata* (Högberg *et al.*, 1976). All these are presumably antibiotics. Similar compounds have been found in marine animals which eat these algae, such as the worm *Balanoglossus* and the mollusc *Aplysia* (Faulkner and Anderson, 1974).

Corals, such as *Gorgonia*, and horny sponges such as *Spongia*, can contain up to 4 % bromine in their dry matter, mostly as 3-bromotyrosine or 3,5-dibromotyrosine. These compounds also occur in molluscs such as *Buccinum* which graze on the coral (Roche *et al.*, 1963; Faulkner and Andersen, 1974). The pigment Tyrian purple was once extracted in quantity from molluscs of the genera *Bolinus*, *Purpura* and *Trunculariopsis*.

A few organic bromides have been found in micro-organisms, such as dibromocaldariomycin from the fungus *Caldariomyces* (Neidleman, 1975), tetrabromopyrrole and hexabromo 2,2′-bipyrrole from *Chromobacterium* sp., together with the antibiotic I, which is also known from *Pseudomonas bromoutilis* (Burkholder *et al.*, 1966; Anderson *et al.*, 1974). I is remarkable in containing 70% bromine by weight.

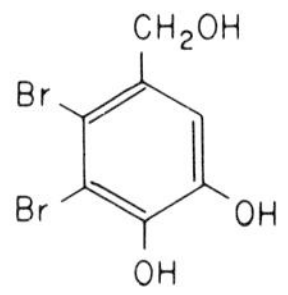

Lanosol

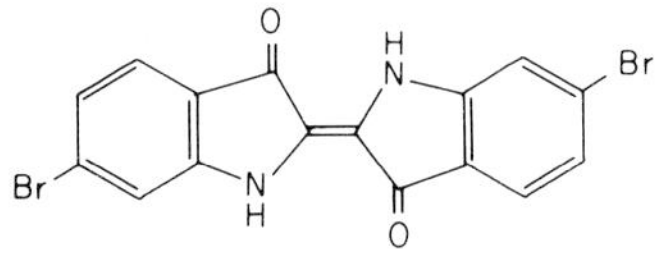

Tyrian purple

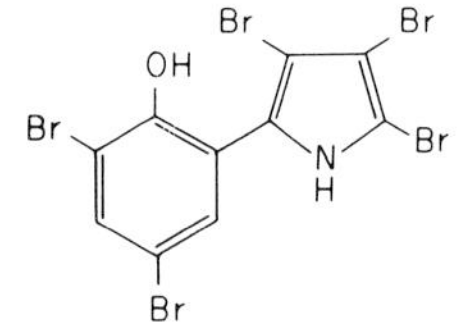

I

ORGANIC IODIDES

Old reports of tri-iodoacetaldehyde and iodotyrosines from marine algae need confirmation (Masuda, 1935; Roche and Yagi, 1952). The red alga *Asparagopsis* contains traces of carbonyl iodide, iodoform, di-iodomethane and di-iodoacetamide as well as mixed halides (Burreson *et al.*, 1976; Woolard *et al.*, 1976). Some of the iodoaminoacids are widespread in the animal kingdom. Thus the scleroproteins of some corals and horny sponges contain 2-iodohistidine, 3-iodotyrosine and 3,5-di-iodotyrosine (Roche *et al.*, 1963). Thyroxine, or 3,3′,5,5′-tetraiodothyronine, occurs in the tunics of Ascidians and is a vital component of the hormone thyroglobulin, which is found in the thyroid gland of all vertebrates. Small amounts of 3-iodotyrosine, 3,5-di-iodotyrosine, 3,3′-di-iodothyronine, 3,3′,5′-tri-iodothyronine and 3,3′,5-tri-iodothyronine also occur in the vertebrate thyroid, together with 4-iodohistidine and 2,4-di-iodohistidine (Wolff and Corelli, 1969). The iodination of histidine and tyrosine is known to be catalysed by chloroperoxidases (Morrison and Schonbaum, 1976).

TRANSPORT COMPLEXES

Small molecules of this type have been discussed in chapter 7; many more doubtless await discovery in plant sap, exudates of micro-organisms etc. Weinberg (1971) has speculated that compounds which sequester iron may be secreted by infectious microbes which have to compete with their host for this metal.

PORPHYRIN COMPLEXES (Falk, 1964; Pratt, 1975)

The porphyrin ring system is a naturally occurring tetradentate ligand which forms complexes with many metals. The basic structure and numbering of the ring system are shown below (A): examples are given in Table 9.1.

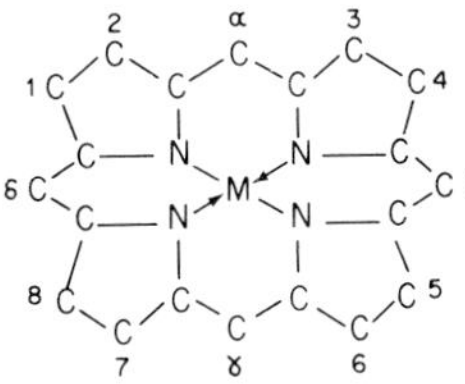

A

The structure below (B) shows the part of the ring system linking C_6 and C_γ in the chlorophylls, while Fig. 9.1 shows the related corrin ring system in cobalamin.

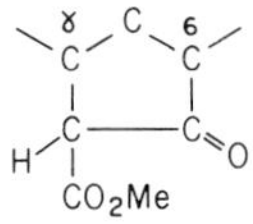

B

The copper porphyrin turacin is a brilliant red pigment found in the feathers of some rare African birds of the Musophagidae (Blumberg and Peisach, 1965).

Iron porphyrins are not normally found in the free state, but are attached to proteins as a prosthetic group. They have at least three vital functions in cells:

(*a*) The cytochromes, which are linked enzyme systems found in all mitochondria, are important in most redox processes in cells; the iron can be present as either Fe(II) or Fe(III) (Lemberg and Barrett, 1973; Nicholls and Elliott, 1974).

(*b*) The red blood cell pigments (haemoglobin in vertebrates, chlorocruorin in some invertebrates), and the muscle pigment myoglobin, are proteins which are able to take up dioxygen reversibly. A related pink pigment, leghaemoglobin, occurs in the root nodules of leguminous plants, but its function is not known (Ellfolk and Sievers, 1971). The ion is normally present as Fe(II), but it may be that porphyrins containing Fe(III) and Fe(IV) occur in some organisms (Aviram *et al.*, 1978).

(*c*) The enzymes peroxidase and chloroperoxidase contain the iron derivative of protoporphyrin IX. During their action the iron atom is reversibly oxidised and reduced, possibly involving the unusual redox reaction (Pratt, 1975);

$$\mathrm{Fe(III)} \rightleftharpoons \mathrm{Fe(V)} + 2e^-$$

The structures and mode of occurrence of chlorophylls has been reviewed

TABLE 9.1 Metalloporphyrins and corrins found in nature

M	Name	1	2	3	4	5	6	7	8	Other	Source
Cu	Uroporphyrin III	A	P	A	P	A	P	P	A		Musophagid feathers
Fe	Protoporphyrin IX	Me	V	Me	V	Me	P	P	Me		Blood, Muscle, Cytochrome b, Leghaemoglobin
Fe	Chlorocruorin	Me	–CHO	Me	V	Me	P	P	Me		Invertebrate blood
Fe	Haem a	Me	–CHMe, OH	Me	V	Me	P	P	–CHO		Cytochrome a
Fe	Haem c	Me	–CHMe, SH	Me	–CHMe, SH	Me	P	P	Me		Cytochrome c
Mg	Chlorophyll a	Me	V	Me	Et	Me	*	Ph	Me	γ*; 7,8-dihydro	Plant chloroplasts
Mg	Chlorophyll b	Me	V	–CHO	Et	Me	*	Ph	Me	γ*; 7,8-dihydro	Plant chloroplasts
Mg	Chlorophyll c	Me	V	Me	Et	Me	*	Ph	Me	*	Algal chloroplasts
Mg	Chlorophyll d	Me	–CHO	Me	Et	Me	*	Ph	Me	γ*; 7,8-dihydro	Red algae
Mg	Bacteriochlorophyll a	Me	–COMe	Me	Et	Me	*	Ph	Me	γ*; 3,4,7,8-tetrahydro	Photosynthetic Bacteria
Mg	Bacteriochlorophyll b	Me	–COMe	Me	=CHMe	Me	*	Ph	Me	γ*; 3,4-dihydro	(Photosynthetic Bacteria)
Co	Cobalamin	Me/Ad	$-CH_2AD$	Me/Ad	$-CH_2Ad$	2 Me	$-CH_2Ad$	Me/Y	Ad	αMe,γMe,δ-Y octahydro-	Vitamin B_{12}

A, $-CH_2CO_2H$; Ad, $-CH_2CONH_2$; Et, $-C_2H_5$; Me, $-CH_3$; P, $-CH_2CH_2CO_2H$; Ph, $-CH_2CH_2CO_2C_{20}H_{39}$; V, $-CH{=}CH_2$; Y, see Fig. 9.1; *see (B), p. 157

by Jackson (1976). Chlorophyll a occurs in the chloroplasts of all green plants. In seed plants, ferns, bryophytes, green algae and Euglenophytes it is accompanied by chlorophyll b; in brown and yellow-green algae this is replaced by chlorophyll c, while in red algae only chlorophylls a and d occur. The bacteriochlorophylls are found in photosynthetic bacteria that do not have chloroplasts, though the chlorophyll molecules are ordered in lamellae. The main function of chlorophyll appears to be its ability to act as a reducing agent when excited by visible light, and there is no obvious reason why magnesium should be necessary for this to occur (see below).

An immense amount of research has been carried out on the structure and reactions of vitamin B_{12}, or cyanocobalamin (Fig. 9.1) (Pratt, 1972; Babior, 1975). It is a red solid containing Co(III) which is synthesised by numerous microbes, and is an essential nutrient for a few bacteria, some diatoms, green and yellow-green algae, and all animals that have been tested, but is not required by fungi or by most green plants. In the human body cyanocobalamin is rare, as it is converted to the more common derivatives methyl cobalamin (dominant in plasma) and adenosyl cobalamin (dominant in erythrocytes, bone marrow and most tissues) (Linnell *et al.*, 1974). Methyl cobalamin is an important biochemical

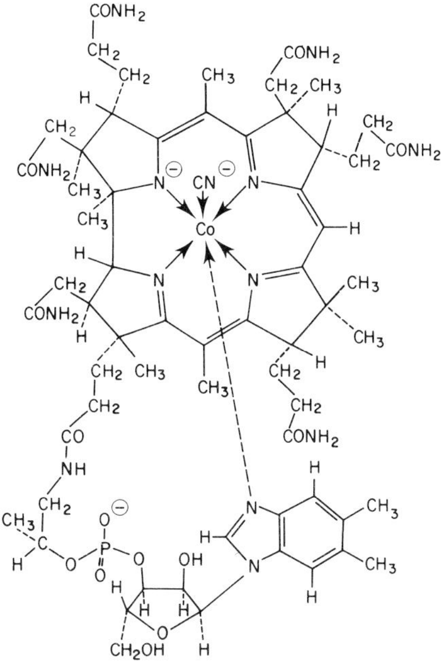

FIG. 9.1 The molecular structure of vitamin B_{12} (cyanocobalamin)

methylating agent. Enzymes sometimes require cobalamins as prosthetic groups, for example the mutases which catalyse hydrogen transfer. Such enzymes include mutases of glutamate, L-β-lysine, ornithine, α-methylene glutarate and methylmalonyl CoA, together with ethanolamine lyase, glycerol dehydrase and ribonucleotide reductase (Ochiai, 1977). The cobalt(III) in cobalamins does not appear to change its valency state in these catalytic reactions.

Other porphyrins which have been reported include coloured cobalt, copper and zinc coproporphyrin(III) in bacteria (*Bacillus*, *Micrococcus*, *Propionibacterium*) and yeast (Townsley and Neilands, 1957; Minagawa, 1958; Sone, 1971; Yamada-Ankei *et al.*, 1977); a distinct iron porphyrin in sulphite reductase (Ochiai, 1977); copper and zinc porphyrins in urine as disease symptoms (Falk, 1964); a possible manganese porphyrin in human blood (Borg and Cotzias, 1958); and nickel and vanadium porphyrins in asphalt (Falk, 1964; Biggs and Swinehart, 1976).

MISCELLANEOUS SMALL MOLECULES

There is no doubt that organisms can synthesise arsenic-carbon bonds, but few organo-arsenic derivatives have been characterised. The fungus *Scopulariopsis* can reduce arsenic(III) to trimethylarsine, with local toxic effects, while *Methanobacterium* can produce dimethylarsine (Waksman, 1952; McBride and Wolfe, 1971). Organo-arsenicals probably occur in algae, molluscs and fish (Lunde, 1970 and 1973; Vaskovsky *et al.*, 1972). Arsenobetaine ($Me_3As^+ . CH_2CO_2^-$) has been isolated from crustacean muscle (Edmonds *et al.*, 1977). Monomethylarsonic acid and dimethylarsinic acid have been found in human urine, river water and sea water; they are less toxic than arsenites (Braman and Forebach, 1973; Andreae, 1977; Shaikh and Tallman, 1978). Mono-, di- and tri-methyl tin derivatives have been found in the same media (Braman and Tompkins, 1979).

Among the volatile metal alkyls produced biologically are tetramethyl lead (Schmidt and Huber, 1976) and dimethyl mercury (Fagerström and Jernelöv, 1972) from bacteria, and unidentified derivatives of cadmium and tellurium. Methyl mercury chloride and ethyl mercury chloride are rapidly synthesised in plants of *Pisum* after exposure to mercury(O) vapour (Fortmann *et al.*, 1977).

The only naturally occurring boron compound appears to be boromycin (Fig. 9.2; Dunitz *et al.*, 1971). This has the formula $C_{45}H_{74}O_{15}NB$, is extracted from *Streptomyces*, and is easily hydrolysed to boric acid. Analytical problems may have prevented the detection of similar compounds in seed plants. At present the function of boron in diatoms and higher plants is unknown, but it seems to play a part in RNA or auxin metabolism, or perhaps in carbohydrate transport (Price *et al.*, 1972; Jackson and Chapman, 1975). Boron deficiency in plants inhibits the ATP-ase activity in the cell membrane as well as uptake of phosphate ions (Pollard *et al.*, 1977).

A 1:1 complex of calcium and sialic acid (*N*-acetylneuramic acid) may be a natural product (Jacques *et al.*, 1977). Much of the magnesium in animal tissues is probably complexed by ADP or ATP. These ligands release Mg^{2+} on acidification, and there is a substantial change in the formation constants of the complexes over the physiological range of pH (Metzler, 1977).

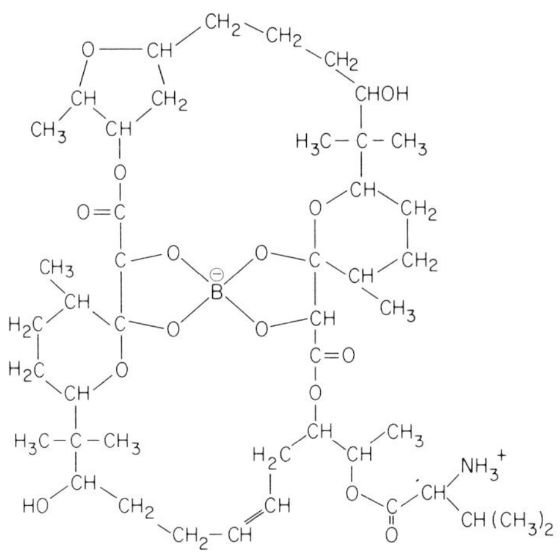

FIG. 9.2 The structure of boromycin; after Dunitz *et al.* (1971)

Several copper and iron compounds which have been isolated from micro-organisms are thought to be antibiotics by Japanese workers. They include brown CuR_2 and purple FeR_3 from a *Pseudomonas*, where R = *N*-methyl-*N*-thioformylhydroxylamine (Shirahata *et al.*, 1970) and a green compound matchamycin ($CuC_{20}H_{13}O_6N_3$) from a *Streptomyces* (Kimura and Nishimura, 1970). FeS is a pigment in *Bacteriodes* (Tracy, 1969).

A number of organo-selenium compounds have been isolated from bacteria, fungi and green plants, and the list in Table 9.2 is condensed from the review by Shrift (1973).

Note that most of these are amino acids, and that only selenomethionine (a methylating agent; Wu and Waksman, 1971) and selenocystine appear to be widely distributed. Other selenium compounds include peptides, adenosine phosphoselenate from bacteria and yeasts, and a plant wax. The only selenium compounds reported from animals appear to be proteins (see

TABLE 9.2 Natural organo-selenium compounds

Formula	Name	Sources
MeSeSeMe	Dimethyl diselenide	Fungi; *Astragalus*
CySeMe	Se-methylselenocysteine	Se accumulators, e.g. *Astragalus*, *Stanleya*
CySeOH	Selenocysteine seleninic acid	*Lolium*, *Trifolium*
$CyCH_2SeMe$	Selenomethionine	Bacteria; fungi; *Lemna*, *Lolium*, *Trifolium*
CySeCy	Selenocystine	
$CyCH_2SeCy$	Selenocystathionine	Se accumulators, e.g. *Astragalus*, *Morinda*
$CyCH_2SeSeCH_2Cy$	Selenohomocystine	One *Astragalus* spp.
$CyCH_2Se(Me){=}O$	Selenomethionine selenoxide	*Lolium*, *Trifolium*
$CySeO.CH{=}CHMe$	Se-propenylselenocysteine selenoxide	*Allium*

Cy, $-CH_2CH(NH_2)CO_2H$; Me, $-CH_3$

below). Clearly some seleno-amino acids are as essential for the formation of specific proteins as are their sulphur analogues. Some of the other selenium compounds may be storage compounds for this easily metabolised and highly toxic element.

Synthetic organo-silicon compounds are now known in large numbers, but the only naturally occurring molecules known appear to be esters of silicic acid, some of which may be artefacts (Voronkov *et al.*, 1971). The complex R_3SiCl, where R = isopropyltropolone, is said to exist in the conifer *Thuja plicata* (Bendz and Lindqvist, 1978). Strongly-bound silicon occurs in alginates from brown algae, pectinates from plants, and mucopolysaccharides such as dermatan, heparan sulphate and hyaluronic acid from animals (Schwarz, 1973; Carlisle, 1974). Reports of compounds containing Si–C and Si–H bonds have been based on infrared spectral evidence which is not unambiguous (Heinen, 1965; Lovering and Engel, 1967).

Amavanadine, whose structure is given below, has been found in the poisonous fly agaric fungus, *Amanita muscaria*, by Kniefel and Bayer (1973). Its function is unknown. Vanadium(III) occurs as a green complex called haemovanadin in the blood cells of some ascidians, where it probably functions as a transport molecule for dioxygen (Biggs and Swinehart, 1976).

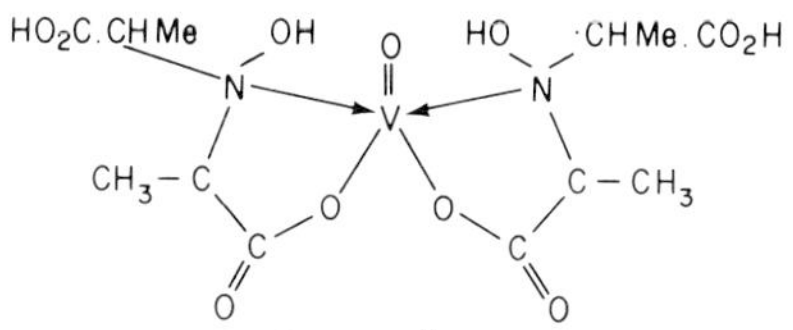

Amavanadine

Large molecules: proteins and enzymes

Three types of compounds involving proteins and an inorganic atom X can be distinguished. *Metal-activated enzymes* are proteins which function as catalysts when X is present, but X is loosely bonded and can frequently be replaced by other elements which may activate the enzyme *in vitro*. During purification of the enzyme, X is frequently lost, but addition of X in solution should reactivate the enzyme. *Metalloproteins* have no enzyme activity, but have X firmly bonded to, or embedded in, the protein in a stoichiometric manner. *Metalloenzymes* are metalloproteins with catalytic properties, and X is usually essential, or at least the enzyme activity is greatly reduced when X is absent or replaced by some other element. During purification, X is not easily removed, and the ratio of [X]/enzyme activity should tend to a constant value or remain constant. Only if X = Se or I does X form an integral part of the polypeptide chain making up the protein. Otherwise either the protein or its prosthetic group functions as a multidentate ligand for X, while X may modify or stabilize the tertiary structure of the protein.

Metal-activated enzymes

A large number of these are now known, but there is no general theory of the function of the metal atom. However, since the majority can be activated by metals with a single valency state, such as Mg^{2+} or K^+, it is reasonable to suppose that an important function of the ion is to hold part of the peptide chain of the protein in a suitable configuration for the desired reaction to take place. It is likely that magnesium may play a part in linking the substrate to the enzyme by interacting with both. Thus Hewitt and Smith (1975) suggest that the true substrate of many enzymes may be bonded through Mg^{2+} or $ADPMg^-$. Some cations may function by polymerising protein subunits, e.g. crayfish haemocyanin depolymerises when calcium is absent, but forms a hexamer when calcium is restored (Jeffrey *et al.*, 1976). Subunits of ribosomes are polymerised by adding Ca^{2+} or Mg^{2+} (Roskoski, 1969) and Zn^{2+} stabilises the dimer of α-amylase (Tellam *et al.*, 1978).

Suelter (1970) and Hewitt and Smith (1975) have listed about 50 enzymes which are activated by singly charged cations, mostly K^+. Most of the reactions catalysed are phosphorylations of $-CO_2H$ or $>C=COH$ groups, or elimination reactions, and are important in protein synthesis and in respiration. Na^+ tends to inhibit respiration, and activates few enzymes apart from Na/K ATPase in the cell membrane.

Magnesium is probably the metal which activates most enzymes, including all known phosphate transferases, many carboxylases and kinases, but not enzymes involved in redox processes. It is essential in glycolysis, the tricarboxylic acid cycle, lipid and protein metabolism etc. (Aikawa, 1963, 1976; Hewitt and Smith, 1975). Yeast enolase (2-phospho-D-glycerate hydrolyase) is unusual in having two tightly bound magnesium atoms per molecule, but requiring two further magnesium atoms for activation; thus it is a metal-activated metalloenzyme (Faller *et al.*, 1977). Magnesium in solution stabilises RNA and DNA (Ochiai, 1977).

Manganese(II) activates many of the same enzymes as magnesium does, e.g. kinases, synthases and a few dehydrogenases. It activates some RNA and DNA polymerases, guanylate cyclase and Na/K ATPase, and is involved in the synthesis of glycoproteins used for cartilage and bone formation (Atkinson *et al.*, 1968; Kimura *et al.*, 1976; Smith and

Gawthorne, 1975). Other valency states of manganese have only been observed in metalloenzymes (see below).

Calcium has been reported to activate about 30 enzymes, but its function is rarely clear (Wyn Jones and Lunt, 1967; Kretsinger, 1976). True calcium activation has been established for some ATPases in cell membranes; at least four mitochondrial enzymes; α-amylase in human saliva; several pancreatic enzymes such as trypsin and DNAase; collagenase (Seltzer *et al.*, 1976); endo-α-1 $\rightarrow$ 6 mannanase (Nakajima *et al.*, 1976); adenylate cyclase (Brostrom *et al.*, 1977); and prothrombin and other enzymes involved in blood clotting. The reversible reaction of calcium ions, aequorin and oxyluciferin in certain Coelenterates is of interest because it takes place when $[Ca^{2+}]$ is only 10 nmol/l. The reverse reaction, in which calcium is set free from its complex with aequorin, gives off visible light with a maximum absorbance at 469 nm, and is the most sensitive method known for determining free calcium ions (Cormier *et al.*, 1974).

Cobalt(II) may be needed for activating catalase and pyrophosphatase in bacteria, phytolysin in fungi, and acid phosphomonoesterase, amino-polypeptidase, arginine desimidase and glycylglycine depeptidase in mammals (Kern and Naef-Roth, 1971; Davies, 1972). There is an old report that chromium(III) activates phosphoglucomutase (Stickland, 1949). Iron activates a few enzymes such as aconitase, L(+) tartrate dehydrase, maleate hydratase, ribotide reductase and *S*-hydroxyanthranilic acid oxidase (Neilands, 1974; Koontz and Shiman, 1976). Zinc is said to activate 14 enzymes (Fiabane and Williams, 1977), but both iron and zinc prefer to form metalloenzymes. Zn^{2+} and Ni^{2+} are able to stabilise the coiled form of ribosomes (Tal, 1968). Lead is usually an enzyme poison, but it stimulates the phosphorylation of protein by microsomes in the absence of sodium (Siegel and Fogt, 1976). Aluminium stimulates 5-aminolaevulinate hydrolyase, an enzyme which requires both K^+ and Zn^{2+} for full activity, and which is poisoned by lead (Meredith *et al.*, 1977).

Metalloproteins and metalloenzymes

Calcium. About 40 calcium-binding proteins have been reported, but few of them have been well characterised (Kretsinger, 1976). Table 9.3 summarises data on molar masses, number of calcium ions and possible functions. When a few calcium ions are bound internally in the coils of polypeptide chains, their function is to stabilise these chains in a fixed configuration. A few specialised proteins containing a high proportion of aspartic acid can bind a large number of calcium ions by electrostatic attraction. These ions are somewhat loosely held, and may act as a store to

TABLE 9.3 Some proteins and enzymes containing tightly bound calcium

Name	Origin	M/1000	*n*(Ca)	Other X	Ref.
Rhodanese	Cow liver	18·5	1		Volini *et al.* (1978)
	Pig gut	12	1		
	Pig parathyroid	15	1		
Ca BP – 1	Pig brain	13			
L – 1	Squid axoplasm	13	2		
Troponin TNC	Muscle	17·8	2–4	+ 2–4 Mg	Kretsinger (1976)
TNC-like protein	Vertebrate brain	17·5	2		
S 100	Glial cells	21	2		
	Chicken gut	28	4		
Concanavalin A	*Canavalia*	25·5	1 ?	+ Mn(II)	Kretsinger (1976); Ochiai (1977)
Thermolysin	*Bacillus* spp.	35	2 + 2	(2 activate)	Kretsinger (1976); Colman *et al.* (1972)
Thermomycolase	*Malbranchea*		1		
Pyrophosphatase	Yeast	67	2		Kretsinger (1976)
Pyruvate kinase	Yeast	175	4		
Pyruvate kinase	Mammals	237	4		
Parvalbumin	Fish muscle	11–15			Drabikowski (1974)
	Wheat	16·6			Kretsinger (1976)
	Spinach chloroplast	33			
Glycoprotein	Mitochondria	30			Drabikowski (1974)
Phosvitin	Egg yolk	40		+ Mg, Fe	Kretsinger (1976)
	Cow plasma	56			Stenflo (1976)
Calsequestrin	Rabbit muscle	44	up to 43		MacLennan and Wong (1971)
Acetylcholine receptor	*Electrophorus* nerve	260	up to 60		Chang and Neumann (1976)

M, molar mass; *n* (Ca), number of Ca atoms per molecule

be drawn on when Ca^{2+} is needed as a messenger; alternatively they may be used to reduce the concentration of calcium ions in the cytoplasm.

Cadmium. The few known Cd-proteins are the metallothioneins from mammalian kidney, which bind both Cd and Zn (Table 9.8), though there are old reports that the mollusc *Pecten* may contain a cadmium protein (Kojima *et al.*, 1976; Vinogradov, 1953). They may function as zinc storage proteins which accidentally capture cadmium, and have low molar masses and a high proportion of cysteine (–SH) groups which act as ligands.

Chlorine. α-Amylase from the pancreas of pigs has a molar mass of 50 000 and one chlorine atom per molecule (Levitzki and Steer, 1974). Chloroplasts also require chlorine for photosynthesis (see below).

Cobalt. Most known cobalt enzymes contain a cobalamin as their prosthetic group, and have been discussed above (p. 160). Cobalt also occurs in transcarboxylase (Table 9.8), which is a large enzyme molecule containing zinc, and there are old reports of a cobalt protein in molluscan liver (Vinogradov, 1953).

Copper. Over thirty copper metalloenzymes are known, and the number is growing steadily (Malkin and Malmstrom, 1970; Ochiai, 1977). A list of the better defined species is given in Table 9.4. While the function of some of these are not known, four distinct functions of these enzymes can be distinguished at present:

1. In electron transport during cellular redox processes, especially in photosynthesis and respiration, where the reversible reaction

$$\mathrm{Cu(II)} + e^- \rightleftharpoons \mathrm{Cu(I)}$$

is involved. Cytochrome oxidase (Cytochrome $a+a_3$) is particularly important because it can reduce dioxygen. It is believed to contain an imidazole group linking Cu(II) and Fe(III) atoms with antiferromagnetically coupled spins (Tweedle *et al.*, 1978).

2. In a number of oxidases, especially amine oxidases, where the same reaction occurs. In some hyperoxide dismutases, Cu and Zn atoms are linked by an imidazole ring (Beem *et al.*, 1977).

3. Haemocyanins are dioxygen carriers in the blood of certain arthropods and molluscs.

4. Ceruloplasmin, from mammalian plasma, not only catalyses the oxidation of Fe(II) to Fe(III) in the presence of dioxygen, but also can release stored Fe(III) from the liver by reducing it to Fe(II) (Smith and Gawthorne, 1975).

Much work is in progress on the nature of the copper binding sites; for example five different types of copper can be distinguished in ceruloplasmin. The reason for the variety of colours in copper proteins is not known; some

TABLE 9.4 Some copper proteins and enzymes

Name	Origin	M/1000	*n* (Cu)	Colour	Remarks	Ref.
Azurin (bacterial)	*Bordetella* spp.	15	1	Blue	e^- transport	Mm
Stellacyanin	*Rhus vernicifera*	20	1	Blue	e^- transport	Mm
Plastocyanin	*Populus*	10·5	1	Blue	Ligands 2 his, cys, met	Z
Plastocyanin	*Chenopodium*	11·5	2	Blue	e^- transport	Mw
Plastocyanin	*Spinachia*	21	2	Blue	Ligands 2 his, cys	Mm, So
Umemycin	*Armoracia*	14·6	1	Blue	e^- transport	Ps
Laccase (fungal)	*Coriolus versicolor*	60	4	Blue		Mo
Laccase	*Rhus vernicifera*	135	6	Blue		Om
Ascorbate oxidase	*Cucurbita*	146	8	Blue		Mm
Copper chelatin	Rat liver	8	6		storage	Wi
Ceruloplasmin	Human plasma	134	6	Blue	Fe utilisation	Rb
Haemocyanin	Crayfish blood	470	12	Blue	O_2 transport; hexamer	Je
Haemocyanin	*Limulus* blood	560	16	Blue	O_2 transport; octomer	Su
Green protein (diatom)	*Phaeodactylum* spp.	850	8	Green	+ Mn; photosynthesis	Ho
Hyperoxide dismutase	Cow erythrocyte	33·6	2	Blue	+ 2 Zn; ligands 4 his	Be
Lysyl oxidase	Chicken heart	60	1	None		Ha
D-galactose oxidase	*Dactylium dendroides*	68	1	None	Cu(II)	Ko
Diamine oxidase	*Pisum sativum*	96		None		Oc
Diamine oxidase	Pig kidney	87	1	Pink		Mm

Name	Origin	M/1000	*n* (Cu)	Colour	Remarks	Ref.
Monoamine oxidase	Cow plasma	255	4	Pink		Mm
Benzylamine oxidase	Pig plasma	195	3	Pink		Bb
Urate oxidase	Pig liver	120	1	Yellow		Mm
Dopamine-β-hydroxylase	Cow adrenals	290	3–6	None		Mm
Quercetinase (fungal)	*Aspergillus*	111	2	Green		Ok
Luciferase	*Diplocardia*	320	1	None		Bc
Albocuprein II	Human brain	14	3 ?	Yellow		Fu
Albocuprein I	Human brain	320	3 ?	Yellow		Fu
Mitochondrocuprein	Infant liver				4 % Cu; storage; cys?	Po
Cytochrome c oxidase	Cow heart	112	2	Red	+2 Fe, 2 haem a	Or
Cytochrome a	Cow heart	93	1	Red	+1 Fe, 1 haem a	Yo
Cytochrome a_3	Cow heart	240		Red	+Fe, haem a	Yo

M, molar mass; *n* (Cu), number of Cu atoms per molecule; cys, cysteine; his, histidine; met, methionine. Copper may also be present in reductases of nitrite, nitric oxide and hyponitrite (Mahler, 1961).

References: Bb, Buffoni and Blaschko (1964); Bc, Bellisario and Cormier (1971); Be, Beem *et al.* (1977); Fu, Fushimi *et al.* (1971); Ha, Harris *et al.* (1974); Ho, Holdsworth and Arshad (1977); Je, Jeffrey *et al.* (1976); Ko, Kosman *et al.* (1977); Mm, Malkin and Malmstrom (1970); Mo, Mosbach (1973); Mw, Milne and Wells (1970); Oc, Ochiai (1977); Ok, Oka and Simpson (1971); Om, Omura (1961); Or, Orii *et al.* (1977); Po, Porter (1971); Ps, Paul and Stigbrand (1970); Rb, Ryden and Björk (1976); So, Solomon *et al.* (1976); Su, Sullivan *et al.* (1976); Wi, Winge *et al.* (1975); Y, Yamanaka and Okunuki (1974); Z, Colman *et al.* (1978)

are intensely blue, with a molar absorptivity more than ten times that of the hydrated copper(II) cation.

Iron. More than seventy iron proteins have been described and several books have been written about them (Jacobs and Worwood, 1974; Neilands, 1974; Brown *et al.*, 1977; Lovenberg, 1977). The number is indefinite because enzymes isolated from different sources, but with the same function, often differ in molar mass or in their state of polymerisation. The bewildering variety of cytochromes reported from microbial sources can be seen in the review by Yamanaka and Okunuki (1974).

Iron proteins can contain porphyrins, flavins or rarely pterins, and occasionally other metal atoms such as molybdenum or copper. They can be broadly classified into those containing labile sulphide groups, the iron-sulphur proteins, and those that do not. Iron-sulphur proteins (Table 9.5) contain clusters of two types, both of which can readily accept an electron with reduction of one iron(III) to iron(II).

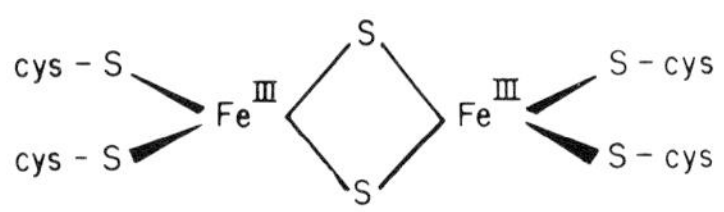

Fe_2S_2 cluster (cys = cysteine)

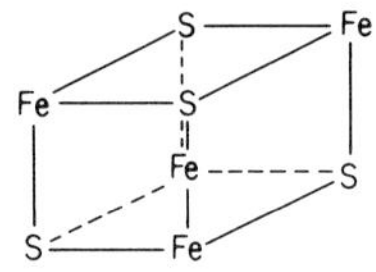

Fe_4S_4 cluster

Other iron proteins are listed in Table 9.6. Some of the main functions of iron proteins are set out below.

1. In electron transfer in cellular redox processes, involving the reaction

$$Fe(III) + e^- \rightleftharpoons Fe(II)$$

e.g. electron transfer at low redox potentials, using ferredoxins and Fe_2S_2 proteins; reductions involving the transfer of 4 or 6 electrons, using Fe_4S_4 flavoproteins, with or without Mo; photosynthesis and respiration, using ferredoxins and the cytochromes containing iron porphyrins.

2. In many mono- and di-oxygenases, where the same reaction is involved.

3. In a few peroxidases, where the suggested reaction mechanism is

$$Fe(III) + H_2O_2 \rightarrow OFe(V) + H_2O$$

but Fe(IV) might be involved (Pratt, 1975).

4. As a dioxygen carrier in blood and muscle, either with or without porphyrins, in haemoglobin, myoglobin and haemerythrin. The dioxygen is a ligand for iron.

5. As iron(III) carriers in blood (transferrin), also in milk and egg white.
6. As the unique storage protein ferritin, which consists of a core of basic iron(III) hydroxide plus some phosphate, surrounded by a thick layer of protein (Brown *et al.*, 1977). Ferritins are polymers of two distinct subunits whose molar masses are 19 000 and 21 000 (Arosio *et al.*, 1978).
7. Pigmentation of red hair and feathers (Flesch, 1968).

Iodine. Thyroglobulin is a thyroid hormone secreted by all vertebrate thyroids (Koutras, 1968), and is also found in the branchial sacs of marine ascidians (Laurens and Commanay, 1969; Suzuki and Kondo, 1971). The iodinated scleroproteins of corals and horny sponges have not been studied in detail (Roche *et al.*, 1963).

Manganese. The properties of twelve manganese proteins are summarised in Table 9.7. Many of them are hyperoxide dismutases, which are widespread in nature, and may be involved in photosynthesis. The reaction mechanism involves

$$\mathrm{Mn(III)} + e^- \rightleftharpoons \mathrm{Mn(II)}$$

and the manganese(III) form is deep pink or red. Arginase, pyruvate carboxylase and concanavalin A contain manganese(II) which does not appear to change its valency during catalysis. The substance pinnaglobin needs re-investigation; the kidney of the mollusc *Pinna nobilis* contains 39 mg Mn/g DM (Ghiretti *et al.*, 1972).

Molybdenum. All known molybdenum enzymes also contain iron, and are included in Table 9.5. The molybdenum in aldehyde oxidase, sulphite oxidase and xanthine oxidase is normally present as Mo(VI), but the redox process

$$\mathrm{Mo(VI)} + 2e^- \rightleftharpoons \mathrm{Mo(IV)}$$

takes place during the oxidation. Despite intensive studies, there is no clear evidence of a valency change at the molybdenum atom in nitrogenase, nor are the ligands for Mo known, and the hypothesis that dinitrogen binds to the Mo atom is only a plausible speculation (Moura and Xavier, 1978).

Nickel. The enzyme urease from *Canavalia* seeds has a molar mass of 105 000 and contains two nickel atoms per molecule (Dixon *et al.*, 1975). Nickel may also be involved in the action of hydrogenase. An α-macroglobulin from rabbit serum, called nickeloplasmin, binds nickel strongly, as do serum albumins (Nomoto *et al.*, 1973). The function of nickel is probably analogous to that of zinc, in maintaining the tertiary structure of the protein without the likelihood of a change in valency.

Selenium. Two well-characterised and two less well-characterised proteins are thought to contain selenium. Glycine reductase from the bacterium *Clostridium* is a small protein with molar mass 12 000, which contains one

TABLE 9.5 Some iron-sulphur and/or iron-flavin proteins, including all iron-molybdenum proteins

Name	Origin	M/1000	*n* (Fe)	*n* (S)	*n* (Mo)	Remarks	Ref.
Adrenodoxin	Pig adrenals	12.5	2	2			Ha
Ferredoxin (plant)	*Spinachia*	10·6	2	2			Ha
Ferredoxin (bacterial)	*Escherichia coli*	12·6	2	2			Ha
EPR protein	*Clostridium*	24	2	2			Ha
Nitrite reductase	*Spinachia*	63	2	2		+1 sirohaem	Ve
Azoferredoxin	*Clostridium*	55	4	4		in nitrogenase	Ha
Ferredoxin (bacterial)	*Desulphovibrio*	6	4	4			Ha
Ferredoxin I	*Rhodospirillum*	8·8	8	8			Ya
Ferredoxin II	*Rhodospirillum*	14·5	4	4			Ya
Ferredoxin III	*Rhodospirillum*	8·5	n	n			Cy
Ferredoxin IV	*Rhodospirillum*	4	8	8			Cy
High potential iron protein	*Chromatium*	9·6	4	4		Redox Potl. +0·35 V	Ha
Hydrogenase	*Clostridium*	60	4/12	4			Ha, Mc
Succinate dehydrogenase 'A'	Mitochondria	27	4	4		+flavin	Oh
Succinate dehydrogenase 'B'	Mitochondria	70	4	4		+FAD	Oh
FeS flavoprotein	Cow heart	63	4	4		+FAD	Ru
Ferredoxin (bacterial)	*Clostridium*	6	8	8		Redox Potl. −0·4 V	Ha
Formate dehydrogenase	*Pseudomonas*	300	5–8	7–8	+	+FMN, Se	Ht
Spermidine dehydrogenase	*Serratia*	76	1?	?		+FAD porphyrin	Sa

Name	Origin	M/1000	*n* (Fe)	*n* (S)	*n* (Mo)	Remarks	Ref.
Trimethylamine dehydrogenase	Bacteria	147	4	4		+ flavin	Z
Sarcosine dehydrogenase	Mitochondria	300	1–2	?		+ flavin	Sa
Malate vitamin K reductase	Mycobacterium		n	?		+ FAD	Kb
Adenylyl sulphate reductase?	*Desulphovibrio*	120	12	12	1		Pe, Zx
NADH dehydrogenase	Mitochondria		23–28	n		+ FMN	Ha
Glutamate synthase	*Escherichia coli*	800	32	32		+ 8 flavin; octomer	Ha
Sulphite reductase	*Escherichia coli*	670	20	12–16		+ 4 FMN, 4 FAD, 4 haem	Si
Sulphite oxidase	Cow liver	110	2	?	2	protoporphyrin IX	Oc
Aldehyde oxidase	Rabbit liver	300	4	?	2	+ 1 FAD	Bw
Xanthine oxidase	Cow milk	275	8	8	2	+ 2 FAD	Ha
Dihydro-orotate dehydrogenase	*Zymobacterium*	120	8	8	2	+ 2 FAD	Ha
Nitrogenase cofactor	*Clostridium*		8	6	1		Sb
Molybdoferredoxin	*Clostridium*	220	20?	20?	2	in nitrogenase	Ha
Molybdoferredoxin	*Azotobacter*	270	32	25	2	in nitrogenase	Sc

M, molar mass; *n* (X), number of atoms of X per molecule; S, 'labile' sulphur; FAD, riboflavin 5′-(trihydrogen diphosphate); FMN, riboflavin 5′-(dihydrogen phosphate)

References: Bw, Bowden (1975); Cy, Carithers *et al.* (1977); Ha, Hall *et al.* (1974); Ht, Höpner and Trautwein (1972); Kb, Kurup and Brodie (1967); Mc, Mortenson and Chen (1974); Oc, Ochiai (1977); Oh, Ohnishi *et al.* (1976); Pe, Peck *et al.* (1972); Ru, Ruzicka and Beinert (1977); Sa, Sagers (1974); Sb, Shah and Brill (1977); Sc, Schneider (1976); Si, Siegel *et al.* (1972); Ve, Vega and Kamin (1977); Ya, Yoch *et al.* (1975); Z, Steenkamp *et al.* (1978); Zx, Moura and Xavier (1978)

TABLE 9.6 Some iron proteins containing neither labile sulphur nor flavins

Name	Origin	M/1000	*n* (Fe)	Remarks	Ref.
Cytochrome c peroxidase	Yeast	34	1	+protoporphyrin IX	Pr
Horseradish peroxidase	*Armoracia*	40	1	+protoporphyrin IX	Pr
Chloroperoxidase (fungal)	*Caldariomyces fumago*	42	1	+protoporphyrin IX	Pr
Rubredoxin (bacterial)	*Clostridium*	6	1	ligands 4 cys; Fe(II)	Ha
Acetate eluted protein	Rat tissues	12	1		Ha
Agavain (proteinase)	*Agave*	56	1	Fe(II)	Ti
Lipoxygenase	*Glycine*	102	1		Pi
Aconitase	Cow liver	89	1	Fe(III); activated by Fe(II)	Oc
Pyrocatechase	*Brevibacterium*	64	1		No
Protocatechuate 4,5-diO_2ase	*Pseudomonas*	150	1		No
Dioxygenase	*Nocardia*	280	1		No
Homogentisate dioxygenase	*Pseudomonas*	380	1	Fe(II)	Oc
Phenylalanine hydroxylase	Rat liver	100	2	Fe(II)+pterin; dimer	Oc
GIBP	Intestinal mucosa	78	2	Fe(III)	Pl
Pyrocatechase	*Pseudomonas*	90	2		No
Hyperoxide dismutase	*Escherichia coli*	39	2	Fe(III)	Sl, Za
Transferrin	Mammal plasma	76	2	Fe(III); ligands his, try, tyr	Mo
Metapyrocatechuase	*Pseudomonas*	140	3		No
Catalase	Horse erythrocytes	250	4		Pr
Lysine 2,3-aminomutase	*Clostridium*	285	4?	Fe(II)	Sa
3,4-diOHphenylacetate 2,3O_2ase	*Pseudomonas*	100	4–5		No

Name	Origin	M/1000	*n* (Fe)	Remarks	Ref.
Protocatechuate 3,4-diO_2ase	*Pseudomonas*	700	8		No
L-serine dehydratase	*Clostridium*	72	n	Fe(II)	Sa
Phospho-transacetylase	*Clostridium*	70?	n		Sa
Dihydro-orotate dehydrogenase	Mammal	61	9 ± 2	+2Zn	Zb
Green protein	Cow erythrocytes	25	1	Fe(III)+haem	Zc
Cytochrome a	Cow heart mitochondria	93	1	+1 Cu, Haem a; dimerises	Yo
Cytochrome b	Cow heart mitochondria	93	1	+protoporphyrin IX	Ne
Cytochrome c	Cow heart mitochondria	13	1	+haem c	Ne
Cytochrome P450	Microsomes	46	1	+haem	No
Tryptophane 2,3-dioxygenase	*Pseudomonas*	120	1	+haem	No
Fatty acid cyclo-oxygenase	Sheep	70	3+	+3 haem	He
Myoglobin	Mammal muscle	16	1	+protoporphyrin IX	Md
Haemoglobin	Mammal erythrocytes	67	4	+4 protoporphyrin IX	Md
Leghaemoglobin	Legume root nodules	16		+protoporphyrin IX	Oc
Phosvitin	Egg yolk	35	47	Fe(III) storage	We
Gastroferrin	Gastric juice	260		Fe(III) storage	We
Ferritin	Mammal liver	440	4500	Fe(III) storage	Br

M, molar mass; *n* (Fe), number of Fe atoms per molecule; cys, cysteine; his, histidine; try, tryptophane; tyr, tyrosine.

References: Br, Brown *et al.* (1977); Ha, Hall *et al.* (1974); He, Hemler *et al.* (1976); Md, Moore and Dubach (1962); Mo, Morgan (1974); Ne, Nicholls and Elliott (1974); No, Nozaki and Ishimura (1974); Oc, Ochiai (1977); Pi, Pistorius *et al.* (1976); Pl, Pollack and Lasky (1977); Pr, Pratt (1975); Sa, Sagers (1974); Sl, Slykhouse and Fee (1976); Ti, Tipton (1965); We, Webb (1975); Yo, Yamanaka and Okunuki (1974); Za, Dougherty *et al.* (1978); Zb, Forman and Kennedy (1978); Zc, De Filippi and Hultquist (1978)

TABLE 9.7 Manganese proteins and enzymes

Name	Origin	M/1000	*n* (Mn)	Remarks	Ref.
Hyperoxide dismutase	*Escherichia coli*	40	2	Mn(III)	Ke
	Porphyridium (red alga)	40	1		Mf
	Pleurotus (fungus)	78	2		Lm
	Yeast	96	4	Mn(III); tetramer	Rf
	Mitochondria	80	4	tetramer	Wf
Manganin	*Arachis* seeds	56	1	hyperoxide dismutase ?	Dr
Concanavalin A	*Canavalia* seeds	26	1	+ Ca; ligands 2 asp, glu, his	Oc
Arginase	Rat liver	118	4	Mn(II)	Hh
Pyruvate carboxylase	Chicken liver	510 ?	2–4	Mn(II) + biotin	Sc
Chloroplast protein	*Zea mays*	25			Ld
Green protein	*Phaeodactylum*	850	0·6–2	+ 8Cu, chlorophyll	Ha
Pinnaglobin (mollusc)	*Pinna* blood			0·35 % Mn	Bo

M, molar mass; *n* (Mn), number of Mn atoms per molecule; asp, aspartate; glu, glutamate; his, histidine

References: Bo, Boeri (1963); Dr, Dickert and Rozacky (1969); Ha, Holdsworth and Arshad (1977); Hh, Hirsch-Holb *et al.* (1971); Ke, Keele *et al.* (1970); Ld, Lagoutte and Duranton (1975); Lm, Lavelle and Michelson (1975); Mf, Misra and Fridovich (1977); Oc, Ochiai (1977); Rf, Ravindranath and Fridovich (1975); Sc, Scrutton *et al.* (1966); Wf, Weisiger and Fridovich (1973)

residue of selenocysteine per molecule (Cone *et al.*, 1976). Glutathione peroxidase, from sheep erythrocytes, has a molar mass of 88 000 and has four selenium atoms per molecule (Oh *et al.*, 1974). A cytochrome c-type enzyme from sheep muscle (M = 10 000), and formate dehydrogenase from *Escherichia coli*, are thought to contain selenium (Whanger *et al.*, 1973; Shum and Murphy, 1972). The administration of selenite to deficient rats increases the activities of two further enzymes, haem oxygenase and δ-aminolaevulinate synthase, but this may be an indirect action (Maines and Kappas, 1976). Labile selenide could exist in liver mitochondria (Caygill *et al.*, 1971).

Zinc. Over 80 enzymes and proteins containing zinc have been reported (Vallee, 1976; Chlebowski and Coleman, 1976). Some of the better characterised ones are listed in Table 9.8. Zinc enzymes are mostly peptidases or anhydrases, which are needed in the synthesis of both nucleic acids and proteins, so that zinc is essential for growth and cell replication. In the presence of oxidising agents such as NAD, some zinc enzymes can catalyse dehydrogenation reactions; this is rather unexpected, since zinc itself does not undergo redox reactions as can Cu, Mn, Mo or Fe. A feature of zinc enzymes is that when the zinc is artificially replaced by cobalt, the enzyme continues to function, often with reduced activity. In the case of the very large molecule transcarboxylase, it is uncertain what the natural proportions of zinc and cobalt are.

Metallothioneins are zinc/cadmium storage proteins found in the mammalian kidney or liver; the protein moiety contains much cysteine. Relatively high zinc concentrations found in the prostate gland and choroid of the eye (Davies, 1972), in the flagella of spermatozoa (Morisawa and Mohri, 1972) and bacteriophage (Kizloff and Lute, 1977), and in the pollen of seed plants (Polar, 1975) await explanation.

Organelles and part organelles

METALS AND THE NITROGENASE ENZYME SYSTEM (Schneider, 1976)

The reduction of dinitrogen to ammonia can be accomplished by a number of specialised micro-organisms using an enzyme system called nitrogenase. All nitrogenases examined consist of a small protein, azoferredoxin, containing 4 iron and 4 labile sulphide atoms per molecule, and a large protein, molybdoferredoxin, containing molybdenum, iron and labile sulphide (Table 9.5). Rawlings *et al.* (1978) have isolated a cofactor containing a new type of cluster compound, with eight iron and six labile

TABLE 9.8 Some zinc metalloenzymes and metalloproteins

Name	Origin	M/1000	*n* (Zn)	Remarks	Ref.
Carbonic anhydrase	Erythrocytes	29	1		Cc
Carboxypeptidase	Cow pancreas	35	1	Ligands; glu, 2 his	Cc
Thermolysin (proteinase)	*Bacillus thermoproteolyticus*	35	1	Ligands; glu, 2 his	Co
Dipeptidase	Pig kidney	47	1		Cc
Neutral protease	Snake venom	22·5	1	+2 Ca	Cc
5-Nucleotidase	*Escherichia coli*	52	1		Cc
Mercaptopyruvate *S*-transferase	*Escherichia coli*	24	1		Cc
Phosphomannose isomerase	*Saccharomyces* (yeast)	450	1		Cc
Aldolase	*Saccharomyces* (yeast)	80	2		Cc
β-lactamase	*Bacillus cereus*	22	2	Ligands; 3his, cys	Sb
Gustin	Human saliva	37	2		Rh
Alkaline phosphatase	*Escherichia coli*	80	2		Cc
DNA polymerase	*Escherichia coli*	109	2		Cc
RNA polymerase	*Escherichia coli*	700	2		Fa
Reverse transcriptase	Bird virus	180	2		Au
5′AMP aminohydrolase	Rat muscle	290	2		Cc
Pyruvate carboxylase	*Saccharomyces*	600	4	+4 biotin	Cc
Aminopeptidase	Pig kidney	300	6		Cc
Aspartate transcarbamylase	*Escherichia coli*	300	6	hexamer	Cc
Transcarboxylase	*Propionibacterium shermanii*	790	6–12	+6 biotin, Co	Cc
Malate dehydrogenase	Cow heart	40	1	+1 NAD	Cc
Lactate dehydrogenase	Liver/muscle	140	1–2	+NAD	Cc, Oc
D-glyceraldehyde $3PO_4$ dehydrogenase	Pig muscle	137	3	+3 NAD	Cc
Alcohol dehydrogenase	Horse liver	80	4	+2 NAD; Ligands; 2 cys, his	Cc
D-lactate cytochrome reductase	*Saccharomyces*	50	4–6	+1 FAD	Cc
Glutamate dehydrogenase	Cow liver	1000	2–4	+NAD	Cc
Hyperoxide dismutase	Cow erythrocyte	33·6	2	+2 Cu; Ligands; asp, 3 his	Be
α-Macroglobulin	Human serum	725	2?	transport?	Ad
Metallothionein	Cow liver	6·8	≤7	+Cd; Ligands; cys	Ko
Metallothionein	Horse kidney	10	3	+5 Cd	Cc

M, molar mass; *n* (Zn), number of zinc atoms per molecule; asp, aspartate; cys, cysteine; glu, glutamate; his, histidine; FAD, riboflavin 5′-(trihydrogen diphosphate); NAD, see p. *xiv*.

References: Ad, Adham *et al.* (1977); Au, Auld *et al.* (1974); Be, Beem *et al.* (1977); Cc, Chlebowski and Coleman (1976); Co, Colman *et al.* (1972); Fa, Falchuk *et al.* (1976); Ko, Kojima *et al.* (1976); Oc, Ochiai (1977); Rh, Henkin (1978); Sb, Baldwin *et al.* (1978). See also Williams and Da Silva (1978)

sulphide atoms per molybdenum atom. The reduction involves the transfer of six hydrogen atoms, and requires ATP:

$$N_2 + 6RH + nMgATP^{2-} \xrightarrow{\text{Nitrogenase}} 2NH_3 + 6R + nMgADP^- + nP^-$$

where RH is a reductant such as dihydrogen or pyruvate, and n is a large number, probably between 12 and 18. Zinc may be present in some molybdoferredoxins.

It seems that $MgATP^{2-}$ binds to the azoferredoxin, which can then reduce the molybdoferredoxin to a form which can bind dinitrogen and reduce it. The site where the dinitrogen is bound is uncertain, but could involve the Mo and/or Fe atoms. Molybdenum can be replaced by vanadium *in vitro* without much loss of enzyme activity.

METALS IN MITOCHONDRIA (Metzler, 1977)

Mitochondria are subcellular organelles which are important in respiration, which is coupled to the phosphorylation of ADP to ATP in a way that is not yet understood. Mitochondria from cow heart contain the molar ratio of 2Cu:5 non haem Fe: $\geqslant 4$ haem Fe. Most of the copper and iron is probably in the cytochrome system of enzymes, such as the terminal oxidase cytochrome oxidase (cytochrome $a + a_3$), which can reduce dioxygen to water in a four electron transfer:

$$O_2 + 4RH \xrightarrow{\text{cytochrome oxidase}} 2H_2O + 4R$$

The oxidised form of cytochrome oxidase is reduced by a chain reaction in which hydrogen atoms provided by a reducing agent such as $NADH_2$ or succinate pass down a chain which is believed to involve several cytochromes, two Fe–S proteins, and the purely organic molecule ubiquinone, as follows:

$$NADH_2 \xrightarrow{\text{Fe–S flavoprotein}} \xrightarrow{\text{Cyt } b_k} \xrightarrow{\text{Ubiquinone}} \xrightarrow{\text{Cyt } b_T} \xrightarrow{\text{Fe–S protein}} \xrightarrow{\text{Cyt } c_1} \xrightarrow{\text{Cyt c}} \xrightarrow{\text{Cyt a}}$$

Note that 7/8 of the steps involve a protein containing iron(II) which can be reversibly oxidised. The energy liberated is used both for the oxidative phosphorylation of ADP and to accumulate ions such as Ca^{2+}, Mg^{2+}, Mn^{2+} and K^+ within the mitochondrial membrane, but the latter process takes precedence over the former (Hunter *et al.*, 1976). The process has some analogy to the charging of a battery. It often leads to the precipitation of granules of calcium phosphate within the mitochondrion (Borle and Anderson, 1976). Both Ca^{2+} and Sr^{2+} stimulate respiration.

METALS IN CHLOROPLASTS (Krogmann, 1973; Metzler, 1977; Renger, 1977)

Chloroplasts are large subcellular organelles, possibly ancient symbionts, which are responsible for the bulk of global photosynthesis. They appear to contain large photosynthetic units of molar mass 930 000, containing about 160 chlorophyll a, 70 chlorophyll b (=230 Mg), 12 Fe (including 5 ferredoxin Fe), 6 Cu (including 1 plastocyanin Cu) and 2 Mn. Photosynthesis is incompletely understood, but the following steps can be distinguished:

1. Water is oxidised to dioxygen by an unknown system involving Mn(III), chlorophyll P682 and the participation of chloride ion (Kono *et al.*, 1976; Hind *et al.*, 1969).

$$R + H_2O \rightarrow RH_2 + \tfrac{1}{2}O_2$$

2. Photosystem II. Chlorophyll P682, or its dimer, absorbs infrared light of wavelength 700 nm and in its excited state is able to reduce an unknown electron acceptor A. The chlorophyll radical ion formed is then reduced by an electron donor D

$$\begin{aligned} Ch^* + A &\rightarrow Ch.^+ + A.^- \\ Ch.^+ + D &\rightarrow Ch + D.^+ \end{aligned}$$

3. $A.^-$ initiates a chain of reversible reductions involving the following reducible substances: plastoquinone, cytochrome b559, cytochrome f, plastocyanin and chlorophyll P700.

4. Photosystem I. Chlorophyll P700 absorbs red light of wavelength about 650 nm and its excited state is able to reduce an electron acceptor, which is probably an Fe–S protein.

5. This electron acceptor, in its reduced form, reduces a ferredoxin which in turn reduces NADP to $NADPH_2$.

6. $NADPH_2$ and ATP are substrates for the synthesis of glucose from carbon dioxide and water, by the overall process:

$$\begin{aligned} 6CO_2 + 12NADPH_2 + 18ATP^{4-} \rightarrow C_6H_{12}O_6 + 12NADP + 6H_2O \\ + 18ADP^{3-} + 18P^- \end{aligned}$$

Among the many features that are not yet clear is why illumination of chloroplasts involves the intake of H^+ and Mg^{2+}, and the excretion of K^+ and Cl^- (Nobel, 1975). The zinc in chloroplasts is probably present as carbonic anhydrase, and could function as a permease exchanging HCO_3^- for OH^- across the chloroplast membrane.

10 *Biological Fractionation of Isotopes*

The majority of elements possess more than one stable isotope, and among the essential elements only Co, F, I, Mn and Na are monoisotopic. While stable isotopes closely resemble one another in their behaviour, there are small differences in mass which can lead to partial fractionation in both kinetic and equilibrium systems. The technique of mass spectrometry has made it possible to measure the extent of fractionation with great precision, and it has long been known that isotopes can be separated by biochemical and geochemical processes. Some of the results in this field, which is still developing rapidly, are described below.

Theory

The theory of isotopic fractionation in chemical reactions, the so-called *isotope effect*, was worked out by Bigeleisen in 1949 (see also Melander, 1960). A light isotope moves faster than a heavy one, because it has less inertia. Isotopic substitution changes the zero point energy of a molecule, which is related to the sum of the normal modes of vibration. This has several consequences.

1. Chemical bond energies are altered by isotopic substitution; C–D are stronger than C–H bonds by about 5–6 kJ mol^{-1}, and their cleavage is slower.

2. In a reaction where an atom X is transferred, the rate constant k depends on the mass of X. For two isotopes X_1 and X_2 with masses m_1 and m_2, the ratio of rate constants is given by

$$\frac{k_1}{k_2} = \sqrt{\frac{m_2}{m_1}} \cdot f(T)$$

where $f(T)$ is a ratio of partition fractions of the molecules taking part.

3. Since the rate constants are affected, the equilibrium constant will usually be affected too.

4. The theory predicts that isotope effects will decrease as the temperature is raised.

In Table 10.1 we compare $\sqrt{(m_2/m_1)}$ for a number of elements of biological interest.

These data show the unique position occupied by the isotopes of hydrogen with regard to fractionation processes. For all the elements except hydrogen, isotope effects in chemical reactions are of the order of a few per cent at most.

TABLE 10.1 $\sqrt{(m_2/m_1)}$ for isotopes of biological interest

Element	m_1	m_2	$\sqrt{(m_2/m_1)}$
H	1	2	1·4142
Li	6	7	1·0801
C	12	13	1·0409
N	14	15	1·0351
O	16	18	1·0606
S	32	34	1·030

Northrop (1975, 1977) has extended the theory to the kinetics of enzymic reactions. Consider an enzyme E reacting with a substrate Su and producing a product Pr

$$Su + E \underset{k2}{\overset{k1}{\rightleftharpoons}} SuE \overset{k3}{\rightarrow} EPr \overset{k5}{\rightarrow} E + Pr$$

where for simplicity the latter two reactions are assumed to be irreversible. It is possible to measure the maximal velocity V of the reaction and conventional to calculate a Michaelis constant

$$K = \frac{k_5(k_2+k_3)}{k_1(k_3+k_5)}.$$

If this is done for reactions where the substrate is labelled with H, D and T, Northrop was able to predict that

$$\frac{L-1}{L^{1\cdot 44}-1} = \left[\frac{(V/K)_{\mathrm{H}}}{(V/K)_{\mathrm{D}}} - 1\right] \bigg/ \left[\frac{(V/K)_{\mathrm{H}}}{(V/K)_{\mathrm{T}}} - 1\right]$$

where $L = k_{3\mathrm{H}}/k_{3\mathrm{D}}$, in good agreement with experiment. The only assumption made is that $k_{3\mathrm{H}}/k_{3T} = (k_{3_\mathrm{H}}/k_{3_\mathrm{D}})^{1\cdot 44}$, which is true between 0 °C and 100 °C (Swain *et al.*, 1958). K, V and V/K may all show isotope effects, but none of them can be interpreted simply, as can the calculated value of L. The maximum value of L found in chemical reactions is 15, but it is usually 6–10 for enzymic reactions.

In cyclic biological processes, the possibility of enhanced enrichment by repeated fractionation of isotopes has been shown by Keeling (1961), but the predicted enrichments are only a few per cent at most except in the case of hydrogen.

Schowen (1977) has provided a useful rule covering isotope exchange reactions – the lighter isotope accumulates where the bonding is weaker. Thus for exchange of hydrogen isotopes with water, protium will accumulate in thiols (RSH) and protonated ethers (R_2OH^+); deuterium will accumulate in α-hydroxyethers ($R_2C(OMe)OH$); while little or no fractionation takes place with alcohols, amines and carboxylic acids.

A NOTE ON STANDARDS

In view of the natural variations in isotope ratios, mass spectrometric measurements are usually reported relative to the ratios in an arbitrary standard, e.g. sea water for isotopes of H and O, atmospheric dinitrogen for isotopes of N, and a fossil limestone belemnite for isotopes of C (and sometimes O). Natural variations occur even among the heavy elements, notably for lead and uranium. The isotopic composition and atomic weight of lead varies according to its source as a radioactive decay product of thorium and/or uranium. Uranium itself varies in isotopic composition (Osmund and Cowart, 1976); in Oklo, central Africa, the ^{235}U content of local ores has been depleted by prehistoric nuclear fission.

Expression of the isotopic abundance data

Table 10.2 gives the normal isotopic composition of six light elements which undergo isotopic fractionation in biological processes. The data given here are fractional abundances. There are several other ways of expressing isotopic composition, of which two of the commonest are (*a*) as a ratio of fractional abundances of two isotopes and (*b*) as delta values, where

$$\delta = K\left[\left(\frac{A_X}{B_X}\right)_{\text{samp}} \times \left(\frac{B_X}{A_X}\right)_{\text{std}} - 1\right]$$

TABLE 10.2 Fractional abundances of the isotopes of normal H, Li, C, N, O and S

Sea water	^{1}H 0·999844	^{2}H 0·000156		
Lithium ores	^{6}Li 0·9248	^{7}Li 0·0752		
Fossil Belemnite	^{12}C 0·98892	^{13}C 0·01108		
Atmospheric N_2	^{14}N 0·99635	^{15}N 0·00365		
Sea water	^{16}O 0·99759	^{17}O 0·000374	^{18}O 0·002039	
Native sulphur	^{32}S 0·95	^{33}S 0·0076	^{34}S 0·0422	^{36}S 0·00014

here A and B are the mass numbers of two isotopes of an element X, samp = sample, std = standard, and K is 1 if δ is a fraction, 100 if δ is a percentage, and 1000 if δ is per thousand. The advantage of using δ values is that they minimise systematic errors of measurement; their disadvantage is that not all workers quoting δ values use the same standard.

Enzymic fractionation of isotopes

Soon after the discovery of deuterium it was shown that succinic dehydrogenase oxidises deuterosuccinate more slowly than it oxidises succinate (Erlenmayer *et al.*, 1936). On the other hand fumarate hydrogenase uses H_2 and D_2 at the same rate (Franke and Mönche, 1948). The current interpretation of these reactions suggests that in the first reaction the oxidation of a C–H bond is rate-limiting, while in the second reaction the formation of a C–H bond is not a rate-limiting step. More recent studies have been able to distinguish between relatively large, primary isotope effects for an atom X, which occur when a C–X bond is involved in the reaction, and small, secondary isotope effects which may occur when the reaction involves other parts of the molecule. Primary hydrogen isotope effects have been found in many redox and some hydrolytic reactions, e.g. oxidation of formate by catalase (Frei and Aebi, 1958); methanol oxidase (Kraus and Simon, 1975); alcohol dehydrogenase (Klinman, 1975); hydrolysis of chitin by lysozyme (Smith *et al.*, 1973). In all cases hydrogen is used more rapidly than deuterium.

Most enzymes act more slowly in a medium of deuterium oxide than in water, for example formic dehydrogenase (Aebi and Frei, 1957) and fumarase (Thomson, 1960). However Nilsson and Kearns (1973) have found that the photochemical inactivation of both trypsin and alcohol dehydrogenase occurs about ten times faster in deuterium oxide than in water. They ascribe this to the increased lifetime of the inactivating species, singlet oxygen, which has a lifetime of about 20 μs in D_2O but only 2 μs in H_2O.

Primary isotope effects have also been found for carbon, nitrogen and oxygen in enzymic reactions, but their magnitude is much smaller than in the case of hydrogen. In order to obtain reproducible results it is necessary to control both temperature and pH, as was done in recent work by O'Leary (1977), who found for example:

X	Enzymic reaction	Ratio of rate constants for stable isotopes
C	Decarboxylation of carboxylates	k_{12}/k_{13} = 1·03 to 1·06
N	Hydrolysis of amides by papain	k_{14}/k_{15} = 1·021
O	Hydrolysis of esters by chymotrypsin	k_{16}/k_{18} = 1·007 to 1·018

In the majority of cases the heavy isotope reacts more slowly than the light one, but Huebner *et al.* (1975) claim that thiamine 4-amino(^{15}N)pyrophosphate reacts faster than the unlabelled compound with pyruvate decarboxylase.

Isotope fractionation by unicellular organisms and lower plants

Substantial isotope effects can occur when hydrogen is metabolised. For example Augood (1954) found that when the green alga *Chlorella* was grown in a nutrient solution it absorbed ^1H_2O preferentially, leaving the mother liquor enriched in deuterium. A marine bacterium, *Pseudomonas*, can produce dihydrogen containing only 30 ppm D from sea water containing 146 ppm D (Cloud *et al.*, 1958). Hydrogen sulphide produced by anoxic bacteria contains about 90 ppm D (Roth, 1956). On the other hand the utilisation of dihydrogen for bacterial reductions involves negligible isotopic fractionation (Linday and Syrett, 1960).

In a study involving the incubation of yeast (*Saccharomyces cerevisiae*) in a nutrient solution containing lithium citrate, it was found that $^6Li^+$ enters the cells slightly faster than does $^7Li^+$. When the lithium-adapted yeast was tranferred to a solution containing potassium citrate, the lighter isotope of lithium left the cells faster than the heavy one, as expected (Bowen, 1956b).

In the case of carbon, both algae (Abelson and Hoering, 1961; Deuser *et al.*, 1968) and the photosynthetic bacterium *Ectothiorhodospira* (Gogotova *et al.*, 1973) have their organic biomass depleted in ^{13}C by 1·3–1·7% with respect to the source of carbon dioxide: that is, they take up $^{12}CO_2$ faster than $^{13}CO_2$. Their biomass is not uniformly depleted in ^{13}C, as the carboxylate carbon from algal amino acids has δ (^{13}C) = $-0{\cdot}31$%, while the rest of the molecule has δ (^{13}C) = $-1{\cdot}46$%.

Calder and Parker (1973) found that the extent of ^{13}C depletion varied from zero to $-1{\cdot}5$% depending on the concentration of carbon dioxide in the feed gas. On the other hand carbon dioxide produced by respiration is usually enriched in ^{13}C, for example the gas produced by fungi respiring in soil (Matschke *et al.*, 1972). In sea water, carbon dioxide has a δ (^{13}C) which varies with depth, with a maximum value of $+0{\cdot}19$% in surface water falling to a constant value of $+0{\cdot}11$% below a depth of 1·7 km (Fig. 10.1; Kroopnick *et al.*, 1972). Algal carbonates also contain carbon which is somewhat enriched in ^{13}C, with δ (^{13}C) $\leqslant +0{\cdot}63$% for calcites and $\leqslant +0{\cdot}33$% for aragonites (Borowitzka, 1977). However, marine phytoplankton have their organic carbon depleted in ^{13}C by 2·2–3·6% (Wong and Sackett, 1978). Bacteria which oxidise carbon monoxide utilise ^{12}CO preferentially by a factor of about 1·9% (Zyakun *et al.*, 1976). The methane produced by

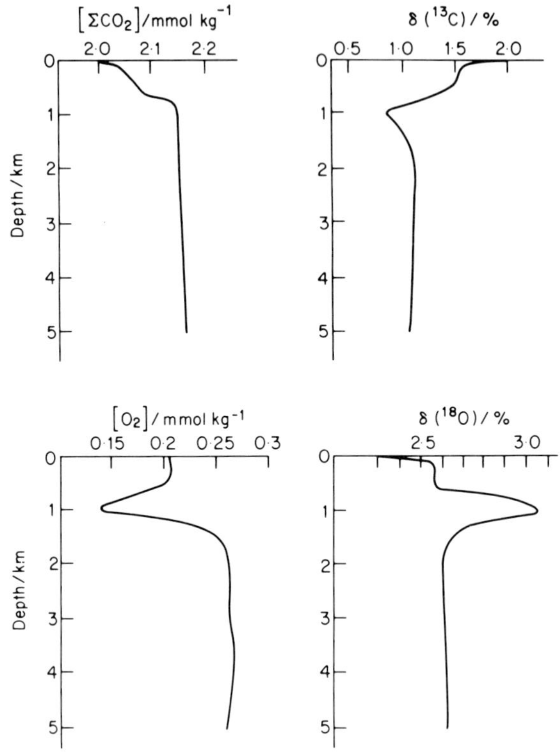

FIG. 10.1 Variation of $[\Sigma CO_2]$, $\delta(^{13}C)$, $[O_2]$ and $\delta(^{18}O)$ with depth in the ocean. Data from Kroopnick *et al.* (1972)

anoxic bacteria is enriched in ^{12}C by an even larger factor with respect to atmospheric carbon dioxide, which has been reported as 6·8 ‰ (Deevey *et al.*, 1963), 4·3 ‰ (Deuser *et al.*, 1973), 4–8 ‰ (Fuex, 1977) and 2·5–6·1 ‰ (Games *et al.*, 1978).

Isotopic fractionation of nitrogen seems to be less extensive than with carbon. There is little or no isotope effect in the fixation of atmospheric dinitrogen by *Azotobacter* on land (Hoering and Ford, 1960; Delwiche and Steyn, 1970) or by *Trichodesmium* in the sea (Wada and Hattori, 1976). However, both $^{14}NO_3^-$ and $^{14}NO_2^-$ are preferentially reduced to dinitrogen by *Pseudomonas* spp. in soils, by a factor of about 2 ‰ (Wellman *et al.*, 1968; Delwiche and Steyn, 1970). $^{14}NH_3$ is preferentially oxidised to nitrite by *Nitrosomonas* in soils by a factor of 2·6 ‰ (Delwiche and Steyn, 1970), or by marine organisms by a factor of 0·5 to 2·1 ‰ (Miyake and Wada, 1971).

The isotopic composition of environmental oxygen has been reviewed by Wagener (1975). The $^{18}O/^{16}O$ ratio is found experimentally to be constant in sea water H_2O, but variable in freshwater H_2O; there is less ^{18}O in polar ice than in equatorial freshwater. Atmospheric dioxygen is enriched in ^{18}O with respect to sea water by a factor of 1·0235, while for dioxygen dissolved in the ocean the ratio of $^{18}O/^{16}O$ depends on the depth (Fig. 10.1; Kroopnick *et al.*, 1972). It is not possible to make a consistent budget of the oxygen-18 cycle using presently available, rather imprecise data, but the following points have been established:

1. Photosynthesis by marine phytoplankton should lead to dioxygen having the same $^{18}O/^{16}O$ ratio as in sea water H_2O, $\pm 0{\cdot}1\,\%$.
2. Photosynthesis by land plants should lead to dioxygen having the same $^{18}O/^{16}O$ ratio as in plant water.
3. Respiration by marine organisms preferentially consumes $^{16}O_2$ by a factor of 1·4% (Kroopnick *et al.*, 1972).
4. Respiration by terrestrial organisms preferentially consumes $^{16}O_2$ by a factor of about 1·7% (Wagener, 1975).

Bacterial transformations of sulphur often show isotope effects and these have been extensively investigated. The equilibrium constant for the inorganic reaction

$$H_2{}^{34}S + {}^{32}SO_4{}^{2-} = H_2{}^{32}S + {}^{34}SO_4{}^{2-}$$

is 1·074 at 25 °C, so it is not surprising that natural sulphides tend to be enriched in ^{32}S while sulphates are enriched in ^{34}S. Thus the oxidation of hydrogen sulphide to sulphate by *Thiobacillus concretivorus* is accompanied by an enrichment of 1·1 to 1·8% ^{34}S in the sulphate formed (Kaplan and Rittenberg, 1962). The reverse reaction involves the enrichment of ^{32}S in the sulphide formed by a factor of 0·5 to 2·7%. This was first pointed out by Thode *et al.* (1951), and subsequently confirmed by others, e.g. McCready *et al.* (1974) using the yeast *Saccharomyces*, and Chambers *et al.* (1976) using bacteria of the genera *Desulphovibrio* and *Desulphomaculum*. The bulk of the isotopic fractionation takes place in the rate-limiting step, which is the reduction of sulphate to sulphite. When sulphites themselves are reduced to sulphides, the fractionation of ^{32}S from ^{34}S depends on the organisms used, e.g.

Organism	$\delta(^{34}S)$ when $SO_3{}^{2-} \rightarrow S^{2-}$	Reference
Desulphovibrio	−0·17 to −2·55%	Kemp and Thode (1968)
Saccharomyces	−3 to −5%	McCready *et al.* (1974)
Clostridium	+4·7 to −1·5%	McCready *et al.* (1975)

The exceptional behaviour of *Clostridium* is contrary to theoretical predictions. During bacterial reduction of sulphates in sea water, the ratio $\delta\,(^{34}S)/\delta\,(^{18}O)$ in residual sulphate remains constant at a value of 4·0 (Mizutani and Rafter, 1969).

The oxidation of elementary sulphur to sulphates by *Thiobacillus thiooxidans*, and the reduction of elementary sulphur to hydrogen sulphide by yeast, show negligible isotope effects, probably because the rate-limiting step is the movement of S_8 molecules across cell membranes (Kaplan and Rittenberg, 1962).

Isotope fractionation by higher plants

Higher plants discriminate against deuterium and tritium, and in favour of protium, when they use water as a metabolite, so that the nutrient solution becomes enriched in the heavier isotopes. Thus Carlbom *et al.* (1956) were able to obtain a four-fold enrichment of deuterium in the residual water used to grow sprouting barley seeds. Lucerne (*Medicago sativa*) grown in water labelled with tritium has a specific activity 22% lower than the water (McFarlane, 1976). Ziegler *et al.* (1976) have shown that all plants discriminate against deuterium by factors of −7·5 to −13·2%, and that there are significant differences between plants with C_3, C_4 and CAM-type metabolism (see below). Old measurements suggest that wood and lignin are somewhat depleted in deuterium (141 ppm D), while cellulose is slightly enriched (160 ppm D) compared with natural water (Bokhoven and Theeuwen, 1956). However Wilson and Grinsted (1975) have shown that cellulose in rings of the gymnosperm tree *Pinus radiata* is slightly depleted in deuterium with respect to the nutrient water, and that the depletion depends on temperature. Epstein *et al.* (1977) confirm that

$$\delta\,(D)_{cellulose} = \delta\,(D)_{water} - 2{\cdot}2 \qquad (\delta\,(D)\text{ in }\%)$$

where 2·2 is an average value for many land plants, but there is some controversy as to whether the tree ring measurements can be used as an archaeo-thermometer (Epstein and Yapp, 1976; Yapp and Epstein, 1977). Schiegl and Vogel (1970) found that woody plants had δ (D) from −2·9 to −3·9% less than local rainfall, while marine algae had δ (D) from −5·7 to −10·3% less than sea water. Plant saps were enriched in deuterium, with δ (D) from +1 to +5%, presumably because transpiration favours loss of light water.

Many workers have confirmed the early finding that plants contain less of the heavy isotope ^{13}C than does atmospheric carbon dioxide (Craig, 1953). It now seems that the two metabolically distinct groups of plants with C_3

and C_4 metabolism can be characterised by their ^{13}C contents (Smith and Epstein, 1971; Medina and Troughton, 1974; Deleens and Garnier-Dardart, 1977). C_3 plants fix carbon dioxide by reaction with ribulosediphosphate carboxylase, which discriminates against $^{13}CO_2$, and their carbon has a δ (^{13}C) of −2·4 to −3·4‰. C_4 plants fix some carbon dioxide as oxalsuccinate, with negligible isotopic fractionation, and have δ (^{13}C) in the range −0·6 to −1·9‰. A small group of plants have Crassulean acid metabolism (CAM), using phospho-enol pyruvate carboxylase to form malate from carbon dioxide and pyruvate in the dark: the malate is broken down again in light. This reaction discriminates weakly in favour of ^{13}C, but CAM plants have variable δ (^{13}C), ranging from −1·4 to −3·3‰ (Bender *et al.*, 1973).

Wilson (1978) has measured the ^{13}C in rings from the tree *Pinus aristata* and shown that the value of δ (^{13}C) suddenly changed from −2·05‰ to −2·13‰ about 1870. He attributes this to the extensive forest clearance and destruction of soil humates which occurred in the late nineteenth century, and suggests that before 1870 atmospheric CO_2 had a δ (^{13}C) of −0·6‰, which has been changed to its present value of −0·9‰ by human activities. Further data are given by Williams (1978).

The observed variations in δ (^{15}N) in plants, from +0·6 to −0·6‰, do not allow any general conclusions to be drawn (Hoering, 1956; Parwel *et al.*, 1957). The lighter isotope ^{16}O is preferentially used in respiration by the freshwater plant *Elodea* (Vinogradov *et al.*, 1959). Both plant water and plant cellulose are enriched in the heavy isotope ^{18}O (Wagener, 1975; Epstein *et al.*, 1977). This probably comes about because of the preferential loss of $H_2{}^{16}O$ by transpiration, since there is negligible isotopic fractionation of oxygen during photosynthesis, and these observations could account for the enrichment of ^{18}O in atmospheric dioxygen. Fractionation of the isotopes of sulphur by aquatic plants has been studied by Mekhtiyeva and Pankina (1968); freshwater plants have δ (^{34}S) of −0·06 to −0·42‰ with respect to the sulphate in their nutrient solution, and marine algae have similar δ values with respect to the ocean.

No biological fractionation of the isotopes of Ca, Li, K, Rb or Sr was found during their uptake by wheat plants (Rick *et al.*, 1961). Walker *et al.* (1958) claim to have detected small enrichments of the heavy isotope of copper (^{65}Cu) in a brown alga (*Sargassum*) and the leaves of oak (*Quercus*), but this needs confirmation.

Isotope fractionation by animals

Early work suggests that the soft tissues of animals are more or less enriched in the lighter isotopes of hydrogen and carbon, while more recent studies indicate that δ (^{13}C) in both vertebrates and invertebrates closely parallels the value in the diet (De Niro and Epstein, 1978). Hard tissues of corals or molluscs are enriched in the heavier isotopes of carbon, oxygen and sulphur (Craig, 1953; Thode *et al.*, 1961; Weber *et al.*, 1976). Ernst and Juricskay (1971) claim that frog muscles can discriminate ^{39}K from ^{41}K, but this should be re-investigated. The $^{40}Ca/^{44}Ca$ ratios in mollusc shells and teeth vary by $\pm 0{\cdot}22\%$ (Russell *et al.*, 1978).

Following the demonstration that the ^{18}O content of limestones can be used as a geological thermometer (Urey *et al.*, 1951), many workers have attempted to calibrate and interpret scales of palaeotemperature (e.g. R. Bowen, 1966; Emiliani, 1978). There have also been unconfirmed reports that $^{12}C/^{13}C$ ratios have varied with geological time, and extensive measurements of $^{12}C/^{13}C$ and $^{18}O/^{16}O$ ratios have been carried out for marine invertebrates (Weber, 1968; Degens *et al.*, 1968; Eichmann and Schidlowski, 1975; Weber *et al.*, 1976). The principle of the oxygen isotope thermometer is that the equilibrium constant of the reaction

$$1/3\, C^{16}O_3{}^{2-} + H_2{}^{18}O = 1/3\, C^{18}O_3{}^{2-} + H_2{}^{16}O$$

varies with temperature, and is 1·018 at 0 °C and 1·014 at 25 °C. Hence precise measurements of ^{18}O in carbonate fossils should reflect the temperature at which they were laid down. It has been found empirically that carbonates derived from terrestrial or air-breathing organisms give erratic results. Calcites are mostly suitable, aragonites less so, and dolomites are so altered by exchange as to be unsatisfactory for this technique. The question of whether the original oxygen can be exchanged is still open, and is believed to vitiate the temperatures derived from Pre-Devonian fossils. The precision of the method can be judged from Fig. 10.2, taken from the data of Keith and Weber (1964) for freshwater carbonates.

Toxic effects of heavy isotopes (Katz *et al.*, 1975)

Katz and co-workers have been able to induce certain blue-green and green algae, and also microbes, to grow in 99·7% deuterium oxide, and many interesting deuterated organic compounds have been biosynthesised in this way. For example, fully deuterated succinic dehydrogenase and alkaline phosphatase have been isolated (Katz and Crespi, 1970). It seems that higher organisms cannot adapt themselves to pure heavy water. For the aquatic seed plant *Lemna*, 50% D_2O in the nutrient solution is toxic, while in mammals 10 to 15 mol% D_2O in the blood causes symptoms of toxicity. Heavy isotopes of carbon and nitrogen appear to be much less toxic, but the

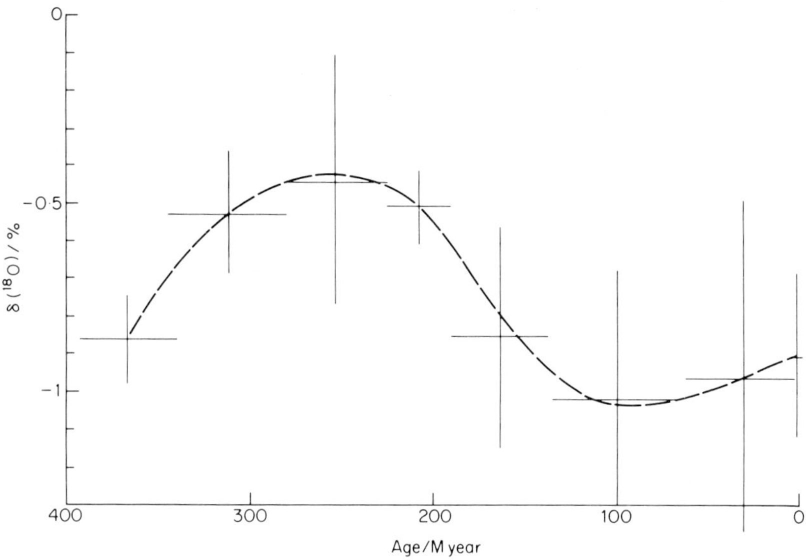

FIG. 10.2 Variation of $\delta(^{18}O)$ with geological age for freshwater carbonates. Data from Keith and Weber (1964)

expense of carrying out experiments with high concentrations of these isotopes has prevented much work from being done. Tobacco plants containing 90% ^{13}C have shown evidence of toxicity in the form of pollen abnormalities.

Environmental effects of biological fractionation of isotopes

Much early information on environmental isotope ratios is collected in two books by Rankama (1954, 1963), while Faure (1977) reviews later studies. In the case of hydrogen, most natural sources contain less deuterium than does sea water, which has 156 ppm D. Fresh waters have 133–153 ppm D, while water believed to be derived from the Earth's mantle has 142–148 ppm D (Taylor, 1978). The methane in natural gas has 122–132 ppm D, while H_2S and H_2 from oil wells have 90 ppm D and 30 ppm D respectively (Schiegl and Vogel, 1970). The two latter molecules are probably produced by bacterial action.

In the case of carbon, a number of ranges of composition can be distinguished, as follows:

1. Biological calcite and sedimentary limestones are enriched in ^{13}C.

They have $89{\cdot}2 > (^{12}C/^{13}C) > 88{\cdot}6$, or $\delta\,(^{13}C)$ between $+0{\cdot}35$ and $-0{\cdot}32\,\%$.

2. Atmospheric carbon dioxide has $\delta\,(^{13}C)$ close to $-0{\cdot}9\,\%$ (Fuex, 1977).

3. Igneous rocks, most shales, terrestrial and marine humates (Nissenbaum and Kaplan, 1972), biological soft tissues and organic carbon, including most petroleum, have isotopic compositions close to that of average crustal carbon, with $91{\cdot}3 > (^{12}C/^{13}C) > 90{\cdot}5$, or $\delta\,(^{13}C)$ between $-1{\cdot}8$ and $-2{\cdot}7\,\%$.

4. Carbon monoxide and some petroleum samples are depleted in ^{13}C and have $94 > (^{12}C/^{13}C) > 91{\cdot}3$, or $\delta\,(^{13}C) < -2{\cdot}7\,\%$ (Stevens *et al.*, 1972; Deuser *et al.*, 1973). Geothermal methane has $\delta\,(^{13}C)$ between $-2{\cdot}5$ and $-3\,\%$ (Fuex, 1977). This topic is reviewed by Stahl (1976).

The small fractionation of nitrogen isotopes found in the soil ($\delta\,(^{15}N)$ from $+2{\cdot}5$ to $-0{\cdot}6\,\%$; Cheng *et al.*, 1964; Rennie and Paul, 1975) and in the ocean ($\delta\,(^{15}N)$ from $-0{\cdot}97$ to $-0{\cdot}17\,\%$: Wada and Hattori, 1976) have not been satisfactorily interpreted. $\delta\,(^{15}N)$ in ammonia and nitrogen dioxide from 'clean' air is about $-1\,\%$, but in air near farmyards the ammonia has a $\delta\,(^{15}N)$ of $+2{\cdot}5\,\%$. Isotopic fractionation probably occurs in the equilibrium between gaseous and condensed phases in the atmosphere (Moore, 1977).

An important feature of the isotopic chemistry of oxygen is the enrichment of atmospheric dioxygen in ^{18}O with respect to sea water. This dioxygen is mostly derived from the water in terrestrial plants, which has been depleted of both $^{1}H_2O$ and $H_2{}^{16}O$ by transpiration and by repeated evaporation and precipitation on land. The oxygen in silicates in the Earth's mantle is thought to have $-0{\cdot}7\,\% > \delta\,(^{18}O) > +0{\cdot}55\,\%$, and

TABLE 10.3 $^{32}S/^{34}S$ ratios in various materials

Material	Range of $^{32}S/^{34}S$	Reference
Sea water $SO_4{}^{2-}$	21·79	Kellogg *et al.* (1972)
Igneous rocks	22·01	Holser and Kaplan (1966)
Volcanic SO_2	21·8–22·4	Kellogg *et al.* (1972)
Volcanic S_8	21·8–22·3	Rankama (1963)
Volcanic H_2S	22·27–22·42	Kellogg *et al.* (1972)
Marine evaporites	21·73	Holser and Kaplan (1966)
Other sediments	22·49	Holser and Kaplan (1966)
Coal	21·7–21·9	Kellogg *et al.* (1972)
Petroleum	21·9–22·45	Kellogg *et al.* (1972)
Terrestrial humates	22·13–22·17	Nissenbaum and Kaplan (1972)
Marine humates	22·3–22·4	Nissenbaum and Kaplan (1972)
Biogenic S_8	22·3–22·7	Rankama (1963)
Biogenic H_2S	22·1–22·7	Kellogg *et al.* (1972)

the same limits are found for the majority of igneous rocks. Where granites are found to have $\delta\,(^{18}O) > 1\,\%$, there is good reason to suspect that they have been formed by remelting sedimentary rocks, which have higher ^{18}O contents, e.g. shales have $+2\,\% > \delta\,(^{18}O) > +1{\cdot}2\,\%$. A few granites have $+0{\cdot}55\,\% > \delta\,(^{18}O) > -0{\cdot}5\,\%$, and these are thought to have interacted strongly with ground water during their magmatic phase; they often have low deuterium contents as well (Taylor, 1978). Plant opal (SiO_2) is strongly enriched in ^{18}O compared to other silica in soils (Dixon and Weed, 1977).

Natural sulphides and elemental sulphur are almost all found to be depleted in heavy ^{34}S, while sulphates are enriched in this isotope. Values for $^{32}S/^{34}S$ in various sources of sulphur are given in Table 10.3. Volcanic sulphur can be distinguished from sulphur produced by bacteria, and stable isotope studies were important in convincing geologists that many massive deposits of elemental sulphur are biogenic (Thode *et al.*, 1954). It does not seem that the $^{32}S/^{34}S$ ratio has varied during geological time, and sulphur bacteria appear to have been functioning in the Pre-Cambrian period more than 2×10^9 years ago, and to have been particularly active during the Devonian era (Ault and Kulp, 1959).

11 *Radioactive Nuclides in the Environment*

Radioactive nuclides have excess energy which they eliminate by giving off ionising radiation. Each radionuclide emits radiation with a characteristic energy and decays with a characteristic half-life. Radioactivity is now measured in becquerels, where 1 Bq = 1 atomic disintegration per second: older units are curies, where 1 Ci = $3{\cdot}7 \times 10^{10}$ Bq, or dpm (disintegrations per minute), where a dpm = 0·0167 Bq. For a long-lived radionuclide of mass number A, the relation between absolute activity (a/Bq), mass (m/g) and half-life ($t_{\frac{1}{2}}$/yr) is

$$a = 1{\cdot}32 \times 10^{16} m/At_{\frac{1}{2}}$$

or

$$m = 7{\cdot}6 \times 10^{-17} At_{\frac{1}{2}} a$$

In order to calculate the specific activity, or activity per g of element, we need to know the fractional abundance of the radioactive isotope.

Natural radionuclides

There are more than sixty natural radionuclides, a number of which are listed in Table 11.1; alpha-emitters are listed separately from beta/gamma emitters since alpha particles have ten times the biological effect that beta particles have. Most of the nuclides listed are primal, and either decay to stable nuclides, e.g.

$$^{40}K \begin{matrix} \nearrow {}^{40}Ar \\ \searrow {}^{40}Ca \end{matrix}$$

or give rise to radioactive decay chains, of which the two most important are those from uranium-238 and thorium-232:

$$^{238}U \xrightarrow{\alpha} {}^{234}Th \rightarrow {}^{234}Pa \rightarrow {}^{234}U \xrightarrow{\alpha} {}^{230}Th \xrightarrow{\alpha} {}^{226}Ra \xrightarrow{\alpha} {}^{222}Rn \xrightarrow{\alpha} {}^{218}Po \xrightarrow{\alpha} {}^{214}Pb \rightarrow {}^{214}Bi \rightarrow {}^{214}Po \xrightarrow{\alpha} {}^{210}Pb \rightarrow {}^{210}Po \rightarrow {}^{210}Bi \xrightarrow{\alpha} {}^{206}Pb$$

$$^{232}Th \xrightarrow{\alpha} {}^{228}Ra \rightarrow {}^{228}Ac \rightarrow {}^{228}Th \xrightarrow{\alpha} {}^{224}Ra \xrightarrow{\alpha} {}^{220}Rn \xrightarrow{\alpha} {}^{216}Po \xrightarrow{\alpha} {}^{212}Pb \rightarrow {}^{212}Bi \rightarrow {}^{212}Po \xrightarrow{\alpha} {}^{208}Tl \rightarrow {}^{208}Pb$$

TABLE 11.1 Natural radionuclides with half-lives less than 10^{12} years

Alpha-emitters			Beta/gamma-emitters		
^{A}X	Half-life/yr	Spec. act./$Bq(gX)^{-1}$	^{A}X	Half-life/yr	Spec. act./$Bq(gX)^{-1}$
^{214}Bi	20 min	b	^{228}Ac	6·1 h	b
^{231}Pa	32 000	$1·8 \times 10^{9}$	^{26}Al	740 000	0·0002 a
^{210}Po	0·38	$1·7 \times 10^{14}$	^{39}Ar	269	$8·4 \times 10^{-5}$ ca
^{212}Po	0·3 μs	b	^{7}Be	0·15	ab
^{214}Po	0·16 μs	b	^{10}Be	$2·7 \times 10^{6}$	26 ca
^{216}Po	0·15 s	b	^{210}Bi	0·014	b
^{218}Po	3·1 min	b	^{212}Bi	61 min	b
^{190}Pt	7×10^{11}	0·013	^{14}C	5760	0·167 ca
^{224}Ra	0·01	b	^{36}Cl	300 000	5×10^{-7} ca
^{226}Ra	1620	$3·7 \times 10^{10}$	^{3}H	12·5	$5·5 \times 10^{-6}$ ca
^{220}Rn	55 s	b	^{40}K	$1·3 \times 10^{9}$	31
^{222}Rn	0·01	$5·5 \times 10^{15}$	^{81}Kr	210 000	0·0045 a
^{147}Sm	$1·1 \times 10^{11}$	120	^{138}La	$1·1 \times 10^{11}$	0·78
^{228}Th	1·9	b	^{22}Na	2·2	ab
^{230}Th	80 000	b	^{32}P	0·038	ab
^{232}Th	$1·4 \times 10^{10}$	4100	^{33}P	0·068	ab
^{234}U	250 000	13 000	^{234}Pa	6·7 h	b
^{235}U	$7·1 \times 10^{8}$	560	^{210}Pb	22	b
^{238}U	$4·5 \times 10^{9}$	12 000	^{212}Pb	10·6 h	b
			^{214}Pb	27 min	b
			^{228}Ra	6·7	b
			^{87}Rb	$4·7 \times 10^{10}$	900
			^{187}Re	7×10^{10}	630
			^{35}S	0·24	ab
			^{32}Si	710	0·00013 ca
			^{234}Th	0·066	b
			^{208}Tl	3·1 min	b

a, produced by cosmic radiation; b, variable; c, in the ocean (Schell, 1970).
For other natural radionuclides, e.g. ^{142}Ce, ^{156}Dy, ^{152}Gd, ^{174}Hf, ^{115}In, ^{176}Lu, ^{144}Nd, ^{204}Pb, ^{190}Pt, ^{146}Sm, ^{148}Sm, ^{180}Ta, ^{123}Te and ^{50}V, see Eisenbud (1973)

TABLE 11.2 Some artificial radionuclides found in the environment

Nuclide	Half-life/yr	Observed in	Source	Amount released/Bq	Ref.
^{110m}Ag	0·69	Marine biota	Reactor waste		
^{241}Am(α)	458	Marine biota, soil	Decay of ^{241}Pu		
^{14}C	5760	Stratosphere etc.	Bomb tests	$3{\cdot}5 \times 10^{17}$	(Fergusson, 1965)
^{45}Ca	0·44	Marine biota	Uncertain		
^{144}Ce	0·79	Marine biota	Bomb tests; fp		
^{242}Cm	0·45	Marine biota	Reactor waste ?		
^{57}Co	0·72	Marine biota	Bomb tests		
^{58}Co	0·20	Marine biota	Bomb tests		
^{60}Co	5·1	Marine biota	Reactor waste		
^{51}Cr	0·07	Estuarine biota	Reactor waste		
^{134}Cs	2·1	Marine biota	Reactor waste		
^{137}Cs	33	Soil, biosphere	Bomb tests; fp	$7{\cdot}2 \times 10^{17}$	(Rice and Wolfe, 1971)
^{55}Fe	2·9	Marine biota etc.	Bomb tests	$1{\cdot}4 \times 10^{18}$	? (Rice and Wolfe, 1971)
^{3}H	12·5	Stratosphere, rain	Bomb tests	$7{\cdot}5 \times 10^{19}$	(Fergusson, 1965)
^{131}I	0·022	Milk, thyroid	Reactor waste; fp		
^{85}Kr	10·3	Atmosphere	Reactor waste; fp	$1{\cdot}3 \times 10^{18}$	(Ehhalt *et al.* 1964)
^{54}Mn	0·82	Marine biota	Bomb tests		
^{63}Ni	92	Marine biota etc.	Bomb tests ?		
^{32}P	0·038	Estuarine biota	Reactor waste		
^{147}Pm	2·6	Ocean	Bomb tests; fp		
^{238}Pu(α)	86·4	Soil, biosphere	Bomb tests	$3{\cdot}2 \times 10^{13}$	
^{239}Pu(α)	24 400	Soil, biosphere	Bomb tests	$7{\cdot}8 \times 10^{15}$	
^{240}Pu(α)	6600	Soil, biosphere	Bomb tests	$5{\cdot}2 \times 10^{15}$	(Holm and Persson, 1977)
^{241}Pu	14·9	Soil, biosphere	Bomb tests	$2{\cdot}2 \times 10^{17}$	
^{106}Ru	1	Marine biota	Reactor waste; fp		
^{46}Sc	0·23	Marine biota	Reactor waste?		
^{90}Sr	28	Soil, biosphere	Bomb tests	$4{\cdot}5 \times 10^{17}$	(Rice and Wolfe, 1971)
^{99}Tc	210 000	Rain	Bomb tests; fp		
^{65}Zn	0·67	Marine biota	Reactor waste		
^{95}Zr/^{95}Nb	0·18/0·096	Marine biota	Reactor waste; fp		

α, alpha-emitter; fp, fission product

The amounts of fission products ^{144}Ce, ^{147}Pm, ^{106}Ru, ^{99}Tc and ^{95}Zr have been estimated from the fission yield of ^{235}U assuming the fission yield of ^{90}Sr to be 5·8%

For information on other artificial radionuclides observed in biota and the environment (^{76}As, ^{108}Ag, ^{140}Ba, ^{207}Bi, ^{109}Cd, ^{113}Cd, ^{115}Cd, ^{141}Ce, ^{152}Eu, ^{155}Eu, ^{59}Fe, ^{203}Hg, ^{140}La, ^{99}Mo, ^{22}Na, ^{239}Np, ^{102}Rh, ^{103}Ru, ^{35}S, ^{124}Sb, ^{125}Sb, ^{89}Sr, ^{132}Te, ^{181}W, ^{185}W, ^{187}W, ^{88}Y and ^{91}Y) see Junge, 1963 (atmosphere); Åberg and Hungate, 1967 (biosphere); Jenkins (1969), Duursma (1972) and Preston (1974) for marine biota

In addition to the primal radionuclides, a number of secondary radionuclides are produced at a more or less constant rate by interactions between cosmic radiation and stable terrestrial nuclides. The secondary species are all beta-emitters with half-lives of less than 10^7 years; the most abundant of them is carbon-14, which is produced by a nuclear reaction in the stratosphere:

$$^{1}n + ^{14}N = ^{14}C + ^{1}p$$

The small flux of neutrons at the surface of the earth, which is about $0{\cdot}33\ m^{-2}\ s^{-1}$ over land and $0{\cdot}11\ m^{-2}\ s^{-1}$ over the sea, produces very low concentrations of radionuclides (Gorshkov *et al.*, 1964). Spontaneous fission of uranium yields very small but detectable quantities of natural fission products in uranium ores (Osmund and Cowart, 1976).

Artificial radionuclides

At least 350 artificial or man-made radionuclides have been observed in the environment, most of them being fission products of uranium or plutonium which were released by nuclear bomb tests in the period 1955–1962. There are also a number of fairly long-lived radionuclides which have been produced by deliberate or accidental activation by neutrons, such as tritium, carbon-14 and iron-55. Almost all the artificial radioisotopes which have been released are beta-emitters, apart from the isotopes of plutonium (plus traces of americium, curium and neptunium), most of which are alpha-emitters. Most of the fission products have half-lives of less than a year, and only those which have stimulated research on their environmental behaviour are given in Table 11.2. Estimates of the amounts released to the environment are not very reliable, but the amounts produced by bomb trials appear to exceed those released by nuclear power plants by a factor of a thousand (Hayes and MacMurdo, 1977).

Radionuclides in the atmosphere

The natural radioactive gases in the atmosphere include $^{14}CO_2$, with a residence time of a few years, together with 55 s ^{220}Rn and 3·8 day ^{222}Rn, whose radioactive half-lives are short compared with their atmospheric residence times. ^{222}Rn is an important source of its decay products, ^{210}Pb and ^{210}Po, in atmospheric dust. Thus Parfenov (1974) estimates that the natural fall-out in rain and dry deposition lies in the range 1·5 to $8{\cdot}9 \times 10^{16}$ Bq ^{210}Pb/yr plus 0·74 to $1{\cdot}4 \times 10^{16}$ Bq ^{210}Po/yr, which greatly exceeds the estimated $2{\cdot}6 \times 10^{13}$ Bq ^{210}Po/yr released by burning coal

(Moore *et al.*, 1976). The residence time of these two nuclides in air is the same as that of atmospheric particles, about 10–25 days (Francis *et al.*, 1970). Schell (1970) gives useful data on the amounts of other natural radionuclides in the stratosphere and troposphere, of which ^{3}H and ^{7}Be are the most abundant.

Artificial radionuclides in the atmosphere include ^{85}Kr which is now a global contaminant. It appears to have been introduced as a by-product of the processing of fission product wastes, as its concentration in air has been rising steadily since the 1950s, without any marked peak around 1962 which characterises the fission products derived from nuclear bomb tests (Ehhalt *et al.*, 1964; Fig. 11.1).

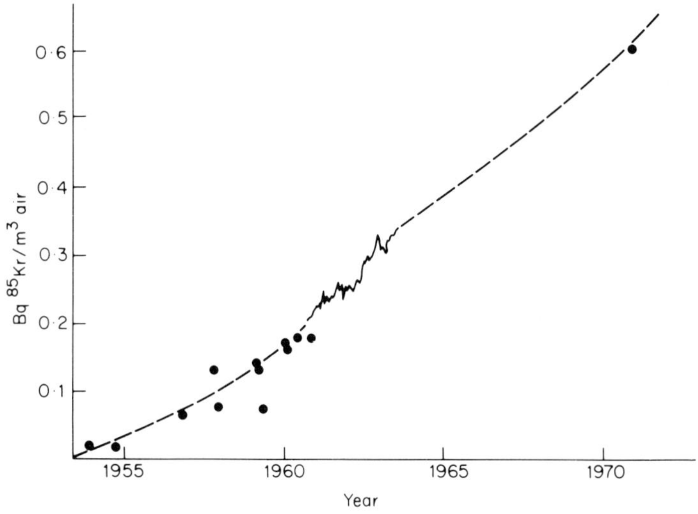

FIG. 11.1 ^{85}Kr in the atmosphere (Ehhalt *et al.*, 1964; Csongor, 1973)

Air-cooled reactors can introduce 2·3 hour ^{41}Ar into the atmosphere, but this is carried away up a tall chimney and decays away before it descends to Earth. Reactor accidents are potential sources of injection of 8 day ^{131}I into the atmosphere, while reactor fuel processing is said to yield gaseous effluents in the following amounts, per MW yr: $1{\cdot}4 \times 10^{13}$ Bq ^{85}Kr, $7{\cdot}4 \times 10^{11}$ Bq ^{3}H and $2{\cdot}1 \times 10^{9}$ Bq ^{222}Rn (Pigford, 1974). These amounts should be compared with the 2×10^{7} Bq ^{226}Ra plus $1{\cdot}3 \times 10^{7}$ Bq ^{228}Ra emitted per MW yr by coal-fired power plants which do not control their emission of dust (Eisenbud and Petrow, 1964). 210 day ^{102}Rh has been deliberately released into the stratosphere as a meteorological tracer (HASP, 1960).

Nuclear bomb tests in the atmosphere have released not only large amounts of fission product nuclides, but also substantial amounts of radionuclides such as ^{14}C, ^{3}H and ^{55}Fe which are formed by reactions of the neutrons produced at the site of the explosion. The fission products are mostly injected into the stratosphere as fine particles, where they have a residence time of the order of years. Once they descend to the troposphere, their residence time is only about thirty days, and most of the bomb debris has now either decayed or descended. The plume of radioactive particles from bomb trials is carried round the earth by prevailing winds, and the resultant deposition is markedly dependent on latitude (Fig. 11.2). It is estimated that bomb tests have injected 2100 kg carbon-14 and 215 kg tritium into the atmosphere, and one of the few useful results of this

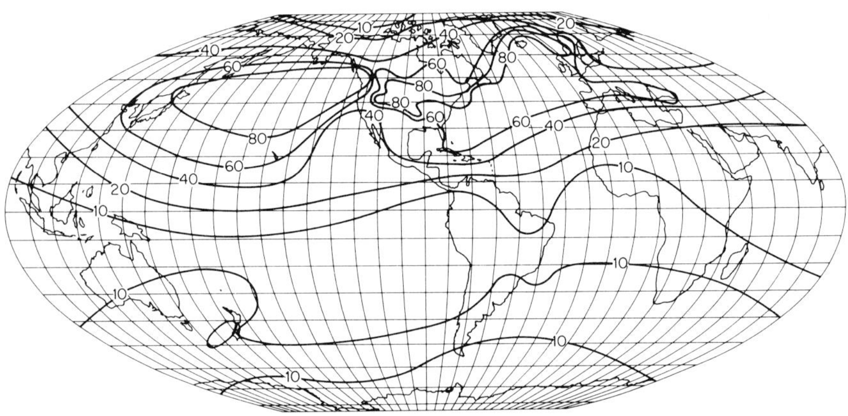

FIG. 11.2 Contours of cumulative deposition of ^{90}Sr. Numbers are mCi $^{90}Sr/km^2$. Data from Hardy *et al.* (1968)

TABLE 11.3 Radioactive gases in the atmosphere at STP

Gas	Source	Half-life	Bq/m^3 in stratosphere	Bq/m^3 in troposphere	Ref.
$^{14}CO_2$	n capture	5760 yr	0·0437*	0·0437*	Rafter and O'Brien (1970)
^{37}Ar	n capture	35 day	0·0041	0·00045	Schell (1970)
^{3}HOH	n capture	12·6 yr	0·13*	0·0015*	Schell (1970)
^{81}Kr	n capture	210 000 yr		0·000018	Schell (1970)
^{85}Kr	fission	10·3 yr	0·62 ?	0·62 (1971)	Csongor (1973)
^{220}Rn	decay of Th	55 s	v. low	0·07–2	Junge (1963)
^{222}Rn	decay of U	3·8 day	low	(0·2–)2–10(–100)	Junge (1963)

* natural level before 1955. 1 m^3 air at STP weighs 1·29 kg

injection has been the determination of residence times of carbon dioxide and water in the atmospheric reservoirs and the hydrosphere (Fergusson, 1965; Nydal and Lövseth, 1970). Both carbon-14 and tritium concentrations rose to a maximum in 1963–4 and have since fallen more or less steadily (Rafter *et al.*, 1970). Table 11.3 gives some information on radioactive gases in the atmosphere. The concentrations of the fission products ^{137}Cs and ^{90}Sr are $\sim 0{\cdot}00004$ Bq/m^3 (Knizhnikov and Barkhudarov, 1975).

Radionuclides in the hydrosphere

Table 11.4 gives some estimates of the amounts of various radionuclides in fresh waters and in the ocean. The radioactivity of most sources of fresh water is very low, of the order of 50–500 Bq/m^3, and is mostly due to potassium-40 and radium-226. Very rarely, natural spring waters contain much larger quantities of radium, and the highest concentration reported seems to be 26 Bq/cm^3 from Japan (Love, 1951). Radioactive contamination of rivers has occurred locally as a result of both uranium mining and the processing of nuclear wastes. For example, Tsivoglou *et al.* (1958) have shown that the radium content in the Colorado river, U.S.A., rose from 11 Bq/m^3 to 3300 Bq/m^3 for a short distance below a uranium mill. Effluents from nuclear processing plants into rivers are controlled by strict legislation. The largest amounts of radioactive wastes known to be discharged to a river are probably those from the Hanford reactors in Washington, U.S.A. The most common radionuclides reaching the estuary of the Columbia river are ^{51}Cr and ^{65}Zn, which are (or were) emitted at rates of 3×10^8 Bq/s and $1{\cdot}3 \times 10^7$ Bq/s respectively (Alexander and Rowland, 1966). This has led to the following activities being recorded from estuarine mud of the Columbia river, in Bq/kg mud (Hubbell and Glenn, 1973): ^{51}Cr 3200; ^{65}Zn 1100; natural ^{40}K 520; other radionuclides (^{60}Co, ^{54}Mn, ^{106}Ru, ^{46}Sc, $^{95}Zr/^{95}Nb$) 250. Contamination of natural waters by tritium from bomb tests is regularly monitored, and the results are often published in TU or tritium units. 1 TU = 1 atom of ^3H per 10^{18} atoms of ^1H, which is equivalent to 120 Bq/m^3 H_2O. Tritium in surface sea water reached a maximum level of 2000 Bq/m^3 about 1963–4, and has since declined to about 800 Bq/m^3 (Dreisigacker and Roether, 1978; Fig. 11.3).

Direct discharge of low-activity nuclear wastes to the ocean is a common practice. The annual effluent from Windscale to the Irish sea is about $2{\cdot}5 \times 10^{15}$ Bq, of which about half is ^{106}Ru, but includes $4{\cdot}8 \times 10^{13}$ Bq ^{137}Cs, $3{\cdot}3 \times 10^{13}$ Bq ^{90}Sr and $3{\cdot}7 \times 10^{12}$ Bq alpha-emitters (Dunster *et al.*, 1964). In this case the critical nuclide which limits the amount that may be

TABLE 11.4 Radionuclides in fresh water and the ocean

Nuclide	Bq/m^3 fresh water (Ref.)	Bq/m^3 sea water (Ref.)	Remarks
^{10}Be	0·008 (Table 2.3)	$1{\cdot}5 \times 10^{-4}$ (Table 2.5)	
^{14}C	1·8 (Table 2.3)	3·3–5·2 (Chipman, 1966)	+3 % from bombs
^{144}Ce		0·1–24 (Burton, 1975)	fp
^{36}Cl		0·01 (Table 2.5)	
^{137}Cs	3·7–37 (Ellis, 1965)	2·1–60 (Duursma, 1972)	fp: max. in N
^{55}Fe	variable	1·5 (Palmer and Beasley, 1967)	
^{3}H	variable	24 (Dreisigacker and Roether, 1978)	max. 2000 in sea
^{40}K	68 (15–300; Table 2.3)	12 000 (Table 2.5)	
^{54}Mn		0·37 (Duursma, 1972)	artificial
^{231}Pa		<0·004 (Cherry and Shannon, 1974)	
^{210}Pb	3–8 (Andrews and Wood, 1974)	0·17 (Andrews and Wood, 1974)	more in deep sea
^{147}Pm		0·2–5 (Burton, 1975)	fp
^{210}Po	0·5–2·6 (Parfenov, 1974)	0·93 (Cherry and Shannon, 1974; Hodge *et al.*, 1974)	
$^{214}Po/^{218}Po$		15 (Cherry and Shannon, 1974)	
^{238}Pu		0·0074 (Cherry and Shannon, 1974)	artificial
$^{239}Pu + ^{240}Pu$	0·02 (Wahlgren *et al.*, 1976)	0·0075–0·21 (Noshkin, 1972; Hodge *et al.*, 1974)	
^{224}Ra		0·24 (Cherry and Shannon, 1974)	
^{226}Ra	4–400 (Love, 1951)	3·3 (Table 2.5)	more in deep sea
^{87}Rb	0·5–8 (Table 2.3)	110 (Table 2.5)	
^{222}Rn	9 (Table 2.3)	7·5 (Cherry and Shannon, 1974)	
^{46}Sc		<0·8 (Duursma, 1972)	artificial
^{90}Sr	7–70 (Ellis, 1965)	1·7–35 (Duursma, 1972)	fp: max. in N
^{99}Tc	0·07–0·63* (Attrep *et al.*, 1971)		fp
^{228}Th		0·06 (Cherry and Shannon, 1974)	
^{230}Th		0·015 (Cherry and Shannon, 1974)	more in deep sea
^{232}Th	0·12 (Table 2.3)	0·004 (Cherry and Shannon, 1974)	
^{234}U	5.2 (Table 2.3)	44 (Cherry and Shannon, 1974)	
^{235}U	0·22 (Table 2.3)	1·8 (Cherry and Shannon, 1974)	
^{238}U	4·8 (Table 2.3)	41 (Cherry and Shannon, 1974)	
$^{95}Zr/^{95}Nb$		<75–300 (Duursma, 1972)	fp

fp, fission product; N, Northern hemisphere; *, in rain

discharged is ^{106}Ru, but the critical nuclide at Hanford, and also at Bradwell, U.K., is ^{65}Zn.

The natural radioactivity of the ocean is dominated by the beta-activity of potassium-40 (Burton, 1975). All other radioactive nuclides present are of minor importance. However useful information can be gained by studying the variation in concentration with depth for both natural and artificial radionuclides. For example, tritium concentrations are negligible in deep ocean water, which must therefore have been isolated from surface ocean water for at least 120 years. On the other hand the natural radionuclides ^{210}Pb and ^{230}Th are both more abundant in the deep ocean than at the surface, and are thought to arise from back-diffusion from sediments. Information on the age of sediments can sometimes be obtained in this way. The recent injection of artificial radionuclides such as ^{137}Cs, ^{3}H and ^{90}Sr has been used to follow ocean currents and to investigate rates of mixing in the ocean (Burton, 1975; Preston, 1974).

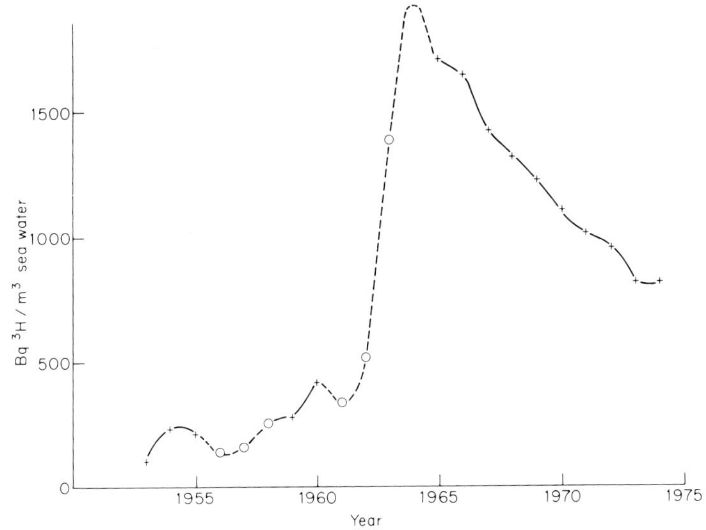

FIG. 11.3 Tritium in sea water from North Atlantic ocean. ×, experimental; ○, calculated. Note the very large injection after 1962. Data from Dreisigacker and Roether (1978)

The distribution of added radioisotopes between true solution and suspended matter is by no means fully understood. Preston (1974) comments that added zinc-65 is still largely present as Zn^{2+} after 100 days in the sea, and has not completely equilibrated with the other zinc species which are known to occur. Nagaya and Nakamura (1974) showed that 99% of the ^{90}Sr and 94% of the ^{137}Cs, but only 20% of the ^{144}Ce, pass through a 20 μm filter. Polyvalent species such as ^{144}Ce, ^{239}Pu and $^{95}Zr/^{95}Nb$ appear to be rapidly absorbed by suspended matter in sea water and then precipitate out. Among natural radionuclides, ^{10}Be, ^{26}Al and ^{32}Si show similar behaviour, as they can be determined in sediments, but their concentrations in sea water are too low for detection (Duursma, 1972). The behaviour of the uranium and thorium decay chains is especially complicated, since certain species (e.g. Ra^{2+}, $UO_2{}^{2+}$) remain in true solution, while others (e.g. Pa(V), Th(IV)) are rapidly absorbed onto sediments. Hence uranium is certainly not in radioactive equilibrium with its decay products in the ocean, nor in marine sediments (Duursma, 1972).

Radionuclides in rocks and soil

The radioactivity of rocks has long been known to be caused almost entirely by potassium-40, rubidium-87 and the decay products of uranium and thorium. Other radioactive species are present in amounts which are usually less than 0·1% of the total activity. For example, it is claimed that ^{244}Pu is present in the mineral bastnasite, in amounts corresponding to 10^{-18} g/g (Hoffman *et al.*, 1972). The major radioactive species are concentrated in the earths crust with respect to the mantle, and their decay is an important source of heat. Igneous rocks usually contain about 2% of potassium, which is fairly uniformly distributed in common minerals. In contrast, uranium and thorium occur at concentrations which are much lower and much more inhomogeneous. This comes about partly because uranium is a soluble element which is readily lost by weathering, and also because both uranium and thorium concentrate in accessory minerals such as monazite, sphene, xenotime or zircon forming hot spots of radioactivity (Rankama, 1963). Sedimentary rocks are usually less radioactive than igneous rocks, since they usually contain less uranium and often less potassium as well. Shales are the most radioactive, and limestone the least. The highest radioactivities in sedimentary rocks in Britain are found in potassium evaporate minerals, and in certain uraniferous nodules (Ponsford, 1955).

The radioactivity of soils is also caused mainly by the nuclides mentioned above, but recently measurable amounts of artificial nuclides have been added from fall-out following nuclear bomb trials. Table 11.5 summarises the available data on radioactive nuclides in soils, and is noteworthy for the importance of some natural nuclides which have not been studied in detail, such as ^{210}Pb, ^{210}Po and ^{230}Th. For natural radionuclides the data are given in Bq/kg, but for artificial radionuclides (and perhaps also for natural decay products of radon) data in Bq/m^2 may be more useful. This is because the latter have only recently been added to the soil and are seldom well-mixed; indeed in natural, unploughed soils they are often confined to the upper layers of the soil, to a depth which rarely exceeds 30 cm and is often much less. Megumi and Mamuro (1977) have shown that ^{238}U is not in equilibrium with its decay products in two soils derived from granite. The ratio $^{230}Th/^{238}U$ was found to be higher than the expected value by a factor of 3 to 5, and more ^{210}Pb and ^{226}Ra was present than radioactive equilibrium would allow. This probably arises through differential leaching of uranium from the soil by rain.

TABLE 11.5 Radionuclides in soils

Nuclide	Source	Bq/kg soil	Bq/m² soil	Ref.
^{14}C	Natural	4 (1·5–6)		Table 4.4
^{137}Cs	Bomb tests	63 (18–90)	4500 (2600–6800)	B, K, R, V
^{55}Fe	Bomb tests		7100	W
^{40}K	Natural	440 (0·2–1200)		Table 4.4; B, V
^{54}Mn	Bomb tests	7·4		V
^{210}Pb	Natural	75–6300*		M
^{210}Po	Natural	8–220		P
^{238}Pu	Bomb tests	0·007–0·07		L, X
$^{239+240}Pu$	Bomb tests	0·05–1·4	76 (44–145)	B, L, No, Ny, X
^{241}Pu	Bomb tests	16		L
^{226}Ra	Natural	30 (7–180)		Table 4.4; B, V
^{87}Rb	Natural	140 (20–560)		Table 4.4
^{90}Sr	Bomb tests	35 (20–55)	19–3000	B, J, K
^{230}Th	Natural	100 (3700–16000)*		M
^{232}Th	Natural	37 (4–78)		Table 4.4; B, V
^{234}U	Natural	26 (9–120)		Table 4.4
^{238}U	Natural	24 (8–110)		Table 4.4

* from one abnormal soil containing 750–3000 Bq ^{238}U/kg.

References: B, de Bortoli and Gaglione (1969); J, Jordan and Kline (1976); K, Klechkovsky *et al.* (1973); L, Livingston *et al.* (1975); M, Megumi and Mamuro (1977); No, Noshkin (1972); Ny, Nyhan *et al.* (1976); P, Parfenov (1974); R, Ritchie *et al.* (1974); V, Vassilaki *et al.* (1966); W, Wrenn and Cohen (1967); X, McLendon *et al.* (1976)

The carbon-14 in soils is not all in equilibrium with atmospheric carbon dioxide, and its specific activity has been used to obtain residence times of various organic fractions in soils, as described in chapter 4.

Radioactivity is added to the soil when fertilisers are applied, especially potassium salts or derivatives of rock phosphate, which is rich in uranium. World usage of potassium and phosphorus as fertilizers is about $1{\cdot}6 \times 10^{10}$ and $1{\cdot}0 \times 10^{10}$ kg/yr respectively. Assuming a U/P ratio of 1/3000 in rock phosphate (Swaine, 1962), about $5{\cdot}2 \times 10^{14}$ Bq ^{40}K and $4{\cdot}1 \times 10^{13}$ Bq ^{238}U are added to arable soils each year. Moore *et al.* (1976) estimate that about $2{\cdot}5 \times 10^{13}$ Bq ^{226}Ra are added to soils each year from phosphatic fertilisers. These figures are an order of magnitude or two lower than the 8×10^{15} Bq/yr of ^{137}Cs or ^{90}Sr added to soils by fall-out in the Northern hemisphere during the past decade (Feeley, 1977). In Britain the recommended application rates for fertilisers on soils used to grow wheat are 3·3 g K/m^2 yr and 1·74 g P/m^2 yr (Cooke, 1975). These involve adding more radioactivity to the soil than fall-out from nuclear bomb tests; each square metre of soil receives 100 Bq ^{40}K + 7 Bq ^{238}U + 4 Bq ^{226}Ra from fertilisers, as against 54 Bq ^{137}Cs + 37 Bq ^{90}Sr from fall-out (Knizhnikov and Barkhudarov, 1975).

The behaviour of important fission products such as ^{137}Cs and ^{90}Sr in soils is now reasonably well known. ^{137}Cs is tenaciously retained by clays and penetrates soil profiles slowly, while ^{90}Sr is somewhat more mobile (Russell, 1966; Ritchie *et al.*, 1974). The behaviour of plutonium is less well understood, but samples deposited 20 years ago in New Mexico have penetrated 20–35 cm down the soil profile. Soils have been contaminated with the alpha-emitters ^{239}Pu and ^{240}Pu, and with larger amounts of the beta-emitting ^{241}Pu; the latter is now giving rise to its alpha-emitting daughter, americium-241 (Schulz *et al.*, 1976; Holm and Persson, 1977). Curium has not yet been reported from natural soils, but its behaviour has been studied by Thomas and Jacobs (1969).

The behaviour of fission products on a geological time scale can be deduced from studies being made at Oklo, central Africa, where uranium ores went critical about $1{\cdot}8 \times 10^9$ years ago. Extinct fission products and other nuclides which are still to be found there include Ag, Bi, Ce, Gd, Nb, Nd, Pd, Ru, Sm, Th, U and Zr. About 90% of the Ba, Cd, Mo and Sr, and about 99% of the Cs, I, Rb and Xe appear to have been lost by leaching or volatilisation (Friedman, 1976).

Radionuclides in marine biota

This subject has been reviewed by Preston (1974), Pentreath (1977) for fish, and by Cherry and Shannon (1974) for alpha-emitters only. Data in the literature is notably variable, and while Table 11.6 summarises the orders of magnitude of reported activities, these can be very different for short-lived artificial radionuclides, depending on the amount which has been released to the local environment and on the time of sampling. The main features of Table 11.6 can be summarised as follows.

1. The dominant natural radionuclides in soft tissues are potassium-40 and polonium-210, though radium-226 is important in some phytoplankton samples. In hard tissues the data is incomplete, but carbon-14, polonium-210 and uranium seem to be important. Hence although the sea is a less radioactive environment than the land, marine organisms are subject to relatively high doses of ionising radiation from internally absorbed alpha-emitters, especially polonium-210.

TABLE 11.6 Radionuclides in marine organisms in Bq/kg DM

Nuclide	Phytoplankton	Algae	Zooplankton	Coral	Molluscs	Mollusc shell	Fish
^{14}C	37 A	53 A	70 A	19 A	67 A	20 A	70 A, Pa
^{137}Cs		0·3–3 P			0·14–1·4 P		0·1–150 J, P, Pa
^{55}Fe	15–1100 Je, Pb	4 Pb	24–220 Pb		5–11 Pb		7–1400 J, Pb
^{40}K	400–1500 A, F	700–4400 A, F	260–370F		250–470 A, F	0·6–8 A	200–600 A, F, J, Pa
^{54}Mn	100–960 F, S	4–280 F, P	15–75 F		7 F		0·3–21 F, J
^{63}Ni					0·0003–0·006 Bh		
^{210}Pb					1·5–14 B		0·25–1·5 B
^{210}Po	30–130 C	4–170 C, H	60–2200 C		3–270 B, C	11 C	0·3–1000 B, C
^{238}Pu		0·027 N					
$^{239+240}Pu$	0·23 C	0·13–16 C, N	0·01–2 C, N	0·07 C	0·014–0·08 C, N	0·015 C, N	0·0001–0·005 C, N, Pa
^{241}Pu		4–27 N					
^{226}Ra	4–1400 C	0·08–8 C	2–20 C		0·3–8 C	0·1–8 C	0·2–8 C, J
^{87}Rb		7A			4·5 A		0·7–5 A
^{106}Ru		3–2800 P			1·4–14 P		0·1–450 P, Pa
^{90}Sr	4–60 S	0·3–3 P			0·01–1·4 P		0·1 ⊃2 P, Pa
^{228}Th	^{23}C		6·7 C		1·4 C	0·037 C	0·27 C, Pa
^{230}Th	2·3 C		0·7–1·4 C				
^{232}Th	1·6 C	0·56 C	0·7–1·4 C				0·01–0·18 J
^{238}U	9 C	0·5–30 C	2–7 C	11–55 C		0·04–22 C	0·01–1 C, Pa
^{65}Zn	8 F	3 F	22 F		2–7 F		6–600 F, J

References: A, calculated from data in chapter 6; B, Beasley and Held (1969); Bh, Beasley and Held (1969); C, Cherry and Shannon (1974); F, Folsom *et al.* (1963); H, Hodge *et al.* (1974); J, Jenkins (1969); Je, Jennings (1977); N, Noshkin (1972); P, Preston (1974); Pa, Pentreath (1977); Pb, Palmer and Beasley (1967); S, Schreiber (1967).

For information on local contamination of marine biota by the following: ^{110m}Ag, ^{241}Am, ^{140}Ba, ^{144}Ce, ^{242}Cm, ^{60}Co, ^{134}Cs, ^{152}Eu, ^{203}Hg, ^{131}I, ^{22}Na, ^{32}P, ^{147}Pm, ^{125}Sb, ^{46}Sc, ^{208}Tl, ^{88}Y, ^{95}Zr, consult Lowman *et al.* (1967), Schreiber (1967), Jenkins (1969), Preston (1974) and Pentreath (1977). Takashima *et al.* (1978a) found 3–7 Bq ^{7}Be/kg DM in 4 algae.

2. Plankton and algae have large concentration factors for a number of artificial radionuclides in sea water, e.g. ^{110m}Ag, ^{144}Ce, ^{60}Co, ^{55}Fe, ^{54}Mn, ^{106}Ru, ^{65}Zn and $^{95}Zr/^{95}Nb$. Animals feeding on these algae and plankton reject most of these elements, except for iron and zinc, which are concentrated still further, and also to some extent caesium, for which the highest activities have been found in fish muscle. Hard tissues accumulate ^{90}Sr and perhaps other nuclides. However, the important long-lived fission products ^{137}Cs and ^{90}Sr are relatively minor contributors to the total radioactivity of marine biota, and are not markedly concentrated by edible organisms.

Radionuclides in terrestial biota

So much research has been carried out on this topic that it is difficult to summarise. Eisenbud (1973) has given a useful overall summary, and Russian work has been described by Klechkovsky *et al.* (1973). Table 11.7 is an attempt to condense the available data in comparable form. The definitive work on human tissues is by Spiers (1968).

The lichens in Table 11.7 are the *Cladonia* species of arctic regions, which cover the soil and absorb their nutrients from rain and dust, including any radionuclides which are present. They are an important source of food for large mammals such as Reindeer and Caribou, which in turn may be eaten by man. Seed plants absorb most of their nutrients from the soil, and reject polyvalent elements such as Ce, Pu and Th at the root surface. Their content of most radionuclides in Bq/kg DM is of the same order of magnitude as the soils on which they grow, apart from thorium and uranium, which plants reject, and carbon-14, most of which they obtain from atmospheric CO_2. There do not seem to be any substantial differences in uptake of radionuclides by woody plants and the vegetable foods used by man. However, grazing mammals have a much greater intake of radioactive contaminants than does man, probably because they ingest fall-out particles with their food while most human food is washed before it is eaten.

The activities found in soft tissues like muscle and hard tissues like bone are very different, and are listed separately in Table 11.7. The main natural radionuclides in terrestrial biota are carbon-14, potassium-40, lead-210 and polonium-210, of which the two last nuclides are notably concentrated in bone. All four of these are ingested predominantly in the diet, with minor amounts in water and inhaled air. Artificial radionuclides which have given cause for concern, especially in the arctic, are caesium-137 and strontium-90, which concentrate in muscle and bone respectively, together with iron-55 and plutonium, whose behaviour is less well known. Comparison of

TABLE 11.7 Radionuclides in some terrestrial biota in Bq/kg DM

X	Lichens	Trees	Vegetables	Mammal muscle	Mammal bone	Human muscle	Human bone
^{14}C	75 A	75 A	75 A	79 A	47 A	79 A	47 A
^{137}Cs	410–3000* H, K		0·5–11 O, Sp	27–450 (–3k*) E, H, W	30–220* H	20–40 Sp	55 Sp
^{55}Fe	240–3600* Ja	100–3300* Ja	1–18* Ja, Wc	35–1500* Ja		1500* Pb	
^{40}K	20–260 A	30–750 A	30–2100 A, O	320 A	46 A	320 A	46 A
^{63}Ni	0·005 Bh						
^{210}Pb	240–2600 K, Km	11–40 Km		<0·1 Km	5–3700* K, Km	0·3–1 Bl	0·5–2·5 Bl
^{210}Po	110–560 P	8–12 Km	0·4–110 P	1–5 (–130*) Km, P	10–80 (–55 k*) Km, P	0·1–2 Bl	1·4 P
^{238}Pu	0·33* Hp		0·0002–0·006 X				
^{239}Pu	5·6* Hp		0·006–0·02 X	0·006* Y	0·015–0·03*	0·09 M	0·003* Fp, Mk
^{241}Pu	37–150* Hp						
^{226}Ra	0·5* H	7 A	0·1–3 (–600) A, C, T	0·08 F	2·7 B	0·004 A	0·75 A
^{228}Ra			3–300 Pc, S		10 Pc	0·03 Sp	0·2 Sp
^{87}Rb	2–22 A	3–9 A	0·5–18 A	21 A	5 A	21 A	5 A
^{106}Ru	240–2000* H		2–37 O				
^{125}Sb	60–84* H		0·4–10 O				
^{90}Sr	50–550* H, K	40–140 J, W	3–50 (–140*) K, Sp	0·005–0·02 (–5*) H, R	12–900 (–5 k*) H, K, R, W		8–54 Bl
^{228}Th	0·25* H		1–220 Pc, T	0·05* H	0·6–3·6 H, Pc		0·33 Ps
^{232}Th	0·4–1 A	0·6–5 A	0·04–2 A, O		0·03 Ps	0·05 C	0·04–0·1 C, Ps
^{238}U	1–3 A	0·06–0·5 A	0·12 A			0·36 A	0·12 A

*, Arctic biota; k, ×1000. ^{239}Pu includes ^{240}Pu. For ^{110m}Ag, ^{144}Ce, ^{60}Co, ^{134}Cs, ^{54}Mn, ^{22}Na, ^{88}Y, ^{65}Zn and ^{95}Zr in arctic biota see Hanson *et al.* (1967). For ^{60}Co and ^{95}Zr in vegetables see Oakes *et al.* (1977). Takashima *et al.* (1978b) found 10–16 Bq ^{7}Be/kg DM in 4 Trees.

References: **A**, calculated from data in chapter 6; **B**, de Bortoli and Gaglione (1972); **Bh**, Beasley and Held (1969); **Bl**, Blanchard (1967); **C**, Clifton *et al.* (1971); **E**, Ellis (1965); **F**, Fowler (1965); **Fp**, Fisenne and Perry (1978); **H**, Hanson *et al.* (1967); **Hp**, Holm and Persson (1977); **J**, Jordan and Kline (1976); **Ja**, Jaakola (1969); **K**, Klechkovsky *et al.* (1973); **Km**, Kauranen and Miettinen (1967); **M**, Moss and Campbell (1972); **Mk**, Magno *et al.* (1967); **O**, Oakes *et al.* (1977); **P**, Parfenov (1974); **Pb**, Palmer and Beasley (1967); **Pc**, Petrow *et al.* (1964); **Ps**, Petrow and Strehlow (1967); **R**, Russell (1966); **S**, Smith and Mercer (1970); **Sp**, Spiers (1968); **T**, Turner *et al.* (1958); **W**, Whicker t al. (1967); **Wc**, Wrenn and Cohen (1967); **X**, McLendon *et al.* (1976); **Y**, Holm and Persson (1976)

Tables 11.5 and 11.7 suggests that biological concentration of radionuclides occurs for ^{210}Po, ^{226}Ra and ^{90}Sr in bone, and possibly for ^{55}Fe in humans. The fact that bone-marrow is an unusually radiosensitive tissue, together with the long residence times for these nuclides in the human body, has inspired extensive research work. However no biological abnormalities have been found which could be attributed to excessive uptake of artificial radionuclides by human populations, even in the Arctic following the major bomb tests in 1962.

Information on the main radionuclides in human diets is rather scanty, but some representative figures are given in Table 11.8. Potassium-40 and carbon-14 are the most important and the dietary content of alpha-emitters is very low. It seems that the perturbations caused by artificial radionuclides, on which so much research effort has been lavished, are of the order of a few per cent at most. They are regularly monitored and have declined steadily since the cessation of major nuclear bomb tests.

TABLE 11.8 Radionuclides in human diets, and maximum permissible body burdens

Nuclide	Source	Bq/day in diet	MPBB/Bq	Ref.
^{14}C	Natural	50	$1{\cdot}5 \times 10^7$	Snyder (1975)
^{137}Cs	Bomb tests	1·25	$1{\cdot}1 \times 10^6$	Eisenbud (1973)
^{40}K	Natural	60–150		Table 8.7; Fowler (1965)
^{210}Pb	Natural	0·9	$1{\cdot}5 \times 10^4$	Magno *et al.* (1970)
^{210}Po	Natural (α)	0·15		Parfenov (1974)
$^{239+240}$Pu	Bomb tests (α)	0·00026	$1{\cdot}5 \times 10^3$	Magno *et al.* (1967)
^{226}Ra	Natural (α)	0·085	$3{\cdot}7 \times 10^3$	Snyder (1975); Fowler (1965)
^{87}Rb	Natural	2·5		Table 8.7
^{90}Sr	Bomb tests	0·37–0·93	$7{\cdot}4 \times 10^4$	Fujita *et al.* (1969); Ellis (1965)
^{232}Th	Natural (α)	0·00022	$1{\cdot}5 \times 10^3$	Clifton *et al.* (1971)
^{238}U	Natural (α)	0·012–0·024	$1{\cdot}9 \times 10^2$	Table 8.7

α, alpha-emitter; MPBB, maximum permissible body burden

Biological effects of radiation and maximum permissible body burdens

Biological effects of ionising radiation are not simply caused by deposition of energy in the tissues. If this were so, it would be a fairly simple matter to calculate the energy deposition per unit weight to obtain the radiation dose in grays, where 1 gray(Gy) = 1 J/kg FM tissue. (Note that 1 Gy = 100 rad, where the rad is a much used non-SI unit.) In practice 1 Gy of alpha radiation has the same biological effect as 10 Gy of beta or gamma radiation, and this must be allowed for when calculating internal radiation doses.

The natural external radiation dose-rate or background depends on altitude and rock type, but typical values at sea level are as follows:

Igneous rock	0·6–3·5 mGy/yr
Sedimentary rock	0·2–0·9 mGy/yr
Ocean	0·45 mGy/yr

to which must be added 0·2–0·35 mGy/yr from internal radiation in man. A few localities in Brazil and southern India have much higher radiation backgrounds of up to 0·44 Gy/yr, but these are exceptional (Eisenbud, 1973).

Maximum permissible body burdens of radionuclides in man are calculated on the basis of the amount which gives an internal radiation dose of 1 mGy/yr (or 0·1 mGy/yr for alpha emitters; Spiers, 1968). Typical values are given in Table 11·8, and a full list is provided by the International Commission for Radiation Protection (ICRP, 1964, 1966). Knizhnikov and Barkhudarov (1975) have given an objective account of natural radiation dose rates and their perturbation by human activities.

12 *Environmental Effects of Human Activities*

The *environment* is taken to consist of the atmosphere, hydrosphere and lithosphere. When a part of the environment changes its chemical composition as a result of human activities, without any obvious biological consequences, we call it *contaminated*. In the dictionary contamination is made a synonym of pollution, literally to make foul or filthy. However, in this work pollution will be taken to mean that adverse biological effects have arisen from contamination. The discussion will be restricted to environmental contamination by inorganic matter, mostly by small molecules and elemental ions, with a brief summary of human depletion of natural resources. A broad view will be presented, and numerous instances of local contamination have been omitted, while local pollution is referred to with the minimum of detail.

Air pollution (Junge, 1963; Stern, 1968; Spedding, 1974)

Most air pollution has arisen from the burning of fossil fuels, which has increased dramatically during this century (Fig. 12.1). The main gaseous pollutants formed are carbon dioxide, carbon monoxide, oxides of nitrogen and sulphur dioxide, together with particulate smoke. Table 12.1 gives the average amounts of these gases produced by burning various fuels, and Table 12.2 gives estimates of the global emissions per year. The gases are discussed individually below.

CARBON DIOXIDE

This is by far the most abundant contaminant added to the atmosphere by combustion, but industrial emissions are only about 7% of global plant respiration. Substantial amounts have been added to the atmosphere as a result of deforestation and oxidation of soil humates during the last century (Stumm, 1977; Williams, 1978). Since low concentrations of the gas are harmless to animals and an essential plant food, it cannot be called a pollutant. Figure 12.2 shows that concentrations of carbon dioxide in the

TABLE 12.1 Gases emitted when various fuels are burned (Altman and Dittmer, 1966)

Gas	g/kg Coal	g/kg Fuel oil	g/kg Natural gas	g/kg Town refuse	g/l Petrol	g/l Diesel oil
CO_2	2950	3150	2750	1000	2000	2300
CO	0·05–22	0·005–0·2	0·07	0·3	240	60
NO	3·5–9	4–12	3–7	0·2–1	18	100
SO_2	17	18	0·01	0·4–0·9	9	40

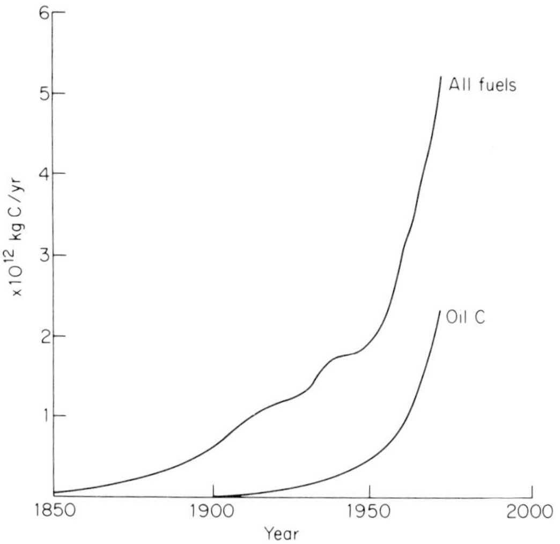

FIG. 12.1 Rate of burning carbonaceous fuels since 1850

TABLE 12.2 Annual global emissions to the atmosphere, 10^{12} kg/year

Substance	Natural emission rate	Industrial emission rate
CO_2	330	10·5
CO	3·2 (1·5–4)	0·64
NO	0·032	0·05 ?
SO_2	(0·24)*	0·092
Smoke	1·7 ?	0·08

See chapter 5 for references. *This figure refers to the SO_2 which is believed to be produced in the atmosphere by natural emissions of H_2S

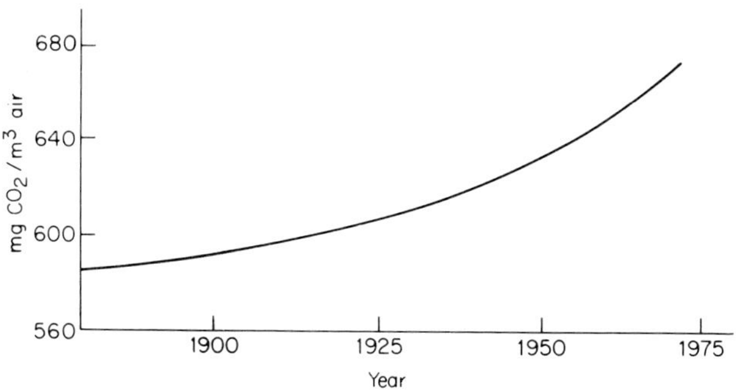

FIG. 12.2 Increase in concentration of carbon dioxide in the atmosphere. There are no data before 1870, when much CO_2 may have been released by agricultural changes

atmosphere have risen by about 12% since 1890. One consequence is that the isotopic composition of atmospheric CO_2 has altered, as the fossil fuel input contains no radioactive carbon-14: this is the Suess effect. Another suggested consequence is that the contaminated atmosphere should absorb more infrared radiation from the sun, and so raise the mean temperature of the earth. This is possible, but short-term changes in the mean temperature of the planet seem to be more dependent on the amount of particulate matter in the atmosphere than on the concentration of carbon dioxide (Lamb, 1972).

CARBON MONOXIDE

This is a global contaminant which is toxic to vertebrates but scarcely toxic to plants; industrial emissions are probably about 20% of natural emissions. The 150% increase in concentration of this gas in the Northern hemisphere, where industrial emissions are highest, as compared to the Southern hemisphere, is not known to have had any biological effects (Weinstock and Chang, 1976). Very locally, in canyon-like city streets or road tunnels, concentrations of carbon monoxide can approach or exceed those known to cause unpleasant symptoms such as headaches in man. Since exhaust fumes from cars are major sources of the gas, and it can be catalytically oxidised to carbon dioxide, legislation has been used to reduce emissions from vehicles in the U.S.A.

OXIDES OF NITROGEN

These are local, rather than global, pollutants because of their short residence times in air. Industrial emissions probably exceed natural emissions but estimates of both quantities are imprecise. Nitrogen dioxide is the only gaseous contaminant which can be seen, on occasion, as a layer of brown air trapped along the length of a city street. Both nitrogen dioxide, and its precursor nitric oxide, are toxic to plants and animals. Nitrogen dioxide is the more toxic of the two, and has both acidic and bleaching properties. Concentrations of the order of 750 $\mu g\ NO_2/m^3$, which can prevent photosynthesis in some plants, are rarely found in the atmosphere; the lower limit for detection by the human nose is about twice this value. However, in the presence of hydrocarbon vapours, NO_2 can form peroxyacetyl nitrate and its analogues, and the latter are present in Los Angeles-type smog. Peroxyacetyl nitrate has serious effects on vegetation at very low concentrations, and is a lachrymator which is highly toxic to mammals. Legislation has been used in the U.S.A. to reduce the emissions of NO_2 by vehicles.

SULPHUR DIOXIDE

Sulphur dioxide is another local, and also seasonal, pollutant with a short residence time in air. Industrial emissions greatly exceed natural emissions from volcanoes, and are thought to be about 40% of the amount produced in the atmosphere by oxidation of natural emissions of hydrogen sulphide. Sulphur dioxide concentrations of 50–200 $\mu g\ SO_2/m^3$, sometimes rising to 1 mg SO_2/m^3, are found in the atmosphere of large cities, especially in the winter months, and also in the neighbourhood of coal-fired power stations and ore smelters. Sulphur dioxide has bleaching properties, and its oxidation product sulphuric acid corrodes metals, attacks limestone and damages paper and textiles. It is somewhat toxic to man, but scarcely so at the concentrations mentioned above, which are seldom detectable by the human nose. It is much more toxic to plants, especially to certain species of lichens, which have now become extinct in the industrialised regions of many countries; about half of the area of the British Isles has been affected in this way (Ferry *et al.*, 1973). Other plant species which are affected are evergreens, such as gymnosperms, which continue to photosynthesise during the winter when pollution is worst. Dramatic decreases in yields of agricultural grasses have been found when they were grown artificially in concentrations of SO_2 similar to those recorded for industrial areas in winter. Fortunately most temperate crops are grown in summer, when SO_2 concentrations are relatively low. There is evidence of genetic selection for resistance to the gas in grasses from Lancashire, England, in a chronically polluted region.

OTHER GASES

Ammonia is a local contaminant from combustion, sewage works and farms. Its effects on the biosphere include changing the lichen community in the immediate neighbourhood of the emission.

Chlorofluorocarbons are recent contaminants from their use as spray propellants and refrigerants. The two most common are monochlorotrifluoromethane and dichlorodifluoromethane, whose cumulative world production up to 1975 has been estimated as $3{\cdot}3 \times 10^9$ kg and $4{\cdot}9 \times 10^9$ kg respectively. Most of this has been released to the atmosphere, leading to measurable concentrations (Table 1.1); the residence time of these gases is not known, but could be about 50 years. Fears have been expressed that their photolysis products could interfere with the ozone cycle in the stratosphere, which has led to a reduction in their usage (Sandalls and Hatton, 1977; Singh *et al.*, 1977).

Hydrogen fluoride is another local contaminant emitted from brickworks, glass factories, aluminium works and some iron smelters. It is extremely toxic and corrosive, and has killed lichen communities as well as causing damage to other plants (Scurfield, 1960). High fluoride concentrations in the atmosphere of Hamilton, Ontario can be correlated with an increase in lung cancer, but other elements may be responsible (Cecilioni, 1972).

Ozone is a local contaminant of the troposphere produced by photochemical reactions, as in Los Angeles smog. It is the most poisonous of all known gases to mammals, and can reach concentrations in the atmosphere which are known to damage plants, as well as exposed metals, rubber and asphalt (Spedding, 1974).

Mercury is emitted locally by cinnabar smelters, coal burning (Joensuu, 1971; Goldwater and Stopford, 1977) and some sewage works (Soldano *et al.*, 1975). Although occupational exposure is a hazard to laboratory workers and dentists, no effects of atmospheric mercury on the biosphere have been observed.

SMOKE

As described in chapter 1, atmospheric particulates are highly heterogeneous, but smoke particles set free by combustion processes usually have a high carbon content and are coloured black. These particles are local pollutants and have adsorbent properties. They act inimically to plant growth, both by reducing the amount of light available for photosynthesis and by physically blocking the stomata of plant leaves. They are selectively trapped by lung tissue and are a potential source of lung cancer in man. Medical statistics clearly show that inversions causing fogs in cities with high rates of emitting smoke are associated with substantial numbers of excess deaths, especially of asthmatic and bronchitic subjects. One result has been legislation to reduce the emission of black smoke, which has been successful in Britain.

OTHER PARTICULATE MATTER

Ammonium sulphate is a major constituent of many industrial aerosols, which acidifies the soil when it descends.

Lead is added to petrol in amounts of the order of 3×10^8 kg/year, and about 75% of this is emitted in particulate form from car exhausts (Huntzicker *et al.*, 1975). Deleterious effects on laboratory rats have been found when the animals were breathing air containing 6 μg Pb/m^3: urban air in Britain contains from 0·2 to 7 μg Pb/m^3. Whether the inhalation of lead-contaminated air, other than in lead factories, has any effect on human populations is a controversial question (Bethea and Bethea, 1975). Lead-free petrols are now required by law in several countries.

Fly ash is finely divided ash from combustion processes which readily becomes airborne. When coal is burned in power stations, the combustion products are gases, fly ash and bottom ash. From an elemental standpoint, most of the Br, C, Cl, F(?), H, Hg, I, N, S and about 32% of the Se in the coal are volatilised as gases; As, Cd, Cu, Ga, Mo, Pb, Sb, Se and Zn are relatively volatile elements which are markedly concentrated in the fly ash with respect to the bottom ash; while most of the other elements are not or scarcely fractionated between fly ash and bottom ash (Klein *et al.*, 1975; Kaakinen *et al.*, 1975; Block and Dams, 1975; Andren *et al.*, 1975; Heinrichs, 1977). Estimates of the effect of combustion of fossil fuels on the elemental burden of the atmosphere have been made by Bertine and Goldberg (1971). They used data for the amounts of fuel burned in 1967, and assumed 5% transfer of fly ash to the atmosphere. The results of a similar calculation are given in Table 12.3. We have used the amounts of fuels burned in 1975, the elemental compositions of Table 3·7, and assumed a 10% transfer of fly ash to the atmosphere (32% for Se, and 100% for Br, C, Cl, Hg, I, N and S). Lantzy and Mackenzie (1979) give rather larger estimates of transfer rates.

The elemental content of aerosols from clean and dirty air have been tabulated and discussed in chapter 1. Of the elements which are strongly enriched in dirty air, Be, Cr and Ni are known to cause lung cancer in factory workers who inhale abnormal amounts. Table 12.3 suggests that coal burning is responsible for large perturbations in the natural cycles of C, Hg, N, S and Se; burning fuel oil for similar perturbations in the cycles of C, N, S and V; and burning 'leaded' petrol for large perturbations in atmospheric Br and Pb.

Preventive measures

Air pollution poses many serious problems, but technical means of prevention are available, and have been applied locally, to reduce emissions of CO, NO_2 and smoke. Lead could be removed from petrol, as has been done in the U.S.A. and in Russia. It is technically impossible to reduce domestic emissions of SO_2, but sulphur could be removed from fuel oil before burning it, at an added cost. It has not proved possible to remove sulphur and pyrite from coal, but several methods of removing SO_2 from the flue gases of power stations have been studied. None of them are cheap, and current policy in Britain is to build higher chimneys so as to disperse the gas into a greater volume.

TABLE 12.3 Annual transfers of the elements in 10^6 kg/year

X	Weathering	River output	Mining	Combustion	X	Weathering	River output	Mining	Combustion
Ag	0·28	11	14·3	0·01	Eu	8·4	0·22		0·14
Al	330 000	11 000	134 000	2800	F	3800	3700	2000	220[b]
As	6	19	47	1·5	Fe	160 000	19 000	680 000	2200
Au	0·004	0·074	1·3	0·0002	Ga	72	3·3	0·03	1·4
B	40	550	2440	20	Ge	7·2		~0·08	1·1
Ba	2000	370	3000	56	Hf	21	0·37	0·027	0·25
Be	10	11	1·1	0·28	Hg	0·2	3·7	8·4	8·4[b]
Bi	0·19	0·74	3·5	0·014	I	0·56	74	~2	4·8[b]
Br	1·5	520	266[b,c]	175[b,c]	In	0·2		0·058	0·0056
C	1900	410 000	5 950 000	5 950 000[a,b]	K	84 000	81 000	18 500	840
Ca	160 000	550 000	~830 000	420	La	128	3·7	~0·3	1·4
Cd	0·4	3·7	7·7	0·065	Li	80	74	4·4	2·8
Ce	270	7·4		3·4	Lu	2	0·11		0·02
Cl	520	260 000	81 000	1500[b]	Mg	92 000	150 000	242	560
Co	80	7·4	22·7	1·1	Mn	3800	300	24 600	14
Cr	400	37	~2500	2·8	Mo	6	19	76	0·8
Cs	12	0·74		0·084	N	100	1900	44 000	84 000[a]
Cu	200	110	6190	4·6	Na	92 000	220 000	53 000	110

X	Weathering	River output	Mining	Combustion	X	Weathering	River output	Mining	Combustion
Nb	80		13	2·8	Sr	1500	2600		42
Nd	150	5·5		2·5	Ta	8	<0·07	0·43	0·054
Ni	320	19	560	5·6[a]	Tb	4·4	0·11		0·062
P	4000	740	11 200	36	Te	0·02		0·22	0·028
Pb	56	110	3340	183[c,d]	Th	48	1·1	0·27	0·56
Pd	0·002		0·1	0·0014	Ti	22 000	190	~1000	140
Rb	360	37		5·6	Tl	2·4		0·03	0·054
S	1000	140 000	51 700	84 000[a]	U	10	15	23	0·28
Sb	0·8	7·4	75	0·28	V	640	19	17·6	20[a]
Sc	64	0·37		1·4	W	5·2	1·1	38·4	0·14
Se	0·2	7·4	1·04	3[b]	Yb	13	0·37		0·1
Si	1 100 000	260 000	~380 000	8400	Zn	300	550	4040	22
Sm	32	4·5		0·34	Zr	760	30	255	14
Sn	8·8	1·9	178	0·54					

Weathering: assuming 4×10^{12} kg crustal rock weathers per year, with composition as in Table 3.3. *River output;* assuming $3{\cdot}7 \times 10^{16}$ kg river water reaches the sea each year, with composition as in Table 2.3. *Mining:* values from Mining Annual Review, 1977, *Combustion:* assuming combustion rates of Table 5.3, composition of coal and oil as in Table 3.7, and 10% transfer of ash to atmosphere except for values marked [b].

[a] more than 50% from burning fuel oil; otherwise mainly from coal. [b] assuming 100% transfer to atmosphere, or 32% for Se. [c] mainly from additives to petrol. [d], Bazell, 1971. [e], McDonald and Merriman (1977)

Water pollution

From a global point of view, it would be of great interest to compare the amounts of elements added to the annual water cycle by Man's activities with the amounts which are cycled naturally. Unfortunately, the amounts concerned are rarely available, and we have to be content with estimates involving arbitrary assumptions, such as those of Table 12.3. Estimates of rock weathering (column 1) are notably uncertain. Barth (1961) estimated that about 4×10^{12} kg of the Earth's crust weathers each year, and we have used this figure to simplify the calculation. However Garrels and Mackenzie (1971) state that whereas the mean rate of weathering over the last 600 million years is about 3×10^{12} kg/year, the present rate may be eight times this value. They also point out that different types of sedimentary rocks weather at very different rates, which should be allowed for in more rigorous calculations. The estimates of river transfer are often based on fragmentary data, and no allowance has been made for the large annual transfer of suspended matter by rivers. These estimates (column 2) include data for many rivers known to be contaminated by human activities. Column 3, giving the rates of mining in 1976, is subject to uncertainties such as the difficulty of obtaining statistics from China and Russia. No estimates have been made for the mobilisation of elements mined incidentally, such as cadmium from zinc ores, or from the weathering of spoil heaps near mines. The rates of mining of most elements are increasing exponentially, as shown in Fig. 12.3, and no steady state has been reached. How much of the annual production of each metal is lost to the environment by corrosion or wastage is poorly known, though estimates have been made for silver (see below). The uncertainties in column 4 have been discussed in the last section.

Inspection of the data in Table 12.3 suggests that rates of mining equal or exceed natural rates of cycling for the following elements; when the natural rate of cycling is exceeded by a factor of ten or more the element is in italic, as it must be a potential pollutant:

Ag, As, *Au*, B, Bi, *C*, *Cd*, Co, *Cr*, *Cu*, Fe, *Hg*, *Mn*, Mo, *N*, Ni, P, *Pb*, *Sb*, *Sn*, U, *W*, *Zn*

A substantial part of the river output is unaccounted for, and could be made up by contamination from human activities, for the following elements:

Ag, As, Au, B, Bi, Cd, Hg, Mo, N, Pb, S, Sb, Se, Zn

Note that some, and possibly all of these elements could be partly cycled through the atmosphere.

These considerations have some value in predicting which elements are

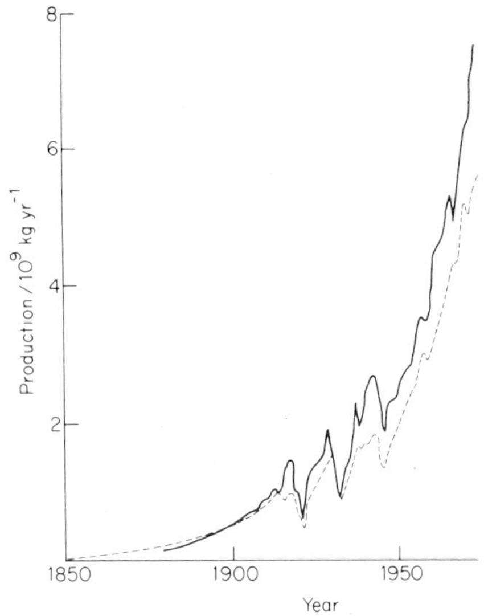

FIG. 12.3 Annual world production of copper (———), and zinc (-----)

most likely to be involved in contamination and pollution of freshwater. They include some elements which have been studied during the last decade with an intensity which at times appears neurotic (Cd, Hg, Pb), and others which have been neglected (Ag, Au, Sb, Sn). The approximate dates at which transfers by human activities exceeded natural transfer rates have been discussed by Bowen (1977b).

FRESHWATER POLLUTION

Local pollution of lakes and rivers has occurred from a variety of sources, such as sewage effluents, agricultural wastes and waters used for mineral processing and industrial works. In the case of large lakes, the time scale is relatively slow, both for the biological effects to appear and for readjustments if and when the source of pollution is eliminated. In the case of rivers, effects such as widespread killing of fish are both rapid and obvious, but self-purification is equally rapid if the pollution is acute rather than chronic. Lakes in river systems function as gigantic sink traps for heavy metals and phosphates, which are partly retained in their sediments. If a river traverses no lakes, any sparingly soluble or particulate contaminants which it carries become concentrated in the estuary and its mud. The estuarine flora and fauna are adapted to resist moderate contamination, and adverse effects on them may pass unnoticed.

Contamination of streams and well-waters by nitrates has been observed in many of the drier parts of the U.S.A. (Keller and Smith, 1967; Winton *et al.*, 1971). Nitrates are scarcely toxic, but are readily reduced to nitrites, which can be poisonous to infants at concentrations exceeding 10 mg N/l Nitrate fertilisers are a possible source of the contamination, as the ion is highly mobile in soils, and attempts have been made to replace them by other forms of nitrogen such as ammonia or urea. All these, and other forms of nitrogen, can be converted to nitrates quite rapidly by soil bacteria if conditions are suitable, and nitrates can accumulate in cultivated soils which are not fertilised at all. The remedy seems to involve the avoidance of discharging excess nitrogenous wastes to local waters, together with careful calculation of the correct rates of application of nitrogenous fertilisers so as to balance inputs and outputs (Stewart, 1970).

Nitrates, together with phosphates, are a major cause of eutrophication in lakes (Owens and Wood, 1968). The sources of these anions may be human sewage, animal manure or artificial fertilisers, and they are both important plant foods. Phosphate is frequently the limiting factor for biological productivity in fresh waters. The symptoms of eutrophication involve an increase in the mass of aquatic vegetation, often with transient blooms of algae which later decay and give off offensive by-products. Salmonid fish avoid eutrophic waters and are replaced by coarser fish such as cyprinids. Normal sewage treatment removes less than 30% of the phosphate from waste waters, and more efficient removal requires special treatment with alum and lime (de Haan and Zwerman, 1976). There is no effective chemical method for removing excess nitrates from lake inputs, though bacterial denitrification might be used.

Cyanides are highly toxic contaminants found in the effluents of some mineral processing operations. They can be rendered less harmful by oxidation with dichlorine before discharge.

Both acute and chronic contamination by heavy metals is often found in individual lakes and rivers. The most frequent cases of serious pollution appear to be due to cadmium, copper, mercury and zinc. Cadmium pollution has caused painful illness and death among Japanese living near the estuary of the river Jintsu (Friberg *et al.*, 1971). Acute copper pollution killed much of the local flora and fauna in the river Churnet in 1938 (Butcher, 1946). Chronic pollution by the same metal caused the flesh of oysters to turn green in Restronguet Creek, Cornwall (Bryan, 1976). Salmon are known to avoid streams contaminated by copper (Sprague *et al.*, 1965). Chronic mercury pollution of the Minamata estuary in Japan has affected between 1000 and 3000 people and resulted in 108 deaths (Tsubaki and Irukayama, 1977). Holden (1972) has reviewed other reports of mercury contamination of the hydrosphere. Some streams draining lead/zinc mines have reduced algal growth and practically no invertebrate fauna (Jones, 1940; Griffiths *et al.*, 1975). The only known preventive measure is to remove the source of pollution, but once metals are trapped in estuarine muds they can contaminate the local flora and fauna, and render them unsuitable for human consumption for a very long time. Heavy metals can be removed from most factory effluents, and to some extent from mine wastes, but not from entire rivers. Improved legislation and inspection could prevent further tragedies such as those of the Jintsu and Minamata. Current legislation has laid down criteria for the maximum permissible concentrations of a number of elements in drinking water (Leh and Lak, 1974).

OCEANIC POLLUTION

In an excellent review of marine pollution, Bryan (1976) considers that contamination of the upper layers of the open ocean has only been demonstrated for lead (Chow and Patterson, 1966), and possibly also for mercury (Chester *et al.*, 1973). No biological effects have been reported from this contamination.

Coastal and estuarine contamination is of frequent occurrence, especially from As, Cd, Cu, Hg, Pb, Sn and Zn, but instances of pollution are rare (Abdullah *et al.*, 1972; Thornton *et al.*, 1975; Bryan, 1976). Although all the elements mentioned are concentrated to a notable extent by marine biota, there is little or no progressive concentration up food chains, except in the case of mercury (see below). There is evidence that some top carnivores, such as seals and sea eagles, have been affected by ingesting mercury from contaminated fish (Bryan, 1976).

Statistical information on waste disposal to the ocean is not readily available. In 1968, the U.S.A. disposed of $3{\cdot}8 \times 10^{10}$ kg dredgings plus 10^{10} kg sewage sludges and industrial wastes to the ocean. Global rates of cropping include $5{\cdot}3 \times 10^{10}$ kg FM fish and 6×10^{8} kg FM marine algae (Hood and McRoy, 1971).

Soil contamination and pollution

Aspects of this subject have been reviewed recently by Bowen (1975b), de Haan and Zwerman (1976) and Purves (1977). The soil is an important habitat for both producers (green plants) and decomposers (bacteria and fungi). While air and water are both self-purifying systems with regard to most inorganic contaminants, the soil is a sink which both receives fall-out from the atmosphere and absorbs or filters material from natural waters.

Most soils have a considerable buffering capacity with respect to contaminants, and may themselves be self-purifying systems, but the time scales involved are probably of the order of kiloyears. We would therefore expect biological effects of contamination to appear slowly, and to take many years to correct.

Under natural conditions the soil is a dynamic system with inputs (rock weathering; decay of plant litter; fall-out from rain or dust) and outputs (production of plant biomass; drainage; loss of volatile substances), so that in principle one could calculate residence times for its constituents. Cultivated soils have additional inputs such as irrigation water, fertilisers (sometimes including metal slags and sewage sludges) and pesticides, while the harvested crop is an extra output. Estimates of all these inputs and outputs vary with local circumstances, and few reliable data are available. The following sources of data have been used here:

Rock weathering. Estimates for the global situation are given in Table 12.3. As most soils are derived from their underlying or neighbouring rocks, local variations will be considerable. The rates of formation of soils have been discussed in chapter 4.

Decay of plant litter. This gives rise to short-term cycling of many inorganic species between the litter layer of the soil and surface-rooting plants.

Fall-out from rain or dust. Atmospheric deposition of plant nutrients has been reviewed by Cooke (1975), while many elements have been studied by Cawse (1977) in Britain and by Lazrus *et al.* (1970) in the U.S.A. Deposition rates vary greatly with rainfall and proximity to industrial areas, volcanoes and the ocean. Very high rates are found in large cities (Volchok and Bogen, 1973).

Irrigation water. Inputs are usually so small as to be negligible, but in hot, dry countries such as Iraq and Pakistan large areas have been rendered infertile by excessive irrigation with saline water (Cantor, 1967).

Fertilisers. World production of fertilisers in 1975 was $4{\cdot}4 \times 10^{13}$ g N, $1{\cdot}85 \times 10^{13}$ g K and $1{\cdot}12 \times 10^{13}$ g P. Application rates vary greatly in different parts of the world and for different crops. Typical rates for British soils are given by Cooke (1975). The elemental composition of fertilisers is given by Swaine (1962) in an extensive monograph; limited data for sewage sludges can be found in papers by Berrow and Weber (1972), Furr *et al.* (1976) and a review by de Haan and Zwerman (1976).

Pesticides. Inputs of inorganic pesticides are small and declining. They involve mainly copper, as a fungicide in vineyards, and mercury as a fungicidal seed dressing. Arsenic, lead, manganese and zinc derivatives are used occasionally (de Ong, 1956; Martin, 1972).

Production of plant biomass and cropping. Production has been discussed in chapter 5. Cooke (1975) and Nye and Tinker (1977) give data on nutrient removal by cropping. Some crops, such as rubber and cereal grains, remove relatively little in the way of nutrients, while others, such as tobacco and cabbages, remove a lot.

Drainage. Outputs can be computed by subtracting the annual mass of water lost by evaporation and transpiration from the rainfall. The composition of drainage water is usually close to that of local river water.

Volatilisation. There is no satisfactory way to measure these outputs, but they are believed to be important for very few elements (Br, C, Cl, Hg, I, N, S and possibly Se). Estimates for volatilisation losses of C, N and S are given in chapter 5.

With all these uncertainties it is difficult to make quantitative estimates of the rates of elemental cycling in soils. An attempt has been made to compare the inputs and outputs of a number of elements in Tables 12.4 and 12.5. Table 12.4 assumes a natural, uncontaminated environment and an undisturbed soil bearing climax vegetation. Table 12.5 assumes a moderately contaminated environment and a soil under permanent cultivation. For both situations temperate conditions, a rainfall of 66 cm/year (the world mean) and an average productivity have been assumed.

From the data in Table 12.4, uncertain and imprecise as it is, two features appear to characterise natural uncultivated soils. Inputs and outputs balance, at least within an order of magnitude, for virtually all the elements listed, though input from rock weathering appears to exceed losses by drainage for Co, Cr, Fe, Ni, P and V (the data for Hg and Se are particularly uncertain). Thus the table provides semi-quantitative evidence for a steady state which has usually been assumed intuitively. Secondly the rate of biological cycling quantitatively exceeds the rate of geological cycling in soils for eleven of the elements listed, namely B, Ca, Cd, Cu, K, Mg, Mn, N, P, S, and Zn, and all these are important essential elements except for cadmium.

The data in Table 12·5 for the moderately contaminated, cultivated soil has several further points of interest. Inputs and outputs roughly balance for Ca, Cl, Co, Cr, Cu, Hg, Mg, Mn, Na, Ni and Zn, and inputs from contaminated atmospheres are a major factor in maintaining the copper and zinc status of cropped soils. Outputs may sometimes exceed inputs, but are not far off balance, for B, K, Mo, N, P and S. It is not surprising that N, P and K occur in this group, since they have long been known to be the most important elements for agricultural fertilisers. Inputs of nitrogen and sulphur, and perhaps also of B and Mo, from contaminated atmospheres

TABLE 12.4 Estimated inputs and outputs in mg X/m² yr for an uncontaminated, uncultivated soil

X	Rock weathering[a] input	Rain and dust[b] input	Litter decay[c] input = Biomass production[d] output	Drainage[e] output	Volatilisation[f] output
As	0·04	0·1–0·5	0·2	0·12	0
B	0·27	4–8	55	3·5	0
Ca	1100	40–80	8000	500–28 000	0
Cd	0·003	0·013[g]	0·15	0·023	0
Cl	3·5	170[h]	4000	250–8000	?
Co	0·54	0·04	0·2	0·046	0
Cr	2·7	0·4	0·23	0·23	0
Cu	1·3	1·5[j]	8	0·7	0
F	26	<3	4	23	0
Fe	1100	40	80	120	0
Hg	0·0013	<0·006	0·08	0·02	0·1?
K	570	30–120	8000	510	0
Mg	620	20[h]	1500	920	0
Mn	26	0·8[j]	400	2	0
Mo	0·04	<0·3	0·15	0·12	0
N	0·67	150[h]	20 000	12	100–2500
Na	620	110[h]	200	150–6000	0
Ni	2·2	0·3[j]	1·2	0·1	0
P	27	20[h]	1500	5	0
Pb	0·37	0·1–2	1·5	0·7	0
S	7	<200[h]	3000	200–8000	≤440
Se	0·0013	0·005	0·02	0·05 ?	?
V	4·3	0·1	0·4	0·12	0
Zn	2	2–6	40	3	0

[a] assuming 27 g crustal rock weathers per year per m² soil. [b] assuming a rainfall on land of 660 kg/m² yr; estimates from data of Table 1.2 and Cawse (1977). [c] assumed to equal biomass production. [d] assuming mean net land production of 0·78 kg DM/m² year; elemental composition as in chapters 5 and 6. [e] assuming 35% of rainfall is run off; elemental composition from Table 2.3. [f] estimated values; see chapter 5 for N and S. [g] Lagerwerff, 1972. [h] Cooke, 1975. [j] Lazrus *et al.*, 1970

TABLE 12.5 Inputs and outputs in mg X/m² yr for a cultivated soil in a moderately contaminated environment, neglecting uncropped biomass and litter decay

X	Rock weathering[a] input	Fertiliser[b] input	Rain and dust[c] input	Cropping[d] output	Drainage[e] output	Volatilisation[f] output
As	0·04	2·4	1–3·2	0·03–1	0·12	0
B	0·27	1·5	4–8	20[h] (1–65)	3·5	0
Ca	1100	0–20 000	800–1600	1000–20 000	500–28 000	0
Cd	0·003	0·02	0·2–1[g]	0·03–0·3	0·023	0
Cl	3·5	5000	2500–5000[h]	1000–7000	250–8000	?
Co	0·54	0·05	0·2–0·4	0·06[m] (0·01–0·5)	0·046	0
Cr	2·7	3·2	1–5	0·1–9	0·23	0
Cu	1·3	0·16	12–23	3–15	0·7	0
F	26	300	3–800	2–13	23	0
Fe	1100	700	150–460	60[h] (35–170)	120	0
Hg	0·0013	0·15[l]	0·05[l]	0·009–0·08	0·01[l]	0·05[l]
K	570	3000–8000	300–1000	4000–30 000	500	0
Mg	620	0–10 000	400[h]–900	500–4500	920	0
Mn	26	5	20	5–60[h]	2	0
Mo	0·04	0·07	<0·5	1[h] (0·4–3)	0·12	0
N	0·67	0–20 000	500–1600[h]	8000–25 000[h]	12	100–2500
Na	620	200 ?	1300–2700[h]	80–1000	150–6000	0
Ni	2·2	0·15	3·3	0·015–2·7	0·1	0
P	27	1700–4400	20[h]–400	1000–6500	5	0
Pb	0·37	1·5	21	0·3–13	0·7	0
S	7	800	900–2000	400–10 000	200–8000	≤440
Se	0·0013	0·05	0·46	0·02–0·07	0·05	?
V	4·3	0·6	5·2	0·2–3	0·12	0
Zn	2	3	50–140	10–110	3	0

[a] assuming 27 g crustal rock weathers per year per m² soil. [b] assuming N, K and P are added to the soil at rates of 10, 6 and 3 g/m² year respectively, using the elemental composition of fertilizers given by Swaine (1962). [c] mostly data from Cawse (1977). [d] assuming a yield of 0·67 kg DM/m² year, elemental composition as in Table 6.2. [e] assuming 35% of rainfall is run off; elemental composition from Table 2.3. [f] estimated values; see chapter 5 for N and S. [g] Lagerwerff (1972). [h] Cooke (1975). [l] Madsen and Jensen (1978). [m] Chichilo and Whittaker (1961)

make a useful contribution to the soil nutrient pool. Potential pollutants, whose inputs exceed their outputs, include As, Cd, F, Fe, Pb, Se and V. For all of these the atmosphere is a major source of contamination, while fertilisers contribute substantial amounts of arsenic and fluorine from their impurities. Apart from iron, which is scarcely toxic, these are the elements whose concentrations in well-characterised soils should be monitored at regular intervals, especially in industrialised countries.

Many authors have pointed out the dangers of repeatedly treating soils with metallurgical slags (Swaine, 1962) or sewage sludges (Berrow and Webber, 1972; Pike *et al.*, 1975; Furr *et al.*, 1976), because of the possible build-up of elements to toxic concentrations. An analogous problem arises from the repeated use of saline water for irrigation. In assessing the hazards of these treatments, some kind of balance sheet should be drawn up for the soil concerned, so as to ensure that inputs do not grossly exceed outputs. Sewage sludges usually contain abnormal amounts of Ag, Ba, Cd, Cr, Cu, Hg, Ni, Pb, Sb, Sn and Zn, and sometimes also Au, B, Be, Bi, Co, Ge, In, Mo and W. In the majority of cases, toxic effects on plants are most likely to arise from the high zinc content of sludge, which is of the order of 2 mg/g DM. Although salt pollution can be overcome quite quickly by washing a soil with clean water, there are no remedial measures for heavy metal pollution of soils other than disposal to the ocean or as landfill.

Local instances of soil pollution

Arsenic. High concentrations of arsenic occur naturally in a few places such as S.W. Britain (Colbourn *et al.*, 1975; Porter and Peterson, 1975), Switzerland and New Zealand (Kipling, 1977). Smelters frequently contaminate soils with arsenic (Crecelius *et al.*, 1974). So much lead arsenate has been applied to certain orchards in the U.S.A. that the soil has become phytotoxic (Steevens *et al.*, 1972). Cows have sometimes died after breaking into potato fields which have been sprayed with arsenates (Kipling, 1977). The element is strongly absorbed by soils and its effects persist for many years.

Boron. The concentration of boron is elevated in urban soils (Purves, 1966) and can become toxic to plants grown in some irrigated soils (Kubota and Allaway, 1972). Excess is readily removed by leaching with clean water.

Cadmium. The concentrations of cadmium in soil near zinc smelters can be very high, and these soils have been abandoned for cultivation (Buchauer, 1973; Fleischer *et al.*, 1974). In Japan, polluted soils containing more than 50 mg Cd/kg have been found in rice-growing districts where Itai itai disease has occurred (Yamagata and Shigematsu, 1970).

Chromium. Chromium(VI) is mobile in soils and toxic to plants, but is rapidly reduced to the less toxic, immobile chromium(III). Plant damage has been observed in waste tips containing chromates from industry (Breeze, 1973).

Copper. High concentrations of copper are found in soils near smelters and on spoil heaps from mines. Such soils may show reduced or negligible enzyme activity (Tyler, 1974, 1976). Copper toxicity to plants has been found in vineyards which have been repeatedly sprayed with Bordeaux mixture. Agricultural usage involves about 7×10^7 kg copper per year.

Fluorine. Fluoride discharges from brickworks in Britain have caused bone damage to sheep and cattle grazing in nearby fields (Allcroft *et al.*, 1965).

Lead. Soils have been contaminated globally by emissions from car exhausts, and the concentrations of lead in all soils examined are rising rapidly (Murozumi *et al.*, 1969; Page and Gange, 1970; Bethea and Bethea, 1975). Very high local contamination is found on spoil heaps and near lead smelters (Little and Martin, 1972; Jennett *et al.*, 1975) and near major highways (Höll and Hampp, 1975). Lead is immobilised in soils, mostly as $PbSO_4$ (Olson and Skogerboe, 1975). High concentrations of lead reduce the bacterial population (Griffiths *et al.*, 1975), but are not very toxic to plants. Plant uptake of lead from soils can be reduced by liming, or by adding phosphates or sulphates (Lagerwerff, 1972). High lead in soils does not appear to harm the invertebrate fauna (Williamson and Evans, 1972, 1973), but lead poisoning is a common cause of death in calves (Todd, 1962) and horses have died from eating hay contaminated with lead (Schmitt *et al.*, 1971). Lead is also said to be responsible for the deaths of many urban dogs (Zook *et al.*, 1972) and of animals in urban zoos (Bazell, 1971). It has numerous toxic effects on man, and high levels in the environment are held responsible for hyperactivity in city children (David, 1974).

Mercury. High concentrations of mercury have been found in the soils of Dutch bulb fields, and in the greens of golf courses, which have been repeatedly treated with mercurial fungicides. Many seed-eating birds and their predators have been killed in districts where methyl mercury chloride has been used as a seed dressing (Rissanen and Miettinen, 1972). Methyl mercurials have now mostly been replaced by aryl mercurials, which are less toxic to wild life. Mercury can be both methylated and demethylated by soil bacteria. Methylation requires anoxic conditions and is inhibited by adding nitrates. Mercury is undoubtedly lost from soils by volatilisation, but the rate of loss is uncertain.

Manganese. Manganese toxicity to plants has been occasionally found in acid soils, and in steam-sterilised soils, but can be corrected by adding lime to increase the pH (Kubota and Allaway, 1972). It has once been noted in soils on which dry batteries had been dumped (J. Patterson, 1971).

Molybdenum. Some soils contain sufficient molybdenum to cause illness in cows grazing the local vegetation: the condition can often be remedied by adding a controlled amount of copper to the soil (Underwood, 1977). One case of molybdenum pollution has been reported from aerial emissions by Fawley oil refinery in Britain (Gardner and Hall-Patch, 1962).

Nickel. This very toxic element is probably phytotoxic in some spoil heaps, but plant damage from sewage sludges and industrial wastes is rare (J. Patterson, 1971).

Selenium. Selenium occurs in high concentrations in soils in many parts of the world, but is very rare as a pollutant; toxic effects on plants have been found near a rectifier factory in Britain (J. Patterson, 1971).

Zinc. Zinc is the element most frequently implicated in causing plant damage from industrial emissions, e.g. on mine wastes, near smelters, sewage sludges, river dredgings, near galvanised steel buildings and where rubber tyres have been burnt (J. Patterson, 1971; Berrow and Webber, 1972). Excess zinc inactivates soil enzymes (Tyler, 1974, 1976), and reduces the bacterial population (Griffiths *et al.*, 1975), but does not have marked effects on the fauna.

Biomagnification in food chains

Several reports have suggested that biological concentration factors may increase as the food chain is ascended, and the term *biomagnification* has been coined to describe this phenomenon (Kneip and Lauer, 1973). As far as metals are concerned, all are concentrated by organisms as described in chapter 7, often by large factors. In aquatic habitats, where carnivorous fish are usually the top predator, these fish have been shown to contain lower concentrations of Cd, Cs, Cu, Mn and Pu than the plankton or algae of the lowest trophic level (Kneip and Lauer, 1973). Only in the case of mercury is there good evidence of biomagnification. Thus in a newly made reservoir, Potter *et al.* (1975) found 0·01 ng Hg/g water, 28–34 ng Hg/g FM in algae and aquatic plants, and 230–500 ng Hg/g FM fish muscle. Berg *et al.* (1966) have shown biomagnification of methyl mercurial compounds, which has

TABLE 12.6 Radioactivity of ^{137}Cs and ^{22}Na in an arctic food chain (Hanson *et al.*, 1967)

Material	Bq ^{137}Cs/kg FM	Bq ^{22}Na/kg FM
Lichen	210–300	0·33–0·55
Caribou flesh	395–730	10·8
Wolf flesh	1260	10·2

led to toxic effects on raptorial birds. Much useful data on this topic can be found in a symposium edited by Aberg and Hungate (1967). In the arctic environment, Hanson *et al.* (1967) have shown clear evidence for biomagnification in the food chain Lichen → Caribou → Wolf, as shown in Table 12.6 for ^{137}Cs and ^{22}Na. In the same food chain, both caribou and wolf flesh had much lower activities than did lichen for the following nuclides: ^{144}Ce, ^{60}Co, ^{54}Mn, ^{106}Ru, ^{125}Sb, ^{88}Y and ^{95}Zr.

Exhaustion of natural resources

A great deal of discussion, often polemical, has been generated concerning the exhaustion of minerals by human activities. An extensive review of this topic has been edited by Govett and Govett (1976), and a few general points are worth making in concluding this chapter.

No stocks of any element or nuclide (apart from ^{235}U) are being depleted as a result of human activities. The chemical form of many mineral deposits has been changed, and many elements are being dispersed from concentrated ores into less concentrated wastes. Virtually all minerals are being exploited at an exponentially increasing rate as the world population expands.

Natural resources are of two kinds; biological products such as coal, oil, limestone, most sulphur, dioxygen and dinitrogen, and geological products which include all ore deposits. In the case of biological products, it is generally believed that resources of coal and oil will last a few centuries at most. Large stocks will probably remain in the Earth's crust, but the cost of extracting these will be too high for exploitation. As mentioned in chapter 5, very little research has been devoted to measuring or even estimating the natural rates of formation of biological products. A few estimates are given in Table 12.7. In an ideal world rates of usage would not exceed rates of formation, but in the real world it appears that reserves of coal, oil, natural gas and biogenetic sulphur are all being used up much faster than natural means of replacement. Stocks of dinitrogen and limestone are in no danger of exhaustion. Difficulties arise in any attempt to predict when coal and oil will become exhausted. Current *reserves*, which are exploitable at an economic price, are thought to be in the range $9–15 \times 10^{15}$ kg coal (Averitt, 1973) and about 6×10^{13} kg oil (Govett and Govett, 1976). Other *resources*, which are low-grade materials such as tar sands which might be exploited with better technical knowledge and/or higher prices, have not been assessed with any precision.

In the case of geological ores, the rates of formation are seldom known at

TABLE 12.7 Estimates of rates of formation and 1976 rates of usage of biological products in 10^9 kg/year

Product	Rate of formation	Rate of usage (1976)
Coal	9 ?	2790
Oil	0·25 ?	2820
Natural Gas	0·15 ?	1700
Limestone	3000 ?	~ 2100
Sulphur	0·00002 ?	10
Dioxygen	67 000	23 000
Dinitrogen	190	44[a] less 42[b]

See chapter 5 for references. [a] fixed by Haber process; [b] produced by burning fossil fuels

all, except for submarine manganese nodules, which are resources rather than reserves. Neither world resources nor reserves are known with any high degree of precision, but there is a rough relation between the estimated mass M of element in reserves (in 10^9 kg) and the concentration c of element in the Earth's crust (in mg/kg).

$$M = Kc$$

Where K is the range 0·68 to 10; this is shown in Fig. 12.4. Although we do

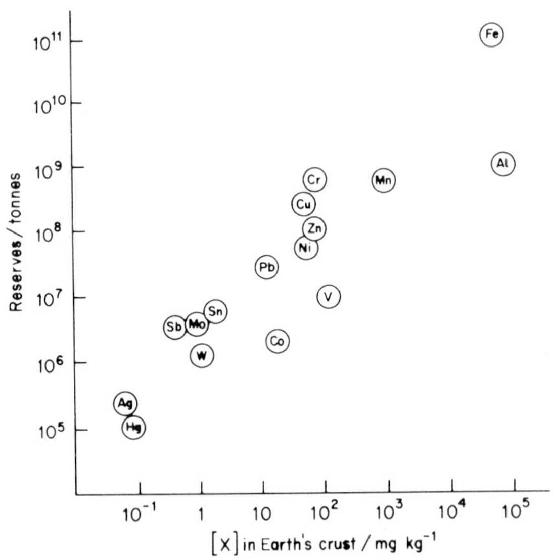

FIG. 12.4 Relation between world reserves of metals and their concentrations in the Earth's crust. After Govett and Govett (1976)

not know whether human activities are reducing ore deposits faster than they are being formed, it seems likely that this is often the case. The best known example is copper. In 1700, ores contained more than 13% Cu, but by 1900 ores containing 2·5% Cu were being mined, and in 1970 ores with as little as 0·25% Cu are being exploited because richer sources have been exhausted.

More information is needed on the rates of weathering of metals such as copper, iron, lead and zinc which human activities have exposed to the atmosphere in large amounts. The rate of weathering of these metals would be a better measure of human perturbations of their cycles than the rate of mining. One complication is that the corrosion of these metals in polluted atmospheres of cities is much faster than in clean air in rural areas. A noteworthy study of the rate of loss of silver from coinage has been carried out by Patterson (1972), who was able to show that silver in coins has a mean residence time of about 40 years, with a range of 25–100 years. Residence times of other metals should be estimated if we are to understand the ultimate effects of mining and smelting activities.

13 *Elements in the Geosphere and the Biosphere*

This chapter summarises information given elsewhere in this book, or in the references quoted, on all the elements from hydrogen to curium, with the exception of francium. It is not a comprehensive survey of all available literature on the subject. The arrangement of elements is alphabetical by symbol. Concentrations are mostly given in mg/kg, or in Bq/kg for some radioactive nuclides. Note the following points:

1. 'Natural radioisotopes' here include truly natural radionuclides, those produced by cosmic radiation, and those introduced into the environment by man.
2. Mean values are given for most concentrations in the geosphere, except for air, where ranges are quoted.
3. 'Annual cycle, natural' refers to the estimated rate of rock weathering and/or the rate of transfer of soluble matter by rivers: units are mostly 10^9 g/yr.
4. 'Mining' is mostly 1976 data; where a substantial contribution comes from the combustion of fossil fuels, this is given here: units are mostly 10^9 g/yr.
5. Ranges of concentration are given for the biosphere, where they have been measured. Some data for mammal muscle comes from Gorski *et al.* (1974).

Ac (Actinium)

Stable isotopes (fract. abund.) None

Natural radioisotopes ($t_{\frac{1}{2}}$) ^{227}Ac(22 yr), ^{228}Ac(6·1 h)

Geosphere (mg/kg)

Granite: 10^{-12}(^{228}Ac). Basalt: 8×10^{-14}(^{228}Ac). Common minerals: Decay product of ^{235}U/^{232}Th in rocks, see Aly *et al.* (1975).

Biosphere: ^{228}Ac

Accumulators: 15–1000 Bq/kg DM in plants from Th-rich soil (Gopal-Ayengar and Mistry, 1962). Accumulated by liver and bone (ICRP, 1972).

Ag (Silver) At. Wt 107·868
Stable isotopes (fract. abund.) ^{107}Ag(0·514), ^{109}Ag(0·486)
Natural radioisotopes ($t_{\frac{1}{2}}$) ^{108m}Ag(127 yr), ^{110m}Ag(253 days)
Geosphere (mg/kg)
Granite: 0·04. Basalt: 0·1. Soil: 0·05. Shale: 0·07. Limestone: 0·12. Sandstone: 0·25. Sea water: 0·00004. Freshwater: 0·0003. Air: 0·2–7 ng/m^3. Common minerals: Ag_2S. Annual cycle, natural: 0·28–11. Mining: 14 $\times 10^9$ g/yr.
Biosphere (mg/kg DM)
Land plants: 0·01–0·8. Edible vegetables: 0·02–0·6. Mammal muscle: 0·009–0·28. Mammal bone: 0·01–0·44. Marine algae: 0·11–0·28. Marine fish: 0·04–0·1
Accumulators: possibly Lycoperdales and *Eriogonum ovalifolium*. Natural compounds: Ag methyls? Toxicity: Very toxic to fungi and plants; 60 mg/day toxic, 1300 mg/day lethal to Man.
Remarks: Unique metabolism; probably binds to –SH groups in proteins.
Reviews: In plants and soils, Warren and Delavault (1960), Boyle (1968); in animals, Underwood (1977); in environment, Smith and Carson (1978).

Al (Aluminium) At. Wt 26·9815
Stable isotopes (fract. abund.) ^{27}Al(1)
Natural radioisotopes ($t_{\frac{1}{2}}$) ^{26}Al(7·4 $\times$ 10^5 yr)
Geosphere (mg/kg)
Granite: 77 000. Basalt: 87 600. Soil: 71 000. Shale: 88 000. Limestone: 9000. Sandstone: 43 000. Sea water: 0·002. Freshwater: 0·3. Air: 150–3500 ng/m^3. Common minerals: AlO_2H, aluminosilicates, clays. Annual cycle, natural: 11 000–330 000. Mining: 134 000 $\times$ 10^9 g/yr.
Biosphere (mg/kg DM)
Land plants: 90–530. Edible vegetables: 0·1–200. Mammal muscle: 0·7–28. Mammal bone: 4–27. Marine algae: 62? Marine fish: 20.
Accumulators: Many plants in Diapensiaceae, Ericaceae, Melastomataceae, Symplocaceae and Theaceae: siliceous Porifera. Clubmosses. Beneficial to some acidiphilic plants. Toxicity: 0·1–30 mg/l toxic to plants; rarely toxic to animals. Natural compounds: Binds to DNA in *Pisum* etc. (Matsumoto *et al.*, 1977). Stimulates 5-aminolaevulinate hydrolyase (Meredith *et al.*, 1977).
Reviews: In soils, Segalen (1973) and Dixon and Weed (1977); in plants and environment, Sorenson *et al.* (1974), Sorenson (1976); in animals, Underwood (1977); toxicity, Burrows (1977) and Elliott *et al.* (1978).

Am (Americium)
Stable isotopes (fract. abund.) None
Natural radioisotopes ($t_{\frac{1}{2}}$) ^{241}Am(458 yr); decay product of ^{241}Pu
Geosphere (mg/kg)
Freshwater: 0·0009 Bq/m^3 (Wahlgren *et al.*, 1976)
Common minerals: 8000 yr ^{243}Am may exist in molybdenites (Cherdyntsev and Mikhailov, 1963).
Biosphere (mg/kg DM)
Accumulators: Mammal bone (ICRP, 1972).
Reviews: In soil, see Cleveland and Rees (1976); in marine algae and fish, see Hampson and Tennant (1973) and Pentreath (1977).

Ar (Argon) At. Wt 39·95
Stable isotopes (fract. abund.) ^{36}Ar(0·00337), ^{38}Ar(0·00063), ^{40}Ar(0·996)
Natural radioisotopes ($t_{\frac{1}{2}}$) ^{37}Ar(35 d), ^{39}Ar(269 yr), ^{41}Ar(1·8 h)
Geosphere (mg/kg)
Granite: 0·04? Basalt: 0·00003–0·002. Shale: 0·05–0·5. Sea water: 0·45. Freshwater: 0·6. Air: 16 g/m^3. Atmosphere: decay product of ^{40}K.
Mining: 700×10^9 g/yr.
Biosphere (mg/kg DM)
Toxicity: negligible.
Remarks: 0·75 mg/l mammalian blood (Regnard and Schlösung, 1897).

As (Arsenic) At. Wt 74·9216
Stable isotopes (fract. abund.) ^{75}As(1)
Natural radioisotopes ($t_{\frac{1}{2}}$) ^{76}As(1·1 days)
Geosphere (mg/kg)
Granite: 1·5. Basalt: 1·5. Soil: 6. Shale: 13. Limestone: 1. Sandstone: 1. Sea water: 0·0037. Freshwater: 0·0005. Air: 1·5–53 ng/m^3. Common minerals: As_2S_3; As_4S_4; As; $FeAs_2$. Annual cycle, natural: 6–19. Mining: 47 $\times 10^9$ g/yr.
Biosphere (mg/kg DM)
Land plants: 0·02–7 (−40). Edible vegetables: 0·01–1·5. Mammal muscle: 0·007–0·09. Mammal bone: 0·08–1·6. Marine algae: 1–30. Marine fish: 0·2–10.
Accumulators: Mammalian hair and nails; probably brown algae, some marine invertebrates. Essential to: Red algae (1) and mammals. Natural compounds: Me_3As and $HAsMe_2$; arsenobetaine; $MeAsO_2H$, Me_2AsOH and Me_3AsO in urine. Toxicity: 0·02–7·5 mg/l toxic to plants; 5–50 mg/day toxic, 100–300 mg/day lethal to Man. Functions: Improves growth of hair?

Remarks: Probably carcinogenic.
Reviews: Aquatic cycle, Crecelius (1975); in animals, Underwood (1977); in Man, Schroeder and Balassa (1966a); in medicine, Frost (1967); in foods, Lewis (1977); in plants, Girling (1978).

At (Astatine)
Stable isotopes (fract. abund.) None
Natural radioisotopes ($t_{\frac{1}{2}}$) ^{210}At(8·3 h), ^{211}At(7·2 h)
Biosphere
Accumulators: Mammalian thyroid (transient); Durbin (1960).

Au (Gold) At. Wt 196·9665
Stable isotopes (fract. abund.) ^{197}Au(1)
Geosphere (mg/kg)
Granite: 0·0018. Basalt: 0·0005. Soil: 0·001. Shale: 0·0025. Limestone: 0·002. Sandstone: 0·003. Sea water: 0·000004. Freshwater: 0·000002. Air: 0·001–0·006 ng/m^3. Common minerals: Au; $AuTe_2$. Annual cycle, natural: 0·004–0·074. Mining: $1{\cdot}3 \times 10^9$ g/yr.
Biosphere (mg/kg DM)
Land plants: 0·0017 (−0·04). Edible vegetables: ≤0·0016. Mammal muscle: 0·00006–0·0003. Mammal bone: 0·016. Marine algae: 0·03? Marine fish: 0·00012.
Accumulators: Possibly the tree *Tectona grandis* (Anon, 1968); mammal liver; fossil plant *Thuchomyces* (Hallbauer, 1978). Toxicity: Scarcely toxic. Natural compounds: Methyl Au compounds.
Remarks: ^{198}Au is used in medicine (Williams and Da Silva, 1978).
Reviews: In plants, Minski *et al.* (1977); Girling (1978).

B (Boron) At. Wt 10·81
Stable isotopes (fract. abund.) ^{10}B(0·196), ^{11}B(0·804)
Geosphere (mg/kg)
Granite: 12. Basalt: 8. Soil: 20. Shale: 130. Limestone: 20. Sandstone: 30. Sea water: 4·44. Freshwater: 0·015. Air: 4 ng/m^3. Common minerals: borax; tourmaline. Annual cycle, natural: 40–550. Mining: 2440×10^9 g/yr.
Biosphere (mg/kg DM)
Land plants: 11–140. Edible vegetables: 8–200. Mammal muscle: 0·33–1. Mammal bone: 1·1–3·3. Marine algae: 100–160. Marine fish: 20.
Accumulators: ±many Cruciferae and Leguminosae (plants). Essential to: Blue-green algae, diatoms, seed plants. Natural compounds: Boromycin. Toxicity: 1–5 mg/l toxic to plants; 4 g/day toxic to Man. Functions: Antibiotic. Involved in auxin and/or RNA metabolism?

Remarks: A useful indicator of salinity.
Reviews: B cycle, Gregor (1965); in plants, Price *et al.* (1972), Jackson and Chapman (1975); in animals, Underwood (1977).

Ba (Barium) At. Wt 137·34
Stable isotopes (fract. abund.) ^{130}Ba(0·00101), ^{131}Ba(0·00097), ^{134}Ba(0·0242), ^{135}Ba(0·0659), ^{137}Ba(0·113), ^{138}Ba(0·7166)
Natural radioisotopes ($t_{\frac{1}{2}}$) ^{140}Ba(12·8 days) f.p.
Geosphere (mg/kg)
Granite: 730. Basalt: 250. Soil: 500. Shale: 550. Limestone: 90. Sandstone: 320. Sea water: 0·013. Freshwater: 0·01. Air: 0·2–90 ng/m^3. Common minerals: $BaCO_3$, $BaSO_4$. Annual cycle, natural: 370–2000. Mining: 3000 $\times 10^9$ g/yr.
Biosphere (mg/kg DM)
Land plants: 8–150. Edible vegetables: 1–150. Mammal muscle: 0·09. Mammal bone: 3–30. Marine algae: 1·5–50. Marine fish: 0·017.
Accumulators: Tree *Bertholletia excelsa*; Protozoan *Xenophyophora*; Plankton *Chaetoceros curvisetus* and *Rhizosolenia calcar-avis*. Perhaps essential to mammals. Natural compounds: $BaSO_4$ (rare). Toxicity: 500 mg/l toxic to plants; 200 mg/day toxic to Man. Functions: Structural and gravity sensor (Kreger and Boeré, 1969; Hubert *et al.*, 1975).
Reviews: In biota, Schroeder *et al.* (1972); in animals, Underwood (1977).

Be (Beryllium) At. Wt 9·0122
Stable isotopes (fract. abund.) ^{9}Be(1)
Natural radioisotopes ($t_{\frac{1}{2}}$) ^{7}Be(54 day), ^{10}Be(2·5 $\times 10^6$ yr)
Geosphere (mg/kg)
Granite: 5. Basalt: 0·3. Soil: 0·3. Shale: 3. Limestone: <1. Sandstone: <1. Sea water: 0·0000056. Freshwater: 0·0003. Air: 0·1–4 ng/m^3. Common minerals: Beryl. Annual cycle, natural: 10. Mining: 1·1 $\times 10^9$ g/yr.
Biosphere (mg/kg DM)
Land plants: 0·04? Edible vegetables: 0·0006? Mammal muscle: 0·00075. Mammal bone: 0·003. Marine algae: 0·003? Marine fish: 0·002–0·2.
Accumulators: Mammal bone. Toxicity: 0·5 mg/l toxic to plants; very toxic to mammals (Tepper *et al.*, 1961).
Reviews: Meehan and Smythe (1967); Reeves (1965) for metabolism by rat.

Bi (Bismuth) At. Wt 208·9804
Stable isotopes (fract. abund.) ^{209}Bi(1)
Natural radioisotopes ($t_{\frac{1}{2}}$) short-lived ^{211}Bi, ^{212}Bi, ^{215}Bi and ^{207}Bi(30 yr)

Geosphere (mg/kg)
Granite: 0·065. Basalt: 0·031. Soil: 0·2? Shale: 0·48. Limestone: 0·17. Sandstone: 0·18. Sea water: 0·00002. Freshwater: 0·00002? Air: 0·2–3 ng/m^3. Common minerals: Bi, Bi_2O_3, Bi_2S_3. Annual cycle, natural: 0·2–0·7. Mining: $3{\cdot}5 \times 10^9$ g/yr.
Biosphere (mg/kg DM)
Land plants: <0·02. Edible vegetables: 0·06? Mammal muscle: 0·032. Mammal bone: <0·2. Marine fish: 0·04.
Accumulators: Mammal kidney. Toxicity: 27 mg/l toxic to plants; 1·5 mg/day toxic, 160 mg/day lethal to rats.

Br (Bromine) At. Wt 79·904
Stable isotopes (fract. abund.) ^{79}Br(0·505), ^{81}Br(0·495)
Geosphere (mg/kg)
Granite: 0·3. Basalt: 0·44. Soil: 10. Shale: 24. Limestone: 6. Sandstone: 1. Sea water: 67·3. Freshwater: 0·014. Air: 0·6–2500 ng/m^3. Common minerals: Salt lake evaporates. Annual cycle, natural: 1–520. Mining: 266 $\times 10^9$ g/yr.
Biosphere (mg/kg DM)
Land plants: 15? Edible vegetables: 0·05–26. Mammal muscle: 8. Mammal bone: 7. Marine algae: 220–3300. Marine fish: 400.
Accumulators: Red and some brown algae; some Porifera; many corals; a few molluscs (*Aplysia* and Muricidae); fungus *Amanita* spp. Possibly essential to: Red algae and mammals. Natural compounds: >50, mostly bromophenols (Miller, 1973). Toxicity: 15–600 mg/l toxic to plants; 3 g/day toxic, 35–350 g/day lethal to Man. Functions: Antibiotic: Pigment (Tyrian Purple).
Reviews: In marine biota, Craigie and Gruenig (1967); in diet and Man, Cross *et al.* (1978a); in animals, Underwood (1977).

C (Carbon) At. Wt 12·011
Stable isotopes (fract. abund.) ^{12}C(0·9889), ^{13}C(0·0111)
Natural radioisotopes ($t_{\frac{1}{2}}$) ^{14}C(5760 yr)
Geosphere (mg/kg)
Granite: 360. Basalt: 610. Soil: 20 000. Shale: 14 000. Limestone: 116 000. Sandstone: 16 000. Sea water: 28. Freshwater: 11. Air: 164 mg/m^3. Common minerals: $CaCO_3$, coal, petroleum. Annual cycle, natural: 1900–410 000. Mining: 6 000 000 $\times 10^9$ g/yr.
Biosphere (mg/kg DM)
Land plants: 450 000. Edible vegetables: 450 000. Mammal muscle: 670 000. Mammal bone: 360 000. Marine algae: 280 000–350 000. Marine fish: 475 000.

Accumulators: All organisms. Essential to: All organisms. Natural compounds: Very numerous. Toxicity: CO is very toxic to mammals, and CN^- is very toxic to all organisms. Functions: Structural, biochemical and metabolic.
Remarks: Extensive isotopic fractionation occurs in metabolic processes.
Reviews: C cycle, Pytkowicz (1973).

Ca (Calcium) At. Wt 40·08
Stable isotopes (fract. abund.) ^{40}Ca(0·970), ^{42}Ca(0·0064), ^{43}Ca(0·00145), ^{44}Ca(0·0206), ^{46}Ca(0·000033), ^{48}Ca(0·00185)
Geosphere (mg/kg)
Granite: 15 800. Basalt: 67 200. Soil: 15 000. Shale: 16 000. Limestone: 340 000. Sandstone: 31 000. Sea water: 412. Freshwater: 15. Air: 0·5–7000 ng/m^3. Common minerals: $CaCO_3$, $CaSO_4$, feldspar. Annual cycle, natural: 160–550. Mining: 830×10^{12} g/yr.
Biosphere (mg/kg DM)
Land plants: 3000–14 000. Edible vegetables: 400–50 000. Mammal muscle: 140–700. Mammal bone: 170 000. Marine algae: 4000–24 000. Marine fish: 76–20 000.
Accumulators: Some red algae; shells of many invertebrates; mammal bone. Essential to: All organisms (except some fungi?) Natural compounds: $CaCO_3$, apatite, Ca alginate, Ca pectate, enzymes. Toxicity: relatively harmless to all organisms. Functions: Hard tissues; cell walls; holding proteins together; catalytic; electrochemical; messenger in nerve and muscle response: Initiates chemotropic response in pollen.
Reviews: In diet, Nordin (1976); in proteins, Kretsinger (1976), Wasserman *et al.* (1977); in calcitic tissues, Lowenstam (1974), Darley (1974a, b); in bone, Simkiss (1975); in plants, Wyn Jones and Lunt (1967).

Cd (Cadmium) At. Wt 112·4
Stable isotopes (fract. abund.) ^{106}Cd(0·0122), ^{108}Cd(0·0087), ^{110}Cd(0·124), ^{111}Cd(0·127), ^{112}Cd(0·241), ^{113}Cd(0·123), ^{114}Cd(0·289), ^{116}Cd(0·0758)
Natural radioisotopes ($t_{\frac{1}{2}}$) ^{109}Cd(470 days), ^{113m}Cd(14 yr), ^{115m}Cd(43 days)
Geosphere (mg/kg)
Granite: 0·09. Basalt: 0·13. Soil: 0·35. Shale: 0·22. Limestone: 0·028. Sandstone: 0·05. Sea water: 0·00011. Freshwater: 0·0001. Air: 0·5–620 ng/m^3. Common minerals: With ZnS. Annual cycle, natural: 0·4–3·7. Mining: $7{\cdot}7 \times 10^9$ g/yr.
Biosphere (mg/kg DM)
Land plants: 0·1–2·4. Edible vegetables: 0·05–0·9. Mammal muscle: 0·14–3·2. Mammal bone: 1·8. Marine algae: 0·3–2·5. Marine fish: 0·1–3.
Accumulators: *Agaricus* and other fungi; mollusc *Ostrea* spp. and *Pecten*

liver. Mammal kidney. Natural compounds: Metallothioneins; methyl Cd compounds? Toxicity: 0·2–9 mg/l toxic to plants; 3–330 mg/day toxic, 1·5–9 g/day lethal to Man. Inhibits enzymes (Lee, 1977). Absorbed by ferritin (Hazard *et al.*, 1977). Possibly essential to rats (Schwarz and Spallholz, 1978).

Reviews: In soils, Shacklette *et al.* (1973); in plants, Shacklette (1972); in foods, Essing *et al.* (1969), MOAFF (1973); in Man, Friberg *et al.* (1971), Prasad (1976), CEC (1978); biological methylation, Huey *et al.* (1975); Cycle, Van Hook *et al.* (1977).

Ce (Cerium) At. Wt 140·12

Stable isotopes (fract. abund.) ^{136}Ce(0·00193), ^{138}Ce(0·0025), ^{140}Ce(0·885), ^{142}Ce(0·111)

Natural radioisotopes ($t_{\frac{1}{2}}$) ^{141}Ce(33 days), ^{144}Ce(284 days) f.p.

Geosphere (mg/kg)

Granite: 83. Basalt: 59. Soil: 50. Shale: 96. Limestone: 20. Sandstone: 78. Sea water: 0·0000012. Freshwater: 0·0002. Air: 0·002–20 ng/m^3. Common minerals: Monazite ($CePO_4$). Annual cycle, natural: 7·4–270. Mining: 0·3? $\times 10^9$ g/yr.

Biosphere (mg/kg DM)

Land plants: 0·25–16. Edible vegetables: 0·01–0·05. Mammal muscle: <0·01? Mammal bone: 2·7. Marine fish: 0·006–0·03.

Accumulators: Tree *Carya* spp.

Cl (Chlorine) At. Wt 35·453

Stable isotopes (fract. abund.) ^{35}Cl(0·755), ^{37}Cl(0·245)

Natural radioisotopes ($t_{\frac{1}{2}}$) ^{36}Cl(300 000 yr)

Geosphere (mg/kg)

Granite: 200. Basalt: 60. Soil: 100. Shale: 160. Limestone: 76. Sandstone: 510. Sea water: 19 350. Freshwater: 7. Air: 9–11 000 ng/m^3. Common minerals: NaCl. Annual cycle, natural: 520–260 000. Mining: 81 000 $\times 10^9$ g/yr.

Biosphere (mg/kg DM)

Land plants: 700–27 000. Edible vegetables: 200–46 000. Mammal muscle: 2000–5200. Mammal bone: 900. Marine fish: 6000.

Accumulators: Halophytic plants, e.g. Chenopodiaceae, Frankeniaceae, Plumbaginaceae, Rhizophoraceae, some Verbenaceae (*Avicennia*). Probably essential to all organisms. Natural compounds: >100 small molecules mostly chlorophenols, +1 enzyme (Miller, 1973). Toxicity: Relatively harmless to all organisms as Cl^-. Functions: Antibiotic: electrochemical; Catalytic (α-amylase and photosynthesis).

Remarks: Cl_2, ClO^- and ClO_3^- are highly toxic to organisms.

Reviews: Neidleman (1975).

Cm (Curium)
Stable isotopes (fract. abund.) None
Natural radioisotopes ($t_{\frac{1}{2}}$) ^{242}Cm(0·45 yr); possibly ^{247}Cm(250 Myr) (Cherdyntsev and Mikhailov, 1963)
Biosphere
Accumulators: Accumulated by mammal bone (ICRP, 1972).
Remarks: For behaviour in soils and plants, see Thomas and Jacobs (1969); for occurrence in effluents to sea, see Hampson and Tennant (1973); in lichens, Holm and Persson (1978).

Co (Cobalt) At. Wt 58·9332
Stable isotopes (fract. abund.) ^{59}Co(1)
Natural radioisotopes ($t_{\frac{1}{2}}$) ^{60}Co(5·3 yr)
Geosphere (mg/kg)
Granite: 1. Basalt: 35. Soil: 8. Shale: 19. Limestone: 0·1. Sandstone: 0·3. Sea water: 0·00002. Freshwater: 0·0002. Air: 0·0008–37 ng/m^3. Common minerals: CoS_2, $CoAs_2$ (usually with copper). Annual cycle, natural: 7–80. Mining: $22{\cdot}7 \times 10^9$ g/yr.
Biosphere (mg/kg DM)
Land plants: 0·005–1. Edible vegetables: 0·01–4·6. Mammal muscle: 0·005–1. Mammal bone: 0·01–0·04. Marine algae: 0·02–9(?). Marine fish: 0·006–0·05.
Accumulators: Plants *Clethra barbinervis*, *Crotalaria cobalticola*, *Nyssa sylvatica*. Essential to: All organisms, except perhaps green algae and seed plants. Natural compounds: Corrins (vitamin B_{12}), some enzymes. Toxicity: 0·1–3 mg/l toxic to plants: 500 mg/day toxic to Man. Functions: Catalytic.
Remarks: Extensive areas are known where low soil Co affects the health of grazing mammals.
Reviews: In foods, Yamagata *et al.* (1963), Schlettwein-Gsell and Mommsen-Straub (1970b), Thomas *et al.* (1974); in animals, Underwood (1977); in environment, Young (1979); in corrins, Pratt (1972), Babior (1975); in enzymes, Ochiai (1977).

Cr (Chromium) At. Wt 51·996
Stable isotopes (fract. abund.) ^{50}Cr(0·0431), ^{52}Cr(0·838), ^{53}Cr(0·0955), ^{54}Cr(0·0238)
Natural radioisotopes ($t_{\frac{1}{2}}$) ^{51}Cr(28 days)
Geosphere (mg/kg)
Granite: 4. Basalt: 90. Soil: 70. Shale: 90. Limestone: 11. Sandstone: 35. Sea water: 0·0003. Freshwater: 0·001. Air: 0·005–300 ng/m^3. Common

minerals: Chromite ($FeCr_2O_4$), $MgCr_2O_4$. Annual cycle, natural: 37–400. Mining: 2500? × 10^9 g/yr.

Biosphere (mg/kg DM)

Land plants: 0·03–10. Edible vegetables: 0·016–14. Mammal muscle: <0·002–0·84. Mammal bone: 0·1–33. Marine algae: 0·5–13. Marine fish: 0·03–2.

Accumulators: Trees *Leptospermum scoparium* and *Pimelea suteri*. Essential to: Mammals. Natural compounds: $Cr(C_2O_4)_3{}^{3-}$. Toxicity: 0·5–10 mg Cr(VI)/l toxic to plants; 200 mg/day toxic, 3 g/day lethal to Man. Functions: Activates phosphoglucomutase; determines tolerance of glucose in mammals (Mertz, 1969).

Reviews: In soils, Breeze (1973), Bartlett and Kimble (1976); in foods, Thomas *et al.* (1974); in animals, Underwood (1977); in Man, Mertz and Cornatzer (1971), Prasad (1976).

Cs (Caesium) At. Wt 132·9054

Stable isotopes (fract. abund.) ^{133}Cs(1)

Natural radioisotopes ($t_{\frac{1}{2}}$) ^{134}Cs(2·1 yr), ^{137}Cs(30 yr) f.p.

Geosphere (mg/kg)

Granite: 5. Basalt: 1. Soil: 4. Shale: 5·5. Limestone: 0·5. Sandstone: 0·5. Sea water: 0·0003. Freshwater: 0·00002. Air: 0·02–1·5 ng/m^3. Common minerals: Pollucite. Annual cycle, natural: 0·7–12. Mining: ? × 10^9 g/yr.

Biosphere (mg/kg DM)

Land plants: 0·03–0·44. Edible vegetables: <0·001–0·08. Mammal muscle: 0·07–1·6. Mammal bone: 0·013–0·05. Marine algae: 0·04–0·4. Marine fish: 0·04–0·18.

Toxicity: Relatively harmless to all organisms.

Remarks: ^{137}Cs contamination is discussed by Miettinen *et al.* (1963) and Marei *et al.* (1972).

Reviews: Cs cycle, Auerbach *et al.* (1964), Witherspoon (1964), Patten and Witkamp (1967); in Man, Yamagata (1962).

Cu (Copper) At. Wt 63·55

Stable isotopes (fract. abund.) ^{63}Cu(0·691), ^{65}Cu(0·309)

Geosphere (mg/kg)

Granite: 13. Basalt: 90. Soil: 30. Shale: 39. Limestone: 5·5. Sandstone: 30. Sea water: 0·00025. Freshwater: 0·003. Air: 0·036–4900 ng/m^3. Common minerals: Cu_2S, CuS, $CuFeS_2$, CuO, $Cu_2CO_3(OH)_2$. Annual cycle, natural: 110–200. Mining: 6190 × 10^9 g/yr.

Biosphere (mg/kg DM)

Land plants: 5–15. Edible vegetables: 4–20. Mammal muscle: 10. Mammal bone: 1–26. Marine algae: 2–68. Marine fish: 0·7–15.

Accumulators: Bryophytes *Merceya* and *Mielichhoferia* spp: a few Cruciferae, Compositae, Labiatae and Tiliaceae; Poriferan *Chondrilla nucula*; blood of Crustacea and Mollusca. Essential to: All organisms. Natural compounds: 1 Porphyrin and >30 enzymes. Toxicity: 0·5–8 mg/l toxic to plants, 250 mg/day toxic to Man. Functions: Pigment: antibiotic: O_2 transport: catalytic in redox reactions.

Remarks: Natural isotopic fractionation has been claimed but needs confirmation.

Reviews: In foods, Schlettwein-Gsell and Mommsen-Straub (1971c); in mammals, Fisher (1975), Underwood (1977); in Man, Prasad (1976); in biochemistry, Peisach *et al.* (1966); in enzymes, Ochiai (1977); in plants, Jones (1972); in medicine, Hambidge and Nichols (1978).

Dy (Dysprosium) At. Wt 162·5

Stable isotopes (fract. abund.) ^{156}Dy(0·00052), ^{158}Dy(0·0009), ^{160}Dy(0·0229), ^{161}Dy(0·189), ^{162}Dy(0·255), ^{163}Dy(0·25), ^{164}Dy(0·282)

Geosphere (mg/kg)

Granite: 5? Basalt: 6·9. Soil: 5. Shale: 5·8. Limestone: 2·1. Sandstone: 6·9. Sea water: 0·00000091. Freshwater: 0·000005? Air: 0·04 ng/m^3? Common minerals: Xenotime.

Biosphere (mg/kg DM)

Land plants: 0·05–0·6. Mammal bone: <0·36. Marine fish: 0·0014? Accumulators: Tree *Carya* spp. Toxicity: low.

Er (Erbium) At. Wt 167·26

Stable isotopes (fract. abund.) ^{162}Er(0·00136), ^{164}Er(0·00156), ^{166}Er(0·334), ^{167}Er(0·229), ^{168}Er(0·271), ^{170}Er(0·149)

Geosphere (mg/kg)

Granite: 3·7. Basalt: 3·9. Soil: 2. Shale: 4. Limestone: 1·7. Sandstone: 4·9. Sea water: 0·00000087. Air: 0·026 ng/m^3? Common minerals: Xenotime.

Biosphere (mg/kg DM)

Land plants: 0·08–0·38. Mammal bone: <0·28. Accumulators: Tree *Carya* sp. Toxicity: low.

Eu (Europium) At. Wt 151·96

Stable isotopes (fract. abund.) ^{151}Eu(0·478), ^{153}Eu(0·522)

Natural radioisotopes ($t_{\frac{1}{2}}$) ^{152}Eu(12 yr), ^{155}Eu(1·8 yr) f.p.

Geosphere (mg/kg)

Granite: 2. Basalt: 2·2. Soil: 1. Shale: 1·2. Limestone: 0·4. Sandstone: 2? Sea water: 0·00000013. Fresh water: 0·000006. Air: 0·00017–1·7 ng/m^3.

Common minerals: Monazite, and impurity in strontianite and sphene. Annual cycle, natural: 0·2–8·4 × 10^9 g/yr.
Biosphere (mg/kg DM)
Land plants: 0·028–0·13. Edible vegetables: <0·005–0·07. Mammal bone: 0·032. Marine fish: 0·00016–0·0009.
Accumulators: Tree *Carya* spp. Toxicity: low.

F (Fluorine) At. Wt 18·9984
Stable isotopes (fract. abund.) ^{19}F(1)
Geosphere (mg/kg)
Granite: 1400. Basalt: 510. Soil: 200. Shale: 800. Limestone: 220. Sandstone: 180. Sea water: 1·3. Freshwater: 0·1. Air: 1–400 ng/m^3. Common minerals: CaF_2, fluoroapatite. Annual cycle, natural: 3800. Mining: 2000 × 10^9 g/yr.
Biosphere (mg/kg DM)
Land plants: 0·02–24. Edible vegetables: 3–19. Mammal muscle: 0·05. Mammal bone: 2000–12 000. Marine algae: 4·5–11. Marine fish: 1400.
Accumulators: A few plants; *Acacia georginae*, *Dichapetalum* spp., *Gastrolobium grandiflorum*; mammal bone and teeth; Poriferan *Dysidea crawshayi*. Probably essential to: Mammals. Natural compounds: Fluoroacids (rare) and nucleocidin (Miller, 1973). Toxicity: 5 mg/l toxic to plants; 20 mg/day toxic, 2 g/day lethal to Man. Functions: Structural in apatite; antibiotic.
Reviews: In foods, Jones *et al.* (1971) and Kumpulainen and Koivistoinen (1977); in bacteria, Kelly (1965); in animals, Underwood (1977); in Man, Prasad (1976); correlation with heart disease, Taves (1978); in bone, Zipkin (1973): in plants, Hall (1977).

Fe (Iron) At. Wt 55·85
Stable isotopes (fract. abund.) ^{54}Fe(0·0582), ^{56}Fe(0·917), ^{57}Fe(0·0219), ^{58}Fe(0·0033)
Natural radioisotopes ($t_{\frac{1}{2}}$) ^{55}Fe(2·6 yr), ^{59}Fe(45 days)
Geosphere (mg/kg)
Granite: 27 000. Basalt: 56 000. Soil: 40 000. Shale: 48 000. Limestone: 17 000. Sandstone: 29 000. Sea water: 0·002. Freshwater: 0·5. Air: 0·8–14 000 ng/m^3. Common minerals: FeS_2, Fe_3O_4, $FeCO_3$, Fe_2O_3, FeO_2H, Olivine. Annual cycle, natural: 19–160. Mining: 680 × 10^{12} g/yr.
Biosphere (mg/kg DM)
Land plants: 70–700. Edible vegetables: 2–250. Mammal muscle: 180. Mammal bone: 3–380. Marine algae: 13–1900. Marine fish: 9–88.
Accumulators: Iron bacteria; lichen *Acarospora smaragdula*; red blood cells; fish liver. Essential to: All organisms. Natural compounds: Por-

phyrins: >70 metalloproteins and enzymes: ferritin. Toxicity: 10–200 mg/l toxic to plants; 200 mg/day toxic to Man. Functions: Pigments; antibiotics; O_2 transport; catalytic in redox reactions; storage.
Reviews: In soils, Dixon and Weed (1977); in microbes, Neilands (1974); in animals, Underwood (1977); in proteins and enzymes, Jacobs and Worwood (1974), Lovenberg (1977), Ochiai (1977); deficiency in Man, Hallberg *et al.* (1970).

Ga (Gallium) At. Wt 69·72
Stable isotopes (fract. abund.) ^{69}Ga(0·604), ^{71}Ga(0·396)
Geosphere (mg/kg)
Granite: 19. Basalt: 17. Soil: 20. Shale: 23. Limestone: 2·5. Sandstone: 6. Sea water: 0·00003. Freshwater: 0·00009. Air: 0·2–1 ng/m^3? Common minerals: associated with ZnS. Annual cycle, natural: 3–72. Mining: 0·03 $\times 10^9$ g/yr.
Biosphere (mg/kg DM)
Land plants: 0·01–0·2. Edible vegetables: 0·03–2. Mammal muscle: 0·0014. Marine algae: 0·5? Marine fish: 0·15.
Toxicity: 10 mg/day is toxic to rats.
Remarks: For behaviour of Ga in rabbit serum, see Hartman and Hayes (1969).

Gd (Gadolinium) At. Wt 157·25
Stable isotopes (fract. abund.) ^{152}Gd(0·002), ^{154}Gd(0·0215), ^{155}Gd(0·147), ^{156}Gd(0·205), ^{157}Gd(0·157), ^{158}Gd(0·249), ^{160}Gd(0·219)
Geosphere (mg/kg)
Granite: 7·4. Basalt: 8. Soil: 4. Shale: 6. Limestone: 2·7. Sandstone: 7·1. Sea water: 0·0000007. Air: 0·045 ng/m^3? Common minerals: Xenotime.
Biosphere (mg/kg DM)
Land plants: 0·0016–0·5. Mammal bone: <0·4.
Accumulators: Tree *Carya* spp. Toxicity: low.

Ge (Germanium) At. Wt 72·59
Stable isotopes (fract. abund.) ^{70}Ge(0·205), ^{72}Ge(0·274), ^{73}Ge(0·0776), ^{74}Ge(0·365), ^{76}Ge(0·0776)
Geosphere (mg/kg)
Granite: 2. Basalt: 1·6. Soil: 1. Shale: 2. Limestone: 0·31. Sandstone: 1·2. Sea water: 0·00005. Air: 3 ng/m^3? Common minerals: None. Annual cycle, natural: 7. Mining: 0·08? $\times 10^9$ g/yr.
Biosphere (mg/kg DM)
Land plants: 1–2·4. Mammal muscle: 0·14. Marine fish: 0·3.
Toxicity: Scarcely toxic (except as GeH_4).
Reviews: Schroeder and Balassa (1967), Underwood (1977).

H (Hydrogen) At. Wt 1·0079
Stable isotopes (fract. abund.) ^{1}H(0·9998), ^{2}H(0·000156)
Natural radioisotopes ($t_{\frac{1}{2}}$) ^{3}H(12·3 yr)
Geosphere (mg/kg)
Granite: 1400. Soil: Variable. Shale: 5600. Limestone: 860. Sandstone: 1800. Sea water: 110 000. Freshwater: 110 000. Air: 300 μg/m^3 DM. Common minerals: Water. Annual cycle, natural: 4·1 × 10^{18} g/yr.
Biosphere (mg/kg DM)
Land plants: 55 000. Edible vegetables: 55 000. Mammal muscle: 93 000. Mammal bone: 52 000. Marine algae: 41 000–72 000. Marine fish: 68 000. Accumulators: All organisms. Essential to: All organisms. Natural compounds: Very numerous. Toxicity: Negligible. Functions: Biochemical and metabolic.
Remarks: 30% D_2O is toxic to mammals. Extensive isotopic fractionation occurs in natural processes.

He (Helium) At. Wt 4·0026
Stable isotopes (fract. abund.) ^{3}He(0·000001), ^{4}He(1)
Geosphere (mg/kg)
Granite: 0·01? Basalt 3 × 10^{-8}–0·001. Shale 0·0004–0·007.
Sea water: 0·0000069. Air: 920 μg/m^3. Common minerals: Decay product of Th and U minerals; natural gas. Mining: 4·34 × 10^9 g/yr.
Biosphere
Toxicity: Negligible.
Remarks: For submarine source of ^{3}He, see Lupton (1976).

Hf (Hafnium) At. Wt 178·49
Stable isotopes (fract. abund.) ^{174}Hf(0·0018), ^{176}Hf(0·052), ^{177}Hf(0·185), ^{178}Hf(0·271), ^{179}Hf(0·138), ^{180}Hf(0·354)
Natural radioisotopes ($t_{\frac{1}{2}}$) ^{174}Hf(2 × 10^{15} yr)
Geosphere (mg/kg)
Granite: 5·8. Basalt: 4·8. Soil: 6? Shale: 2·8. Limestone: 0·3. Sandstone: 3·9. Sea water: 0·000007. Freshwater: 0·00001. Air: 0·0005–0·3 ng/m^3. Common minerals: None; associated with zircon. Annual cycle, natural: 0·4–21. Mining: 0·027 × 10^9 g/yr.
Biosphere (mg/kg DM)
Edible vegetables: $<$0·001–1. Mammal muscle: 0·04? Mammal bone: 0·12. Toxicity: Scarcely toxic.

Hg (Mercury) At. Wt 200·59
Stable isotopes (fract. abund.) ^{196}Hg(0·00146), ^{198}Hg(0·1), ^{199}Hg(0·168), ^{200}Hg(0·231), ^{201}Hg(0·132), ^{202}Hg(0·298), ^{204}Hg(0·0685)

Natural radioisotopes ($t_{\frac{1}{2}}$) ^{203}Hg(47 days)
Geosphere (mg/kg)
Granite: 0·08. Basalt: 0·012. Soil: 0·06. Shale: 0·18. Limestone: 0·16. Sandstone: 0·29. Sea water: 0·00003. Freshwater: 0·0001. Air: 0·009–38 ng/m^3. Common minerals: Cinnabar (HgS). Annual cycle, natural: 0·2–3·7. Mining: 8·4 × 10^9 g/yr (and 8·4 × 10^9 g/yr from burning fossil fuel).
Biosphere (mg/kg DM)
Land plants: 0·005–0·02. Edible vegetables: 0·013–0·17. Mammal muscle: 0·02–0·7. Mammal bone: 0·45. Marine algae: 0·03. Marine fish: 0·4.
Accumulators: Tree *Betula papyrifera* and *Minuartia setacea*?; mammal kidney. Natural compounds: CH_3HgCl, $Hg(CH_3)_2$ and C_2H_5HgCl. Toxicity: 0·4 mg/day toxic, 150–300 mg/day lethal to Man. Very toxic to fungi and plants. Functions: Reacts with –SH groups and inhibits enzymes.
Reviews: In soil, Shacklette *et al.* (1971); in plants, Shacklette (1970); in foods, MOAFF (1971), Schelenz and Diehl (1973); aquatic cycle, Fagerström and Jernelöv (1972); in environment, D'Itri (1972), Holden (1972); pollution, Tsubaki and Irukayama (1977); in Man, Prasad (1976).

Ho (Holmium) At. Wt 163·934
Stable isotopes (fract. abund.) ^{165}Ho(1)
Geosphere (mg/kg)
Granite: 1·3. Basalt: 1·5. Soil: 0·6. Shale: 1·8. Limestone: 0·28. Sandstone: 2? Sea water: 0·00000022. Air: 0·013 ng/m^3? Common minerals: Xenotime.
Biosphere (mg/kg DM)
Land plants: 0·03–0·11. Mammal bone: <0·08.
Accumulators: Tree *Carya* spp. Toxicity: low.

I (Iodine) At. Wt 126·9045
Stable isotopes (fract. abund.) ^{127}I(1)
Natural radioisotopes ($t_{\frac{1}{2}}$) ^{129}I(16 Myr), ^{131}I(8·1 days) f.p.
Geosphere (mg/kg)
Granite: 0·2. Basalt: 0·08. Soil: 5. Shale: 19. Limestone: 14? Sandstone: 0·1. Sea water: 0·06. Freshwater: 0·002. Air: 0·2–60 ng/m^3. Common minerals: $NaIO_3$ in evaporates. Annual cycle, natural: 0·56–24. Mining: 2? (+ 4·8 from coal) × 10^9 g/yr.
Biosphere (mg/kg DM)
Land plants: 3–5. Edible vegetables: 0·07–10. Mammal muscle: 0·05–0·5. Mammal bone: 21? Marine algae: 90–2500. Marine fish: 1.
Accumulators: Red and brown algae and their marine predators; many Porifera, some Coelenterata; mammalian thyroid. Essential to: corals, sponges, Ascidians, vertebrates (and seed plants?). Natural compounds:

>70 small molecules and proteins (Miller, 1973). Toxicity: 1 mg/l toxic to plants; 2 mg/day toxic, 35–350 g/day lethal to Man. Function: Antibiotic; hormonal.

Reviews: I cycle, Miyake and Tsunogai (1963); in plants, Shacklette and Cuthbert (1967); in foods, Sengupta and Pal (1971); in animals, Underwood (1977); in hormones, Koutras (1968).

In (Indium) At. Wt 114·82
Stable isotopes (fract. abund.) ^{113}In(0·0428), ^{115}In(0·957)
Geosphere (mg/kg)
Granite: 0·04. Basalt: 0·058. Soil: 1? Shale: 0·057. Limestone: 0·009. Sandstone: 0·01. Sea water: 0·00000011. Air: 0·001–1 ng/m^3. Common minerals: None; associated with ZnS. Annual cycle, natural: 0·2. Mining: $0{\cdot}058 \times 10^9$ g/yr.
Biosphere (mg/kg DM)
Edible vegetables: 0·0008. Mammal muscle: 0·016?
Toxicity: 30 mg/day toxic, 200 mg/day lethal to rats; for toxicity to plants, Liebig *et al.* (1943).
Review: Smith and Carson (1978).

Ir (Iridium) At. Wt 192·22
Stable isotopes (fract. abund.) ^{191}Ir(0·373), ^{193}Ir(0·627)
Geosphere (mg/kg)
Granite: 0·000002–0·00002. Basalt: 0·000004–0·0003. Air: 0·0003 ng/m^3? Common minerals: None; IrOs.
Biosphere (mg/kg DM)
Mammal muscle: 0·00002? (Koch and Roesmer, 1962).

K (Potassium) At. Wt 39·098
Stable isotopes (fract. abund.) ^{39}K(0·931), ^{41}K(0·0688)
Natural radioisotopes ($t_{\frac{1}{2}}$) ^{40}K(0·000118: 1·3 Gyr)
Geosphere (mg/kg)
Granite: 33 400. Basalt: 8300. Soil: 14 000. Shale: 24 500. Limestone: 3100. Sandstone: 15 000. Sea water: 399. Freshwater: 2·2. Air: 0·3–40 000 ng/m^3. Common minerals: Feldspars, micas, KCl. Annual cycle, natural: 81. Mining: $18{\cdot}5 \times 10^{12}$ g/yr.
Biosphere (mg/kg DM)
Land plants: 5000–34 000. Edible vegetables: 1000–68 000. Mammal muscle: 16 000. Mammal bone: 2100. Marine algae: 32 000–52 000. Marine fish: 15 000. Essential to: All organisms. Natural compounds: Activates some enzymes. Toxicity: Scarcely toxic, except by intravenous injection to mammals. Functions: Electrochemical; catalytic.

Remarks: Isotopic fractionation has been claimed, but needs confirmation.
Reviews: In enzymes, Suelter (1970), Hewitt and Smith (1975), Ochiai (1977).

Kr (Krypton) At. Wt 83·8
Stable isotopes (fract. abund.) ^{78}Kr(0·00354), ^{80}Kr(0·0227), ^{82}Kr(0·116), ^{83}Kr(0·115), ^{84}Kr(0·569), ^{86}Kr(0·174)
Natural radioisotopes ^{81}Kr(210 kyr), ^{85}Kr(10·6 yr)
Geosphere (mg/kg)
Granite: 0·00001? Basalt: 3×10^{-8}–3×10^{-6}. Shale: 0·0008. Sea water: 0·00021. Air: 4·1 mg/m^3. Common minerals: None. Annual cycle, natural: None. Mining: ~8000 kg/yr.
Biosphere
Toxicity: Negligible.

La (Lanthanum) At. Wt 138·905
Stable isotopes (fract. abund.) ^{138}La(0·00089), ^{139}La(0·999)
Natural radioisotopes ($t_{\frac{1}{2}}$) ^{140}La(40 h) f.p.
Geosphere (mg/kg)
Granite: 43. Basalt: 21. Soil: 40. Shale: 49. Limestone: 10. Sandstone: 42. Sea water: 0·0000034. Freshwater: 0·0001. Air: 0·0005–10 ng/m^3. Common minerals: Monazite. Annual cycle, natural: 4–128. Mining: $0·3? \times 10^9$ g/yr.
Biosphere (mg/kg DM)
Land plants: 0·003–15. Edible vegetables: <0·0004–2. Mammal muscle: 0·0004. Mammal bone: <0·08–9. Marine algae: 10? Marine fish: 0·014. Accumulators: Tree *Carya* spp.; mammal bone. Toxicity: Low; 720 mg/day is lethal to rats.
Reviews: In plants, Yliruokanen (1975b).

Li (Lithium) At. Wt 6·94
Stable isotopes (fract. abund.) ^{6}Li(0·074), ^{7}Li(0·926)
Geosphere (mg/kg)
Granite: 30. Basalt: 10. Soil: 25. Shale: 76. Limestone: 7·5. Sandstone: 38. Sea water: 0·18. Freshwater: 0·002. Air: 2 ng/m^3? Common minerals: None; associated with pegmatites. Annual cycle, natural: 74–80. Mining: $4·4 \times 10^9$ g/yr.
Biosphere (mg/kg DM)
Land plants: 0·5–3·4. Edible vegetables: 0·8–1·3. Mammal muscle: 0·023. Marine algae: 5?
Toxicity: >30 mg/l toxic to plants; 200 mg/day toxic to Man. Functions: induces lethargy in manics.

Remarks: Yeast can fractionate Li isotopes to some extent.
Reviews: In soils, Shacklette *et al.* (1973); in plants, Ezdakova (1964); in animals, Underwood (1977); in medicine, Johnson and Johnson (1978).

Lu (Lutecium) At. Wt 174·97
Stable isotopes (fract. abund.) ^{175}Lu(0·974), ^{176}Lu(0·0259)
Geosphere (mg/kg)
Granite: 0·54. Basalt: 0·49. Soil: 0·4. Shale: 0·8. Limestone: 0·16. Sandstone: 0·8? Sea water: 0·00000015. Freshwater: 0·000003? Air: 0·004–0·03 ng/m^3. Common minerals: Xenotime. Annual cycle, natural: 0·1–2 $\times$ 10^9 g/yr.
Biosphere (mg/kg DM)
Land plants: 0·03. Edible vegetables: 0·0006–0·06. Mammal bone: <0·08. Accumulators: Tree *Carya* spp. Toxicity: Low.

Mg (Magnesium) At. Wt 24·305
Stable isotopes (fract. abund.) ^{24}Mg(0·787), ^{25}Mg(0·101), ^{26}Mg(0·112)
Geosphere (mg/kg)
Granite: 1600. Basalt: 45 000. Soil: 5000. Shale: 16 000. Limestone: 5800. Sandstone: 11 500. Sea water: 1290. Freshwater: 4. Air: 1–11 000 ng/m^3. Common minerals: Olivine, talc, serpentine, dolomite etc. Annual cycle, natural: 92–150. Mining: 0·242 $\times$ 10^{12} g/yr.
Biosphere (mg/kg DM)
Land plants: 1000–9000. Edible vegetables: 700–5600. Mammal muscle: 900. Mammal bone: 700–1800. Marine algae: 6400–20 000. Marine fish: 1200.
Accumulators: Some seeds; calcareous tissues. Essential to: All organisms. Natural compounds: $Mg(OH)_2$ (rare); chlorophylls, MgATP, enzyme activator. Toxicity: Scarcely toxic except by intravenous injection. Functions: Electrochemical; Catalytic; Structural (holding enzyme to substrate).
Reviews: In Man, Nordin (1976), Prasad (1976); in enzymes, Ochiai (1977).

Mn (Manganese) At. Wt 54·938
Stable isotopes (fract. abund.) ^{55}Mn(1)
Natural radioisotopes ($t_{\frac{1}{2}}$) ^{54}Mn(280 days)
Geosphere (mg/kg)
Granite: 400. Basalt: 1500. Soil: 1000. Shale: 850. Limestone: 620. Sandstone: 460. Sea water: 0·0002. Freshwater: 0·008. Air: 0·01–900 ng/m^3. Common minerals: MnO_2, $MnCO_3$, $MnSiO_3$. Annual cycle, natural: 300–3800. Mining: 24 600 $\times$ 10^9 g/yr.

Biosphere (mg/kg DM)
Land plants: 20–700. Edible vegetables: 0·3–1000. Mammal muscle: 0·2–2·3. Mammal bone: 0·2–14. Marine algae: 1–500. Marine fish: 0·3–4·6.
Accumulators: Plants of Ericaceae and Theaceae; Diatom *Coscinodiscus*; Poriferan *Terpios zeteki*; Annelid *Hermione*: kidney of mollusc *Pinna*: Ascidians *Didemnum* and *Halocynthia*. Essential to: All organisms. Natural compounds: Porphyrin ?; 12 Mn proteins or enzymes. Toxicity: 1–100 mg/l toxic to plants; 10–20 mg/day toxic to rats. Functions: Catalytic (redox or not): enzyme activator.
Remarks: Mn^{2+} induces DNA mutations in yeast mitochondria (Putrament *et al.*, 1977).
Reviews: In soils, Dixon and Weed (1977); in plants, Jones (1972); in foods, Schlettwein-Gsell and Mommsen-Straub (1971a); in animals, Underwood (1977); in Man, Prasad (1976); in enzymes, Ochiai (1977).

Mo (Molybdenum) At. Wt 95·94
Stable isotopes (fract. abund.) ^{92}Mo(0·158), ^{94}Mo(0·0904), ^{95}Mo(0·157), ^{96}Mo(0·165), ^{97}Mo(0·0946), ^{98}Mo(0·238), ^{100}Mo(0·0963)
Natural radioisotopes ($t_{\frac{1}{2}}$) ^{99}Mo(67 h)
Geosphere (mg/kg)
Granite: 2. Basalt: 1. Soil: 1·2. Shale: 2·6. Limestone: 0·16. Sandstone: 0·2. Sea water: 0·01. Freshwater: 0·0005. Air: $<0{\cdot}2$–10 ng/m^3. Common minerals: MoS_2. Annual cycle, natural: 6–19. Mining: 76×10^9 g/yr.
Biosphere (mg/kg DM)
Land plants: 0·06–3. Edible vegetables: 0·03–5. Mammal muscle: 0·02–0·07. Mammal bone: $<0{\cdot}7$. Marine algae: 0·2–2. Marine fish: 1.
Accumulators: Composite plant *Grindelia fastigiata*? Essential to: All organisms. Natural compounds: A few Fe enzymes. Toxicity: 0·5–2 mg/l toxic to plants; 5 mg/day toxic, 50 mg/day lethal to rats. Functions: Catalytic (e.g. in N_2 fixation). Excess inhibits reduction of sulphates.
Reviews: In water and soil, Chappell (1975); in the environment, Chappell and Petersen (1976); in plants, Jones (1972); in animals, Underwood (1977); in enzymes, Moura and Xavier (1978).

N (Nitrogen) At. Wt 14·0067
Stable isotopes (fract. abund.) ^{14}N(0·99634), ^{15}N(0·00366)
Geosphere (mg/kg)
Granite: 21. Basalt: 30. Soil: 2000. Shale: 600. Limestone: 78. Sandstone: 120. Sea water: 0·64 + 15·5 (N_2). Freshwater: 0·05 (nitrate). Air: 973 g/m^3. Common minerals: Atmosphere; $NaNO_3$ (rare). Annual cycle, natural: 0·1–1·9. Mining: $44+84$ (combustion) $\times 10^{12}$ g/yr.

Biosphere (mg/kg DM)
Land plants: 12 000–38 000. Edible vegetables: 30 k–75 k. Mammal muscle: 72 000. Mammal bone: 43 000. Marine algae: 18 000–25 000. Marine fish: 80 000–96 000.
Accumulators: All organisms. Essential to: All organisms. Natural compounds: Proteins, amino acids, amines, purines, pyrimidines, nucleic acids etc. Toxicity: Harmless to plants; NH_3 is toxic to fish; nitrites are toxic to infants. Functions: Biochemical and metabolic.
Remarks: Natural isotopic fractionation occurs in some processes.
Reviews: N cycle, Burns and Hardy (1975); in soils, Bartholomew and Clark (1965), Hauck and Bremner (1976); N fixation, Hardy (1977).

Na (Sodium) At. Wt 22·9898
Stable isotopes (fract. abund.) ^{23}Na(1)
Natural radioisotopes ($t_{\frac{1}{2}}$) ^{22}Na(2·58 yr)
Geosphere (mg/kg)
Granite: 27 700. Basalt: 19 400. Soil: 5000. Shale: 5900. Limestone: 1300. Sandstone: 10 400. Sea water: 10 770. Freshwater: 6. Air: 7–7000 ng/m^3. Common minerals: NaCl, albite, nepheline. Annual cycle, natural: 92–220. Mining: 53×10^{12} g/yr.
Biosphere (mg/kg DM)
Land plants: 35–1500. Edible vegetables: 2–2400. Mammal muscle: 2600–7800. Mammal bone: 10 000. Marine algae: 26 000–41 000. Marine fish: 8000.
Accumulators: Halophytic plants. Essential to: some microbes and plants, and all animals. Toxicity: Not very toxic: 320 mg/day is lethal to rats. Functions: Electrochemical: catalytic (activates some enzymes).
Reviews: In enzymes, Suelter (1970), Hewitt and Smith (1975).

Nb (Niobium) At. Wt 92·904
Stable isotopes (fract. abund.) ^{93}Nb(1)
Natural radioisotopes ($t_{\frac{1}{2}}$) ^{95}Nb(35·1 days) f.p.
Geosphere (mg/kg)
Granite: 20. Basalt: 20. Soil: 10? Shale: 18. Limestone: 0·3. Sandstone: 0·05. Sea water: 0·00001. Air: 0·1 ng/m^3? Common minerals: Associated with pegmatites. Annual cycle, natural: 80. Mining: 13×10^9 g/yr.
Biosphere (mg/kg DM)
Land plants: 0·3? Mammal muscle: 0·14. Mammal bone: <0·07.
Accumulators: Ascidian *Molgula manhattensis*. Toxicity: Moderately toxic to mammals by intravenous injection.

Nd (Neodymium) At. Wt 144·24
Stable isotopes (fract. abund.) ^{142}Nd(0·271), ^{143}Nd(0·122), ^{144}Nd(0·238), ^{145}Nd(0·083), ^{146}Nd(0·172), ^{148}Nd(0·0573), ^{150}Nd(0·0562)
Natural radioisotopes ($t_{\frac{1}{2}}$) ^{144}Nd(2·4 × 10^{15} yr)
Geosphere (mg/kg)
Granite: 44. Basalt: 32. Soil: 35. Shale: 41. Limestone: 9. Sandstone: 38. Sea water: 0·0000028. Freshwater: 0·00015. Air: 0·3 ng/m^3? Common minerals: Monazite. Annual cycle, natural: 5–150 × 10^9 g/yr.
Biosphere (mg/kg DM)
Land plants: 0·3–7? Mammal bone: <0·28.
Accumulators: Tree *Carya* spp. Toxicity: low.

Ne (Neon) At. Wt 20·18
Stable isotopes (fract. abund.) ^{20}Ne(0·909), ^{21}Ne(0·00257), ^{22}Ne(0·0882)
Geosphere (mg/kg)
Granite: 0·00007. Basalt: 7 × 10^{-7}–6 × 10^{-5}. Shale: 0·00008–0·0002. Sea water: 0·00012. Air: 16 mg/m^3. Common minerals: None. Annual cycle, natural: None. Mining: ~1000 kg/yr.
Biosphere (mg/kg DM)
Toxicity: Negligible.

Ni (Nickel) At. Wt 58·71
Stable isotopes (fract. abund.) ^{58}Ni(0·679), ^{60}Ni(0·262), ^{61}Ni(0·0119), ^{62}Ni(0·0366), ^{64}Ni(0·0108)
Natural radioisotopes ($t_{\frac{1}{2}}$) ^{63}Ni(92 yr)
Geosphere (mg/kg)
Granite: 0·5? Basalt: 150. Soil: 50. Shale: 68. Limestone: 7. Sandstone: 9. Sea water: 0·00058. Freshwater: 0·0005. Air: <1–120 ng/m^3. Common minerals: NiS, Garnierite; associated with pyrrhotite (FeS). Annual cycle, natural: 19–320. Mining: 560 × 10^9 g/yr.
Biosphere (mg/kg DM)
Land plants: 1–5. Edible vegetables: 0·02–4. Mammal muscle: 1–2. Mammal bone: <0·7. Marine algae: 0·4–5. Marine fish: 0·1–4.
Accumulators: A few plants in Cruciferae, Flacourtiaceae, Loganiaceae, Rubiaceae, Sapotaceae, Thymelaeaceae, Tiliaceae and Violaceae: Poriferan *Dysidea*. Essential to: some microbes and plants and all vertebrates. Natural compounds: Urease; hydrogenase. Toxicity: 0·5–2 mg/l toxic to plants; 50 mg/day toxic to rats; carcinogenic. Functions: Catalytic: stabilises coiled ribosomes.
Remarks: $Ni(CO)_4$ is highly toxic industrial product.
Reviews: In foods, Schlettwein-Gsell and Mommsen-Straub (1971b), Thomas *et al.* (1974); in environment, Schroeder *et al.* (1962); in animals, Underwood (1977); in nutrition, Nielsen (1971).

Np (Neptunium)
Stable isotopes (fract. abund.) none
Natural radioisotopes ($t_{\frac{1}{2}}$) ^{239}Np(2·35 days)
Biosphere
Accumulators: Mammalian bone (ICRP, 1972).
Reviews: Foster (1963).

O (Oxygen) At. Wt 15·9994
Stable isotopes (fract. abund.) ^{16}O(0·99759), ^{17}O(0·000374), ^{18}O(0·002039)
Geosphere (mg/kg)
Granite: 492 000. Basalt: 467 000. Soil: 490 000. Shale: 483 000. Limestone: 494 000. Sandstone: 492 000. Sea water: 857 000. Freshwater: 889 000. Air: 299 g/m^3. Common minerals: Water, aluminosilicates, carbonates, O_2. Annual cycle, natural: 67. Mining: 23 (Combustion) $\times 10^{15}$ g/yr.
Biosphere (mg/kg DM)
Land plants: 400 000–440 000. Edible vegetables: 400 000. Mammal muscle: 160 000. Mammal bone: 285 000. Marine algae: 480 000. Marine fish: 290 000.
Essential to: All organisms. Natural compounds: Very numerous. Toxicity: Harmless as O_2 or H_2O, very toxic as O_3. Functions: Structural (in hard tissues), biochemical and metabolic: O_2 is oxidant.
Remarks: Natural isotopic fractionation occurs in some processes.

Os (Osmium) At. Wt 190·2
Stable isotopes (fract. abund.) ^{184}Os(0·00018), ^{186}Os(0·0159), ^{187}Os(0·0164), ^{188}Os(0·133), ^{189}Os(0·161), ^{190}Os(0·264), ^{192}Os(0·41)
Geosphere (mg/kg)
Basalt: 0·00001? Sandstone: 0·00005? Common minerals: None. OsIr is very rare.
Biosphere
Toxicity: OsO_4 is very toxic to organisms.
Reviews: Smith and Carson (1978).

P (Phosphorus) At. Wt 30·97376
Stable isotopes (fract. abund.) ^{31}P(1)
Natural radioisotopes ($t_{\frac{1}{2}}$) ^{32}P(14 days), ^{33}P(33 days)
Geosphere (mg/kg)
Granite: 700. Basalt: 1400. Soil: 800. Shale: 700. Limestone: 700. Sandstone: 440. Sea water: 0·06. Freshwater: 0·02. Air: $<$180–1100 ng/m^3. Common minerals: Apatite. Annual cycle, natural: 700–4000. Mining: 11 200 $\times 10^9$ g/yr.

Biosphere (mg/kg DM)
Land plants: 120–3000. Edible vegetables: 130–10 k. Mammal muscle: 3000–8500. Mammal bone: 62 k–71 k. Marine algae: 2800–4000. Marine fish: 18 000.
Accumulators: Mammalian bone. Essential to: All organisms. Natural compounds: Apatite; ATP; nucleic acids; organic phosphates. Toxicity: Phosphates are relatively harmless; P_4 and PH_3 are highly toxic to mammals and fish. Functions: Structural in bone: biochemical and metabolic. Natural 2-aminoethyl phosphonate has a C—P bond.
Reviews: Environmental phosphorus, Griffith *et al.* (1973); in soils, Dixon and Weed (1977); P cycle, Lerman *et al.* (1975); in polyphosphates, Janakidevi *et al.* (1965); in mammal metabolism, Nordin (1976); toxicity of P_4 to fish, Zitko *et al.* (1970); in bone, Williams and Da Silva (1978).

Pa (Protoactinium)
Stable isotopes (fract. abund.) none
Natural radioisotopes ($t_{\frac{1}{2}}$) ^{231}Pa(35 kyr), ^{234}Pa(6·7 h)
Geosphere (mg/kg) ^{231}Pa (566 Bq associated with 1 g Uranium at equilibrium)
Sea water: $\leqslant 5 \times 10^{-11}$.
Biosphere
Accumulators: Mammal bone.

Pb (Lead) At. Wt 207·2 (variable)
Stable isotopes (fract. abund.) ^{204}Pb(0·0148), ^{206}Pb(0·236), ^{207}Pb(0·226), ^{208}Pb(0·523)
Natural radioisotopes ($t_{\frac{1}{2}}$) ^{210}Pb(22 yr) and short-lived ^{211}Pb, ^{212}Pb, ^{214}Pb
Geosphere (mg/kg)
Granite: 24. Basalt: 3. Soil: 35 (contam.). Shale: 23. Limestone: 5·7. Sandstone: 10. Sea water: 0·00003. Freshwater: 0·003 (contam.). Air: 0·6–13 200 ng/m^3. Common minerals: PbS, $PbCO_3$, $PbSO_4$. Annual cycle, natural: 56–110. Mining: 3340 + 180 (combustion) $\times 10^9$ g/yr.
Biosphere (mg/kg DM)
Land plants: 1–13. Edible vegetables: 0·2–20. Mammal muscle: 0·23–3·3. Mammal bone: 3·6–30. Marine algae: 2–40. Marine fish: 0·001–15.
Accumulators: Lichen *Stereocaulon pileatum*; plants *Amorpha canescens* and *Minuartia verna*; mammalian bone. Natural compounds: $Pb(CH_3)_4$ (Schmidt and Huber, 1976). Toxicity: 3–20 mg/l toxic to plants; 1 mg/day toxic, 10 g/day lethal to Man. Functions: can stimulate microsomes. Inhibits δ-aminolaevulinate dehydrolyase in mammals (Lee, 1977).

Remarks: Crystals containing lead are precipitated by dictyosomes of *Zea mays* (Malone *et al.*, 1974).

Reviews: In food, Lehnert *et al.* (1969), MOAFF (1972); in environment, Bethea and Bethea (1975); in animals, Underwood (1977); in plants, Lagerwerff (1972); in Man, Prasad (1976); toxicity, Waldron and Stöfen (1974); Cycle, Van Hook *et al.* (1977), Nriagu (1978).

Pd (Palladium) At. Wt 106·4

Stable isotopes (fract. abund.) ^{102}Pd(0·0096), ^{104}Pd(0·11), ^{105}Pd(0·222), ^{106}Pd(0·273), ^{108}Pd(0·267), ^{110}Pd(0·118)

Geosphere (mg/kg)

Granite: 0·0005? Basalt: 0·0008? Air: 0·7 pg/m^3? Common minerals: None. Annual cycle, natural: 0·002. Mining: $0{\cdot}1 \times 10^9$ g/yr.

Biosphere (mg/kg DM)

Edible vegetables: 0·03? Mammal muscle: 0·002? (Koch and Roesmer, 1962).

Natural compounds: Methyl Pd compounds? Toxicity: Very toxic to fungi, and to mammals by intravenous injection.

Reviews: Smith and Carson (1978).

Pm (Promethium)

Stable isotopes (fract. abund.) None

Natural radioisotopes ($t_{\frac{1}{2}}$) ^{147}Pm(2·65 yr)

Biosphere

Accumulators: Mammalian bone and liver.

Remarks: Metabolism in mice, see Gensicke (1966).

Po (Polonium)

Stable isotopes (fract. abund.) None

Natural radioisotopes ($t_{\frac{1}{2}}$) ^{210}Po(138 days) and short-lived ^{212}Po, ^{214}Po, ^{216}Po, ^{218}Po

Geosphere ^{210}Po

Soil: 8–220 Bq/kg. Sea water: 0·93 Bq/m^3. Freshwater: 0·25–2·6 Bq/m^3. Air: $1{\cdot}5 \times 10^{-6}$ Bq/m^3.

Biosphere (Bq/kg DM) ^{210}Po

Land plants: 8–12. Edible vegetables: 0·04–110. Mammal muscle: 0·07–130. Mammal bone: 1–55 000. Marine algae: 4–170. Marine fish: 0·3–1000.

Accumulators: Mammalian bone and hair. Toxicity: Very toxic to mammals.

Reviews: In air, Moore *et al.* (1976); in sea, Cowen *et al.* (1977); in biota, Parfenov (1974).

Pr (Praseodymium) At. Wt 140·9077
Stable isotopes (fract. abund.) ^{141}Pr(1)
Geosphere (mg/kg)
Granite: 11. Basalt: 7·8. Soil: 3–12. Shale: 11. Limestone: 2·5. Sandstone: 11. Sea water: 0·00000064. Air: 0·06 ng/m^3? Common minerals: Monazite.
Biosphere (mg/kg DM)
Land plants: 0·06–0·3. Mammal bone: <0·08. Marine fish: 0·003?
Accumulators: Tree *Carya* spp. Toxicity: Low.

Pt (Platinum) At. Wt 195·09
Stable isotopes (fract. abund.) ^{190}Pt(0·000127), ^{192}Pt(0·0078), ^{194}Pt(0·329), ^{195}Pt(0·338), ^{196}Pt(0·253), ^{198}Pt(0·0721)
Natural radioisotopes ($t_{\frac{1}{2}}$) ^{190}Pt (600 Gyr)
Geosphere (mg/kg)
Granite: <0·0005? Basalt: 0·002? Shale: 0·0001? Air: <10 pg/m^3. Common minerals: None.
Biosphere (mg/kg DM)
Mammal muscle: 0·002? (Koch and Roesmer, 1962).
Toxicity: To plants, see Hamner (1942); moderately toxic to mammals by intravenous injection.
Review: Biological and medical effects of *cis*-dichloro Pt (II) complexes, see Williams and Da Silva (1978).

Pu (Plutonium)
Stable isotopes (fract. abund.) None
Natural radioisotopes ($t_{\frac{1}{2}}$) ^{238}Pu(86 yr), ^{239}Pu(24·4 kyr), ^{240}Pu(6·58 kyr), ^{241}Pu(13 yr)
Geosphere (Bq/kg) ^{239}Pu and ^{240}Pu
Soil: 0·05–1·4. Sea water: 0·0075–0·21. Freshwater: 0·02.
Biosphere (Bq/kg DM) ^{239}Pu and ^{240}Pu
Land plants: 0·006–0·02. Edible vegetables: 0·4–2·2. Mammal muscle: 0·006–0·09. Mammal bone: 0·003–0·03. Marine algae: 0·13–16. Marine fish: 0·0001–0·005.
Toxicity: Very toxic to mammals.
Reviews: In soil and ocean, Noshkin (1972); in Man, Langham (1972); mammalian metabolism, ICRP (1972); in plants and soils, McLendon *et al.* (1976); Cycle, Garten *et al.* (1978).

Ra (Radium) At. Wt 226·054
Stable isotopes (fract. abund.) None
Natural radioisotopes ($t_{\frac{1}{2}}$) ^{226}Ra(1622 yr), ^{228}Ra(6·7 yr)
Geosphere (mg/kg) ^{226}Ra
Granite: 0·7 μ? Basalt: 0·6 μ? Soil: 0·8 μ. Shale: 1·1 μ. Limestone: 0·4 μ. Sandstone: 0·7 μ. Sea water: 8·9 × 10^{-11}. Freshwater: 4 × 10^{-10}. Common minerals: Pitchblende.
Biosphere (mg/kg DM) ^{226}Ra
Land plants: 0·2 μ. Edible vegetables: 0·03 μ–1·6 μ. Mammal muscle: 0·00023 μ. Mammal bone: 0·004 μ. Marine algae: 0·03 μ–0·1 μ. Marine fish: 0·015 μ.
Accumulators: Tree *Bertholletia*; mammalian bone. Toxicity: Very toxic to mammals because of its radioactivity.
Remarks: $\mu = \times 10^{-6}$.
Reviews: In soil, Megumi and Mamuro (1977); in environment and biota, de Bortoli and Gaglione (1972).

Rb (Rubidium) At. Wt 85·468
Stable isotopes (fract. abund.) ^{85}Rb(0·7215), ^{87}Rb(0·2785)
Natural radioisotopes ($t_{\frac{1}{2}}$) ^{87}Rb(47 Gyr)
Geosphere (mg/kg)
Granite: 150. Basalt: 37. Soil: 150. Shale: 160. Limestone: 52? Sandstone: 150. Sea water: 0·12. Freshwater: 0·001. Air: 0·6–20 ng/m^3. Common minerals: None. Annual cycle, natural: 37–360 × 10^9 g/yr.
Biosphere (mg/kg DM)
Land plants: 2–50. Edible vegetables: 0·5–50. Mammal muscle: 20–70. Mammal bone: 0·1–5. Marine algae: 7–47. Marine fish: 0·8–6.
Toxicity: Scarcely toxic to plants; 10 mg/day is toxic to rats.
Remarks: Has been used to replace K in some bacteria, green algae, fungi and eggs of Echinoderms.
Reviews: In soils and plants, Aidinyan (1959), Yamagata *et al.* (1959); in animals, Underwood (1977); in Man, Yamagata (1962).

Re (Rhenium) At. Wt 186·2
Stable isotopes (fract. abund.) ^{185}Re(0·371), ^{187}Re(0·629)
Natural radioisotopes ($t_{\frac{1}{2}}$) ^{187}Re(60 Gyr)
Geosphere (mg/kg)
Granite: 0·00056. Basalt: 0·00025. Sea water: 0·000004. Common minerals: None; associated with MoS_2.
Biosphere (mg/kg DM)
Marine algae: 0·014–0·222. Marine fish: 0·0008.

Accumulators: Plant *Atriplex confertifolia*?; human thyroid? Toxicity: Re(VII) is relatively harmless to mammals.
Reviews: In sea, Scadden (1969); in plants, Myers and Hamilton (1961).

Rh (Rhodium) At. Wt 102·9055
Stable isotopes (fract. abund.) ^{103}Rh(1)
Natural radioisotopes ($t_{\frac{1}{2}}$) ^{102}Rh(210 days) has been used as atmospheric tracer
Geosphere (mg/kg)
Basalt: 0·0002? Air: $\leqslant$0·3 ng/m^3. Common minerals: None; associated with dunite/serpentine.
Reviews: Auerbach and Olson (1963); in medicine, Williams and Da Silva (1978).

Rn (Radon)
Stable isotopes (fract. abund.) None
Natural radioisotopes ($t_{\frac{1}{2}}$) ^{220}Rn(54 s), ^{222}Rn(3·82 days)
Geosphere (mg/kg) ^{222}Rn
Sea water: 6×10^{-16}. Freshwater: $1{\cdot}7 \times 10^{-15}$. Air: $(0{\cdot}04\text{–}20) \times 10^{-15}$ g/m^3. Common minerals: None.
Biosphere
Toxicity: Highly toxic to mammals.
Reviews: Megumi and Mamuro (1972, 1973).

Ru (Ruthenium) At. Wt 101·07
Stable isotopes (fract. abund.) ^{96}Ru(0·0551), ^{98}Ru(0·0187), ^{99}Ru(0·127), ^{100}Ru(0·126), ^{101}Ru(0·171), ^{102}Ru(0·316), ^{104}Ru(0·186)
Natural radioisotopes ($t_{\frac{1}{2}}$) ^{103}Ru(40 days), ^{106}Ru(1 yr) f.p.
Geosphere (mg/kg)
Basalt: 0·001? Sea water: 0·0000007. Air: $\leqslant$0·3 ng/m^3. Common minerals: None; RuS_2 occurs with some sulphide ores.
Biosphere (mg/kg DM)
Edible vegetables: 0·004? Marine algae: <0·001–0·002 (Koda, 1977).
Toxicity: RuO_4 is highly toxic to mammals.
Reviews: In water, Neuberger and Fourcy (1969); in plants, Auerbach and Olson (1963); in animals, Bruce and Carr (1961).

S (Sulphur) At. Wt 32·06
Stable isotopes (fract. abund.) ^{32}S(0·95), ^{33}S(0·0076), ^{34}S(0·0422), ^{36}S(0·00014)
Natural radioisotopes ($t_{\frac{1}{2}}$) ^{35}S(87 days)
Geosphere (mg/kg)
Granite: 270. Basalt: 250. Soil: 700. Shale: 2400. Limestone: 1100. Sandstone: 2300. Sea water: 905. Freshwater: 3·7. Air: 2·3–13 μg/m^3. Common minerals: S_8, FeS_2, $CaSO_4$ etc. Annual cycle, natural: 1–140. Mining: 52(+82 from combustion)$\times 10^{12}$ g/yr.
Biosphere (mg/kg DM)
Land plants: 1000–9000. Edible vegetables: 600–6000. Mammal muscle: 5000–11 000. Mammal bone: 500–2400. Marine algae: 8000–30 000. Marine fish: 4700.
Accumulators: Plants in Cruciferae and *Allium* spp.: vertebrate hair, feathers etc: sulphur bacteria. Essential to: All organisms. Natural compounds: Some amino acids, most proteins, sulphates, organo–S compounds. Toxicity: S_8 is highly toxic to bacteria and fungi, but harmless to green plants and mammals; H_2S is highly toxic to mammals; SO_2 is very toxic to lichens and plants; sulphates are scarcely toxic. Functions: Metabolic and biochemical; digestive fluid in ascidians and cell sap in some red algae is H_2SO_4.
Remarks: Natural isotopic fractionation occurs in many processses.
Reviews: In lakes, Stuiver (1967); in oil and environment, Meyer (1977); in bacteria, Zobell (1963); in plants, Kjaer (1963); in nutrition, Muth and Oldfield (1970), May *et al.* (1973); biochemistry and metabolism, Roy and Trudinger (1970), Greenberg (1975), Jocelyn (1972), Friedman (1973).

Sb (Antimony) At. Wt 121·75
Stable isotopes (fract. abund.) ^{121}Sb(0·572), ^{123}Sb(0·428)
Natural radioisotopes ($t_{\frac{1}{2}}$) ^{124}Sb(60 days), ^{125}Sb(2·7 yr) f.p.
Geosphere (mg/kg)
Granite: 0·2. Basalt: 0·2? Soil: 1. Shale: 1·5. Limestone: 0·3. Sandstone: 0·05. Sea water: 0·00024. Freshwater: 0·0002. Air: 0·0017–63 ng/m^3. Common minerals: Sb_2S_3. Annual cycle, natural: 0·8–7·4. Mining: 75 $\times 10^9$ g/yr.
Biosphere (mg/kg DM)
Land plants: 0·002–0·03. Edible vegetables: 0·0001–0·2. Mammal muscle: 0·0003–0·2. Mammal bone: 0·01–0·6. Marine algae: 0·02–0·9. Marine fish: 0·004–0·2.

Accumulator: seed of *Psoralea corylifolia* (Turel *et al.*, 1978).
Toxicity: 100 mg/day is toxic to Man; 11–75 mg/day is lethal to rats.
Reviews: In environment, Valente (1978).

Sc (Scandium) At. Wt 44·9559
Stable isotopes (fract. abund.) ^{45}Sc(1)
Natural radioisotopes ($t_{\frac{1}{2}}$) ^{46}Sc(84 days)
Geosphere (mg/kg)
Granite: 6·9. Basalt: 25. Soil: 7. Shale: 13. Limestone: 1. Sandstone: 1. Sea water: 0·0000006. Fresh water: 0·00001. Air: 0·00012–3 ng/m^3. Common minerals: Thortveitite ($Sc_2Si_2O_7$; rare). Annual cycle, natural: 0·4–64 × 10^9 g/yr.
Biosphere (mg/kg DM)
Land plants: 0·01–0·03. Edible vegetables: 0·00001–0·2. Mammal muscle: <0·0002. Mammal bone: 0·001–6. Marine algae: 0·07–0·45.
Toxicity: Low.
Reviews: Hutchinson (1943).

Se (Selenium) At. Wt 78·96
Stable isotopes (fract. abund.) ^{74}Se(0·0087), ^{76}Se(0·0902), 77Se(0·0758), ^{78}Se(0·235), ^{80}Se(0·498), ^{82}Se(0·0919)
Geosphere (mg/kg)
Granite: 0·05? Basalt: 0·05. Soil: 0·4. Shale: 0·5. Limestone: 0·03. Sandstone: <0·01. Sea water: 0·0002. Freshwater: 0·0002. Air: 0·0056–30 ng/m^3. Common minerals: Associated with sulphide ores. Annual cycle, natural: 0·2–7·4. Mining: 1 (and 3 from combustion) × 10^9 g/yr.
Biosphere (mg/kg DM)
Land plants: 0·03? Edible vegetables: 0·001–0·5. Mammal muscle: 0·42–1·9. Mammal bone: 1–9. Marine algae: 0·04–0·1. Marine fish: 0·17.
Accumulators: Fungus *Boletus edulis*: plants in Compositae, Lecythidaceae, Leguminosae (e.g. *Astragalus*) and Rubiaceae. Essential to: Some plants and microbes and all vertebrates. Natural compounds: Some amino acids, proteins, 1 plant wax. Toxicity: 1–2 mg Se (IV)/l is toxic to plants; 5 mg/day toxic to Man; 1–2 mg/day lethal to rats. Functions: Catalytic; biochemical and metabolic; possibly antibiotic; protects against Ag, Cd, Hg and Tl (Lee, 1977).
Remarks: Metabolised by *Micrococcus selenicus* etc. (Ehrlich, 1971).
Reviews: In fertilisers, Gissel-Nielsen (1971); in soils, Lewis (1976); in plants, Rosenfeld and Beath (1964); in foods, Cross *et al.* (1978b); in animals, Underwood (1977); in Man, Schroeder *et al.* (1970), Prasad (1976); in biochemistry, Shrift (1972), Zingaro and Cooper (1974); in medicine, Muth (1967).

Si (Silicon) At. Wt 28·086
Stable isotopes (fract. abund.) ^{28}Si(0·922), ^{29}Si(0·047), ^{30}Si(0·031)
Natural radioisotopes ($t_{\frac{1}{2}}$) ^{32}Si(710 yr)
Geosphere (mg/kg)
Granite: 336 000. Basalt: 240 000. Soil: 330 000. Shale: 275 000. Limestone: 32 000. Sandstone: 327 000. Sea water: 2·2 Freshwater: 7. Air: <0·1–63 μg/m^3. Common minerals: SiO_2, numerous silicates and aluminosilicates, clays. Annual cycle, natural: 260–1100. Mining: 380? $\times 10^{12}$ g/yr.
Biosphere (mg/kg DM)
Land plants: 400–60 000. Edible vegetables: 200–8000. Mammal muscle: 100–200. Mammal bone: 17? Marine algae: 1500? Marine fish: 70.
Accumulators: Most Cyperaceae, Gramineae and Juncaceae + *Moquilea* spp.; many Protozoa and Porifera; Oncidiid Mollusca; diatoms. Essential to: Diatoms, sponges and vertebrates; probably also for seed plants. Natural compounds: SiO_2 and esters of silicic acid; R_3SiCl in *Thuja* (Bendz and Lindqvist, 1978). Toxicity: Scarcely toxic, except by inhalation by mammals. Functions: Structural (hard tissues and cell walls) (Williams and Da Silva, 1978).
Reviews: In soils, Dixon and Weed (1977); in trees, Amos (1952); in animals, Underwood (1977); in biota, Voronkov *et al.* (1971).

Sm (Samarium) At. Wt 150·4
Stable isotopes (fract. abund.) ^{144}Sm(0·0309), ^{147}Sm(0·15), ^{148}Sm(0·112), ^{149}Sm(0·138), ^{150}Sm(0·0744), ^{152}Sm(0·267), ^{154}Sm(0·227)
Natural radioisotopes ($t_{\frac{1}{2}}$) ^{147}Sm($6{\cdot}7 \times 10^{11}$ yr), ^{151}Sm(93 yr) f.p.
Geosphere (mg/kg)
Granite: 8·5. Basalt: 7·3. Soil: 4·5. Shale: 7. Limestone: 2·1. Sandstone: 8·4. Sea water: 0·00000045. Freshwater: 0·00006? Air: 0·00006–1 ng/m^3. Common minerals: Monazite. Annual cycle, natural: 4–32 $\times 10^9$ g/yr.
Biosphere (mg/kg DM)
Land plants: 0·1–0·8. Edible vegetables: 0·01–0·1. Mammal bone: 0·23. Accumulators: Tree *Carya* spp. Toxicity: Low.

Sn (Tin) At. Wt 118·69
Stable isotopes (fract. abund.) ^{112}Sn(0·0096), ^{114}Sn(0·0066), ^{115}Sn(0·0035), ^{116}Sn(0·143), ^{117}Sn(0·0761), ^{118}Sn(0·24), ^{119}Sn(0·0858), ^{120}Sn(0·328), ^{122}Sn(0·0472), ^{124}Sn(0·0598)
Geosphere (mg/kg)
Granite: 3·5. Basalt: 1. Soil: 4. Shale: 6. Limestone: 0·5. Sandstone: 0·5. Sea water: 0·000004. Fresh water: 0·000009. Air: <1–800 ng/m^3. Common minerals: Cassiterite, SnO_2. Annual cycle, natural: 2–8·8. Mining: 178 $\times 10^9$ g/yr.

Biosphere (mg/kg DM)
Land plants: 0·2–2. Edible vegetables: 0·2–6·8. Mammal muscle: 0·009–2. Mammal bone: 1·4. Marine algae: 0·3?
Accumulators: Plant *Silene vulgaris*?; Poriferan *Terpios zeteki*. Probably essential to: Vertebrates. Natural compounds: Methyl tin compounds (Braman and Tompkins, 1979). Toxicity: Scarcely toxic to plants; 2 g/day toxic to Man.
Remarks: Triphenyl tin halides are fungicides.
Reviews: In plants, Sarosiek and Klys (1962), Peterson *et al.* (1976); in animals, Schwarz (1971), Underwood (1977); in Man, Schroeder *et al.* (1964).

Sr (Strontium) At. Wt 87·62
Stable isotopes (fract. abund.) ^{84}Sr(0·0056), ^{86}Sr(0·0986), ^{87}Sr(0·0702), ^{88}Sr(0·826)
Natural radioisotopes ($t_{\frac{1}{2}}$) ^{89}Sr(51 days), ^{90}Sr(29 yr) f.p.
Geosphere (mg/kg)
Granite: 285. Basalt: 465. Soil: 250. Shale: 300. Limestone: 610. Sandstone: 20. Sea water: 7·9. Freshwater: 0·07. Air: 2–50 ng/m^3. Common minerals: $SrCO_3$, $SrSO_4$. Annual cycle, natural: 1·5–2·6. Mining: 0·01? × 10^{12} g/yr.
Biosphere (mg/kg DM)
Land plants: 7–90. Edible vegetables: 3–400. Mammal muscle: 0·12–0·35. Mammal bone: 36–140. Marine algae: 60–1600. Marine fish: 0·2–20.
Accumulators: Protozoan *Acanthometra*: all brown algae: vertebrate bone. Possibly essential to: Mammals. Natural compounds: $SrSO_4$ (rare). Toxicity: Scarcely toxic to plants; 8 mg/day toxic, 900 mg/day lethal to rats.
Functions: Structural.
Remarks: Can be made to replace most Ca in some bacteria and fungi.
Reviews: In biota, Schroeder *et al.* (1972); in animals, Underwood (1977).

Ta (Tantalum) At. Wt 180·948
Stable isotopes (fract. abund.) ^{180}Ta(0·000123), ^{181}Ta(0·9999)
Geosphere (mg/kg)
Granite: 3·5? Basalt: 0·5. Soil: 2. Shale: 2. Limestone: 0·05. Sandstone: 0·05. Sea water: 0·000002. Freshwater: <0·000002. Air: 0·01 ng/m^3? Common minerals: Rare, in pegmatites. Annual cycle, natural: <0·07–8. Mining: 0·43 × 10^9 g/yr.
Biosphere (mg/kg DM)
Edible vegetables: <0·001. Mammal bone: 0·03.

Accumulators: The ascidian *Styela plicata*. Toxicity: 300 mg/day lethal to rats.
Remarks: 0·02–0·04 mg/kg DM in marine invertebrates.
Reviews: In marine biota, Kokubu and Hidaka (1965), Burton and Massie (1971). In plants, Renan *et al.* (1979).

Tb (Terbium) At. Wt 158·9254
Stable isotopes (fract. abund.) ^{159}Tb(1)
Geosphere (mg/kg)
Granite: 1. Basalt: 1·2. Soil: 0·7. Shale: 1. Limestone: 0·44. Sandstone: 2? Sea water: 0·00000014. Freshwater: 0·000003. Air: 0·01 ng/m^3? Common minerals: Xenotime. Annual cycle, natural: 0·1–4·4 × 10^9 g/yr.
Biosphere (mg/kg DM)
Land plants: 0·001–0·12. Edible vegetables: 0·001? Mammal bone: 0·05.
Accumulators: Tree *Carya* spp. Toxicity: Low.

Tc (Technetium)
Stable isotopes (fract. abund.) none
Natural radioisotopes ($t_{\frac{1}{2}}$) ^{99}Tc(210 kyr) f.p.
Biosphere
Accumulators: Mammalian thyroid?
Remarks: Human metabolism, see Shimmins *et al.* (1965); rat metabolism, see Papadopoulos *et al.* (1967); plant uptake, see Routson and Cataldo (1978).

Te (Tellurium) At. Wt 127·6
Stable isotopes (fract. abund.) ^{120}Te(0·00089), ^{122}Te(0·0246), ^{123}Te(0·0087), ^{124}Te(0·0461), ^{125}Te(0·0699), ^{126}Te(0·187), ^{128}Te(0·318), ^{130}Te(0·345)
Natural radioisotopes ($t_{\frac{1}{2}}$) ^{132}Te(78 h) f.p.
Geosphere (mg/kg)
Granite: 0·005? Basalt: 0·004? Shale: $<$0·1. Air: $\leqslant$0·2 ng/m^3. Common minerals: Rare; $AuTe_2$, Bi_2Te_3. Annual cycle, natural: 0·02? Mining: 0·22 × 10^9 g/yr.
Biosphere (mg/kg DM)
Edible vegetables: 0·01? Mammal muscle: 0·017?
Natural compounds: $Te(CH_3)_2$. Toxicity: 6 mg/l toxic to plants; 0·25 mg/day toxic, 1–9 mg/day lethal to rats.
Remarks: High values given in the literature are probably errors. 2–25 mg/kg in a few plants from Te-rich soils (Beath *et al.*, 1935).
Reviews: Bacterial metabolism, Silverman and Ehrlich (1964), Nagai (1965), Summers and Jacoby, (1977); rat metabolism, Hollins (1969); toxicology, Cerwenka and Cooper (1961).

Th (Thorium) At. Wt 232·0381
Stable isotopes (fract. abund.) none
Natural radioisotopes ($t_{\frac{1}{2}}$) ^{228}Th(1·9 yr), ^{230}Th(75 kyr), ^{232}Th(13·9 Gyr)
Geosphere (mg/kg) ^{232}Th
Granite: 23. Basalt: 1·6. Soil: 9. Shale: 12. Limestone: 1·7. Sandstone: 3·8. Sea water: 0·000001. Freshwater: 0·00003. Air: 0·00006–1 ng/m^3. Common minerals: Dispersed in pegmatites, zircon, monazite and sphene. Annual cycle, natural: 1–48. Mining: $0{\cdot}27 \times 10^9$ g/yr.
Biosphere (mg/kg DM) ^{232}Th
Land plants: 0·03–1·3. Edible vegetables: <0·008–0·2. Mammal bone: 0·002–0·012. Marine algae: 0·06–1·5. Marine fish: 0·32.
Accumulators: Mammal bone. Toxicity: Moderately toxic by intravenous injection of mammals; radioactive hazard.
Reviews: In marine algae, Strohal and Pinter (1973); mammal metabolism, ICRP (1972); in Man, Clifton *et al.* (1971).

Ti (Titanium) At. Wt 47·9
Stable isotopes (fract. abund.) ^{46}Ti(0·0793), ^{47}Ti(0·0728), ^{48}Ti(0·739), ^{49}Ti(0·0551), ^{50}Ti(0·0534)
Geosphere (mg/kg)
Granite: 2300. Basalt: 9000. Soil: 5000. Shale: 4600. Limestone: 300. Sandstone: 3500. Sea water: 0·001. Freshwater: 0·005. Air: 2–1000 ng/m^3. Common minerals: TiO_2, $FeTiO_3$, $CaTiO_3$, Sphene. Annual cycle, natural: 0·2–22. Mining: $1? \times 10^{12}$ g/yr.
Biosphere (mg/kg DM)
Land plants: 1·5–56. Edible vegetables: <0·02–3. Mammal muscle: 1–2. Marine algae: 3–40. Marine fish: 0·2.
Accumulators: Hard siliceous tissues. Toxicity: Relatively harmless.
Reviews: Underwood (1977); in soils, Dixon and Weed (1977).

Tl (Thallium) At. Wt 204·37
Stable isotopes (fract. abund.) ^{203}Tl(0·295), ^{205}Tl(0·705)
Natural radioisotopes ($t_{\frac{1}{2}}$) ^{208}Tl(3·1 min)
Geosphere (mg/kg)
Granite: 1·1. Basalt: 0·08. Soil: 0·2? Shale: 1·2. Limestone: 0·14. Sandstone: 0·36. Sea water: 0·000019. Air: 0·06–0·22 ng/m^3. Common minerals: Rare; dispersed in potash feldspar and pollucite. Annual cycle, natural: 2·4. Mining: $0{\cdot}03? \times 10^9$ g/yr.
Biosphere (mg/kg DM)
Edible vegetables: 0·03–0·3. Mammal muscle: 0·07. Mammal bone: 0·002. Marine fish: 0·08.

Accumulators: Mammalian hair. Natural compounds: Methyl Tl compounds? Toxicity: 1 mg/l toxic to plants; 600 mg/day lethal to Man. Reviews: As pollutant, Zitko (1975), Smith (1978).

Tm (Thulium) At. Wt 168·9342
Stable isotopes (fract. abund.) ^{169}Tm(1)
Geosphere (mg/kg)
Granite: 0·54. Basalt: 0·42. Soil: 0·6. Shale: 0·6. Limestone: 0·16. Sandstone: 1? Sea water: 0·00000017. Air: 0·004 ng/m^3? Common minerals: Xenotime.
Biosphere (mg/kg DM)
Land plants: 0·004–0·07. Edible vegetables: 0·004. Mammal bone: <0·08. Accumulators: Tree *Carya* spp. Toxicity: Low.

U (Uranium) At. Wt 238·029
Stable isotopes (fract. abund.) none
Natural radioisotopes ($t_{\frac{1}{2}}$) ^{234}U(0·000056: 248 kyr), ^{235}U(0·0072: 0·71 Gyr), ^{238}U(0·9927: 4·5 Gyr)
Geosphere (mg/kg)
Granite: 4·4. Basalt: 0·43. Soil: 2. Shale: 3·7. Limestone: 2·2. Sandstone: 0·45. Sea water: 0·0032. Freshwater: 0·0004. Air: 0·02 ng/m^3? Common minerals: Uraninite (rare). Annual cycle, natural: 10–15. Mining: 23 $\times 10^9$ g/yr.
Biosphere (mg/kg DM)
Land plants: 0·005–0·04. Edible vegetables: 0·01–0·06. Mammal muscle: 0·0009–0·003. Mammal bone: 0·00016–0·07. Marine algae: 0·4–0·9. Marine fish: 0·04–0·08.
Accumulators: Plants *Coprosma arborea* and *Uncinia leptostachya*?; some corals. Toxicity: 36 mg/day lethal to rats; additional radioactive hazard.
Reviews: In water and soil, Osmund and Cowant (1976); biogeochemistry, Kovalsky *et al.* (1968).

V (Vanadium) At. Wt 50·941
Stable isotopes (fract. abund.) ^{50}V(0·0024), ^{51}V(0·9976)
Natural radioisotopes ($t_{\frac{1}{2}}$) ^{50}V(6×10^{15} yr)
Geosphere (mg/kg)
Granite: 72. Basalt: 250. Soil: 90. Shale: 130. Limestone: 45. Sandstone: 20. Sea water: 0·0025. Freshwater: 0·0005. Air: 0·0015–2000 ng/m^3. Common minerals: VS_4, Calcium vanadate. Annual cycle, natural: 19–640. Mining: 17·6 (and 20 from combustion) $\times 10^9$ g/yr.
Biosphere (mg/kg DM)
Land plants: 0·001–0·5. Edible vegetables: 0·001–0·5. Mammal muscle:

0·002–0·02. Mammal bone: 0·003–0·03. Marine algae: 1–16? Marine fish: 0·03.
Accumulators: Fungus *Amanita muscaria*; plants *Astragalus confertiflorus*? and parsley; some Ascidians. Essential to: Some Ascidians, all vertebrates and some microbes and seed plants. Natural compounds: Amavanadine and Haemovanadin. Toxicity: 10–40 mg/l toxic to plants; 0·25 mg/day toxic, 2–4 mg/day lethal to rats. Functions: Antibiotic?; O_2 transport?
Associated with transferrin in blood (Sabbioni *et al.*, 1979).
Reviews: In soils and plants, Cannon (1963); in marine biota, Biggs and Swinehart (1976); in animals, Underwood (1977); in Man and foods, Byrne and Kosta (1978); biochemistry, Hopkins and Mohr (1971); toxicity, Hudson (1964).

W (Tungsten) At. Wt 183·85
Stable isotopes (fract. abund.) ^{180}W(0·00135), ^{182}W(0·264), ^{183}W(0·144), ^{184}W(0·306), ^{186}W(0·284)
Natural radioisotopes ($t_{\frac{1}{2}}$) ^{181}W(30 days), ^{185}W(74 days), ^{187}W(1 day)
Geosphere (mg/kg)
Granite: 1·5. Basalt: 0·36. Soil: 1·5? Shale: 1·9. Limestone: 0·56. Sandstone: 1·6. Sea water: 0·0001. Freshwater: 0·00003. Air: 0·016–18 ng/m^3. Common minerals: $FeWO_4$, $CaWO_4$. Annual cycle, natural: 1·1–5·2. Mining: $38{\cdot}4 \times 10^9$ g/yr.
Biosphere (mg/kg DM)
Land plants: 0·0005–0·15. Edible vegetables: 0·01–0·06. Mammal muscle: 0·005. Mammal bone: 0·00025. Marine algae: 0·035. Marine fish: 0·0014.
Accumulators: Gymnosperm *Pinus cembra*? Toxicity: >10 mg/l toxic to plants; 30–50 mg/day lethal to rats.
Reviews: Radioecology, Romney and Childress (1963); enzyme stimulation, Andreesen and Ljungdahl (1973).

Xe (Xenon) At. Wt 131·3
Stable isotopes (fract. abund.) ^{124}Xe(0·00096), ^{126}Xe(0·0009), ^{128}Xe(0·0192), ^{129}Xe(0·264), ^{130}Xe(0·0408), ^{131}Xe(0·212), ^{132}Xe(0·269), ^{134}Xe(0·104), ^{136}Xe(0·0887)
Geosphere (mg/kg)
Granite: 2 μ–8 μ. Basalt: 0·025 μ–1 μ. Shale: 0·0003–0·001.
Sea water: 0·00005. Air: 500 μg/m^3. Common minerals: None (Atmosphere). Annual cycle, natural: None. Mining: $\sim$600 kg/yr.
Biosphere (mg/kg DM)
Natural compounds: Binds to haemoglobin *in vivo*. Toxicity: Harmless to plants, anaesthetic to mammals (Cullen and Gross, 1951).

Remarks: Xe prevents sporulation in the fungus *Neurospora crassa* (Ne does not); Schreiner (1962).
Reviews: Protein complexes, Schoenborn *et al.* (1965); toxicity of Xe(VIII), Finkel *et al.* (1963).

Y (Yttrium) At. Wt 88·9059
Stable isotopes (fract. abund.) ^{89}Y(1)
Natural radioisotopes ($t_{\frac{1}{2}}$) ^{90}Y(64 h) f.p., ^{91}Y(51 days)
Geosphere (mg/kg)
Granite: 33. Basalt: 27. Soil: 40. Shale: 41. Limestone: 23. Sandstone: 54. Sea water: 0·000013. Air: 0·2–6 ng/m^3. Common minerals: Xenotime.
Biosphere (mg/kg DM)
Land plants: 0·15–0·77. Mammal muscle: 0·02. Mammal bone: 0·07.
Accumulators: Poriferan *Melithoea* spp.; tree *Carya* spp.; calcareous tissues.
Toxicity: Low.
Reviews: Rat metabolism, Magnusson (1963); in Man, Erämetsä *et al.* (1968).

Yb (Ytterbium) At. Wt 173·04
Stable isotopes (fract. abund.) ^{168}Yb(0·00135), ^{170}Yb(0·0303), ^{171}Yb(0·143), ^{172}Yb(0·218), ^{173}Yb(0·161), ^{174}Yb(0·318), ^{176}Yb(0·127)
Geosphere (mg/kg)
Granite: 3·3. Basalt: 3·4. Soil: 3? Shale: 3·9. Limestone: 1·6. Sandstone: 4·4. Sea water: 0·00000082. Freshwater: 0·00001. Air: 0·004–1 ng/m^3. Common minerals: Xenotime. Annual cycle, natural: 0·4–13 × 10^9 g/yr.
Biosphere (mg/kg DM)
Land plants: 0·02–0·6. Edible vegetables: 0·02? Mammal muscle: 0·00012? Mammal bone: <0·28.
Accumulators: Tree *Carya* spp. Toxicity: Low.

Zn (Zinc) At. Wt 65·38
Stable isotopes (fract. abund.) ^{64}Zn(0·489), ^{66}Zn(0·278), ^{67}Zn(0·0411), ^{68}Zn(0·186), ^{70}Zn(0·0062)
Natural radioisotopes ($t_{\frac{1}{2}}$) ^{65}Zn(245 days)
Geosphere (mg/kg)
Granite: 52. Basalt: 100. Soil: 90. Shale: 120. Limestone: 20. Sandstone: 30. Sea water: 0·0049. Freshwater: 0·015. Air: 0·03–16 000 ng/m^3. Common minerals: ZnS. Annual cycle, natural: 300–550. Mining: 4040 × 10^9 g/yr.
Biosphere (mg/kg DM)
Land plants: 20–400. Edible vegetables: 1–160. Mammal muscle: 240. Mammal bone: 75–170. Marine algae: 6–260. Marine fish: 9–80.

Accumulators: A few plants in Caryophyllaceae, Cruciferae, Plumbaginaceae and Violaceae; pollen; flagella; barnacle *Balanus*; molluscs (*Ostrea*, *Pecten*); snake venom; mammal eye and prostate gland. Essential to: All organisms. Natural compounds: Porphyrin?: >80 metalloproteins/enzymes. Toxicity: 60–400 mg/l toxic to plants; 150–600 mg/day toxic to Man. Functions: Catalytic (peptidases and anhydrases); stabilises coiled ribosomes.

Reviews: In plants, Jones (1972); in foods, Schlettwein-Gsell and Mommsen-Straub (1970a); in animals, Underwood (1977); in Man, Schroeder *et al.* (1967); in enzymes, Chlebowski and Coleman (1976), Ochiai (1977); metabolism, Prasad (1976); therapy, Portnoy and Molokhia (1974), Hambidge and Nichols (1978); Cycle, Van Hook *et al.* (1977).

Zr (Zirconium) At. Wt 91·22

Stable isotopes (fract. abund.) ^{90}Zr(0·515), ^{91}Zr(0·112), ^{92}Zr(0·171), ^{94}Zr(0·174), ^{96}Zr(0·028)

Natural radioisotopes ($t_{\frac{1}{2}}$) ^{95}Zr(65 days) f.p.

Geosphere (mg/kg)

Granite: 240. Basalt: 150. Soil: 400. Shale: 160. Limestone: 20. Sandstone: 220. Sea water: 0·00003. Freshwater: 0·0008. Air: <0·7–26 ng/m^3. Common minerals: ZrO_2, $ZrSiO_4$. Annual cycle, natural: 30–760. Mining: 255×10^9 g/yr.

Biosphere (mg/kg DM)

Land plants: 0·3–2. Mammal muscle: 0·08. Mammal bone: <0·1.

Toxicity: 250–700 mg/day lethal to rats.

Reviews: Schroeder and Balassa (1966b); Underwood (1977); Smith and Carson (1978); in soils, Dixon and Weed (1977).

References

Aaronson, S. (1970). "Experimental Microbial Ecology", Academic Press, London, New York

Abbott, D. C. and Tatton, J. O'G. (1970). *Pestic. Sci.* **1**, 99

Abdullah, M. I. and Royle, L. G. (1974). *J. Mar. Biol. Ass. U.K.* **54**, 581

Abdullah, M. I., Royle, L. G. and Morris, A. W. (1972). *Nature, Lond.* **235**, 158

Abedinzadeh, Z., Owlya, A. and Rafii, H. (1978). *Radiochem. Radioanal. Lett.* **32**, 315

Abelson, P. H. and Hoering, T. C. (1961). *Proc. Nat. Acad. Sci. Wash.* **47**, 623

Åberg, B. and Hungate, F. P. (eds) (1967). "Radioecological Concentration Processes", Pergamon Press Ltd., Oxford, New York

Abu-Erreish, G. M., Whitehead, E. I. and Olson, O. E. (1968). *Soil Sci.* **106**, 415

Addy, J. K. and Parker, R. E. (1963). *J. Chem. Soc.* 915

Adham, N. F., Song, M. K. and Rinderknecht, H. (1977). *Biochem. Biophys. Acta,* **495**, 212

Aebi, H. and Frei, E. (1957). *Helv. Chim Acta,* **40**, 1695

Ahl, T. and Jönsson, E. (1972). *Ambio,* **1**, 66

Ahrens, L. H. (ed) (1968). "Origin and Distribution of the Elements", Pergamon Press Ltd., Oxford, New York

Aidinyan, R. K. (1959). *Geochem. USSR,* **4**, 428

Aikawa, J. K. (1963). "The Role of Magnesium in Biologic Processes", C. C. Thomas, Springfield, Ill.

Aikawa, J. K. (1976). "Trace Elements in Health and Disease", Vol. 2, p. 47 (Prasad, A. S., ed), Academic Press, London, New York

Akcetin, S., Ayca, E. and Hoste, J. (1973). *Radiochem. Radioanal. Lett.* **15**, 13

Alexander, G. V. and Rowland, R. H. (1966). *Nature, Lond.* **210**, 155

Alexander, M. L. (1977). "Introduction to Soil Microbiology", 2nd Ed., John Wiley and Sons Ltd., London, New York

Allcroft, R., Burns, K. N. and Herbert, C. N. (1965). "Fluorosis in Cattle", Animal Disease Surv. Rept. no. 2, H.M. Stationery Office

Altman, P. L. and Dittmer, D. S. (1966). Environmental biology. Fed. Amer. Soc. Exper. Biol.

Altman, P. L. and Dittmer, D. S. (1972). Biology Data Book 1,459. Fed. Amer. Soc. Exper. Biol.

Aly, H. F., Herpers, U. and Herr, W. (1975). *J. Radioanal. Chem.* **26**, 279

Amos, G. L. (1952). *Comm. Sci. Indust. Res. Organ. Bull.* **267**, 1

Andelmann, J. B. (1973). "Trace Metals and Metal-organic Interactions in Natural Waters", p. 57 (Singer, P. C., ed), Ann Arbor Science, Ann Arbor, Mich.

Anderson, R. J., Wolfe, M. S. and Faulkner, D. J. (1974). *Mar. Biol.* **27**, 281

Anderson, A. T. (1974). *Bull. Geol. Soc. Amer.* **85**, 1485

Anderssen, A. (1967). *Grundfoerbattring*, **20**, 95

Andreae, M. O. (1977). *Analyt. Chem.* **49**, 820

Andreesen, J. R. and Ljungdahl, L. G. (1973). *J. Bact.* **116**, 867
Andren, A. W., Klein, D. H. and Talmi, Y. (1975). *Envir. Sci. Technol.* **9**, 856
Andrews, J. N. and Wood, D. F. (1974). *Health Phys.* **27**, 307
Anon (1968). THAI AEC-17, Atomic Energy for Peace Office, Bangkok, Thailand
Anspaugh, L. R., Robison, W. L., Martin, W. H. and Lowe, O. A. (1971). U.S. Atomic Energy Comm. Rept. UCRL 51013
Aras, N. K., Zoller, W. H., Gordon, G. E. and Lutz, G. J. (1973). *Analyt. Chem.* **45**, 1481
Aristovskaya, T. V. and Zavarzin, G. A. (1971). "Soil Biochemistry", Vol. 2, p. 385 (McLaren, A. D. and Skujins, J., eds), M. Dekker, New York
Arnon, D. I. (1950). *Lotsya*, **3**, 31
Arnon, D. I. and Stout, P. R. (1939). *Plant Physiol.* **14**, 599
Arnon, D. I. and Wessel, G. (1953). *Nature, Lond.* **172**, 1039
Arnott, H. J. (1973). "Biological Mineralization", p. 609 (Zipkin, I., ed), John Wiley and Sons Ltd., London, New York
Arosio, P., Adelman, T. G. and Drysdale, J. W. (1978). *J. Biol. Chem.* **253**, 4451
Arrhenius, G., Bramlette, M. N. and Picciotto, E. (1957). *Nature, Lond.* **180**, 85
Asher, C. J. and Ozanne, P. G. (1967). *Soil Sci.* **103**, 155
Atkinson, A., Hunt, S. and Lowe, A. G. (1968). *Biochim. Biophys. Acta*, **167**, 469
Attrep, M., Enochs, J. A. and Broz, L. D. (1971). *Envir. Sci. Technol.* **5**, 344
Aubert, H. and Pinta, M. (1977). "Trace Elements in Soils", Elsevier, Amsterdam, London, New York
Auerbach, S. I. and Olson, J. S. (1963). "Radioecology", p. 509 (Schultz, V. and Klement, A. W., eds), Reinhold, New York
Auerbach, S. I., Olson, J. S. and Waller, H. D. (1964). *Nature, Lond.* **201**, 761
Augier, J. and Mastagli, P. (1956). *C.R. Acad. Sci. Paris*, **242**, 190
Augood, D. R. (1954). *Indust. Chem. Manuf.* **30**, 585
Auld, D. S., Kawaguchi, H., Livingston, D. M. and Vallee, B. L. (1974). *Biochem. Biophys. Res. Comm.* **57**, 967
Ault, W. and Kulp, J. L. (1959). *Geochim. Cosmochim. Acta*, **16**, 201
Averitt, P. (1973). *U.S. Geol. Surv. Prof. Paper*, **820**, 133
Aviram, I., Wittenberg, B. A. and Wittenberg, J. B. (1978). *J. Biol. Chem.* **253**, 5685
Babior, B. M. (ed) (1975). "Cobalamin", John Wiley and Sons Ltd., London, New York
Babu, S. R. (ed) (1975). *Advances in Chem.* **141**, 1
Bacso, J., Kovacs, P. and Horvath, S. (1978). *Radiochem. Radioanal. Lett.* **33**, 273
Baker, P. F. (1972). *Progr. Biophys. Molec. Biol.* **24**, 177
Baker, P. F. (1976). *Sympos. Soc. Exp. Biol.* **30**, 67
Baker, P. F. (1977) *Sci. Progr.* **64**, 95
Baker, W. E. (1973). *Geochim. Cosmochim. Acta*, **37**, 269
Baker, W. E. (1978). *Geochim. Cosmochim. Acta*, **42**, 645
Baldwin, G. S., Galdes, A., Hill, R. A. O., Smith, B. E., Waley, S. G. and Abraham, E. P. (1978). *Biochem. J.* **175**, 441
Balevska, P. S. and Kasabova, T. A. (1971). *Dokl. Bolg. Akad. Nauk.* **24**, no. 8, 1133
Banwart, W. L. and Bremner, J. M. (1975). *J. Envir. Quality*, **4**, 363
Barghoorn, E. S. and Tyler, S. A. (1965). "Current Aspects of Exobiology", p. 93 (Mamikunian, G. and Briggs, M. H., eds), Pergamon Press Ltd., Oxford, New York
Barker, J. L. and Anders, E. (1968). *Geochim. Cosmochim. Acta*, **32**, 627

Barnes, R. S. and Schell, W. R. (1973). "Cycling and Control of Metals", p. 45, National Environment Research Center, Cincinnati
Barth, T. F. W. (1961). *Geochim. Cosmochim. Acta*, **23**, 1
Bartholomew, W. V. and Clark, F. E. (1965). "Soil Nitrogen", Madison, Wisc. Amer. Soc. Agron.
Bartlett, R. J. and Kimble, J. M. (1976). *J. Envir. Quality*, **5**, 379
Bate, L. C., Lindberg, S. E. and Andren, A. W. (1976). *J. Radioanal. Chem.* **32**, 125
Batley, G. E. and Florence, T. M. (1975). *J. Electroanal. Chem.* **61**, 205
Baumeister, W. (1958). "Encyclopedia of Plant Physiology", **4**, pp. 5 and 482 (Ruhland, W., ed), Springer-Verlag, Berlin, Heidelberg, New York
Baur, W. H. and Wlotzka, F. (1972). "Handbook of Geochemistry", Vol. II/3 (Wedepohl, K. H., ed), Springer-Verlag, Berlin, Heidelberg, New York
Bazell, R. J. (1971). *Science*, **173**, 130
Bear, F. E. (ed) (1964). "Chemistry of the Soil", 2nd Ed., Chapman and Hall, London
Beasley, T. M. and Held, E. E. (1969). *Science*, **164**, 1161
Beauford, W., Barber, J. and Barringer, A. R. (1975). *Nature, Lond.* **256**, 35
Beavington, F. and Cawse, P. A. (1978). *Sci. Total Envir.* **10**, 239
Beck, A. B. (1956). *Austral. J. Zool.* **4**, 1
Becker, V. J., Bennett, J. H. and Manuel, O. K. (1972). *Chem. Geol.* **9**, 133
Beem, K. M., Richardson, D. C. and Rajagopalan, K. V. (1977). *Biochem.* **16**, 1930
Beers, J. R. (1966). *Limnol. Oceanog.* **11**, 520
Beeson, K. C. (1941). *U.S. Dept. Agric. Misc. Publ.* **369**
Beliles, R. P. (1975). "Toxicology", p. 454 (Casarett, L. J. and Doull, J., eds), Macmillan, London
Bellisario, R. and Cormier, M. J. (1971). *Biochem. Biophys. Res. Comm.* **43**, 800
Belyayeva, N. V. (1973). *Oceanology*, **13**, 247
Bender, M. M., Rouhami, I., Vines, H. M. and Black, C. C. (1973). *Plant Physiol.* **52**, 427
Bendz, G. and Lindqvist, I. (eds) (1978). "Biochemistry of Silicon and Related Problems". Plenum Press, London, New York
Beneš, P., Giessing, E. T. and Steinnes, E. (1976). *Water Res.* **10**, 711
Benninger, L. K., Lewis, D. M. and Turekian, K. K. (1975). "Marine Chemistry in the Coastal Environment", p. 202 (Church, T. M., ed), Amer. Chem. Soc., Washington, DC
Berg, W., Johnels, A., Sjöstrand, B. and Westermark, T. (1966). *Oikos*, **17**, 71
Bergerioux, C. and Zikovsky, L. (1978). *J. Radioanal. Chem.* **45**, 176
Berner, R. A. (1971). *J. Geophys. Res.* **76**, 6597
Berrow, M. L. and Weber, J. (1972). *J. Sci. Food Agric.* **23**, 93
Bertine, K. K., Chan, L. H. and Turekian, K. K. (1970). *Geochim. Cosmochim. Acta*, **34**, 641
Bertine, K. K. and Goldberg, E. D. (1971). *Science*, **173**, 223
Bertine, K. K. and Goldberg, E. D. (1972). *Limnol. Oceanog.* **17**, 877
Bertine, K. K. and Turekian, K. K. (1973). *Geochim. Cosmochim. Acta*, **37**, 1415
Bertrand, D. (1950). *Bull. Amer. Mus. Nat. Hist.* **94**, 409
Bertrand, G. (1905). *C.R. Acad. Sci. Paris*, **124**, 1032
Bertrand, G. (1941). *Annls. Inst. Pasteur*, **81**, 121
Bethea, R. M. and Bethea, N. J. (1975). *Residue Rev.* **54**, 55
Bidstrup, P. L. (1964). "Toxicity of Mercury and its Compounds", Elsevier, Amsterdam, London, New York

Biersteker, K. and Evendijk, J. E. (1976). *Envir. Res.* **12**, 214
Bigeleisen, J. (1949). *J. Chem. Phys.* **17**, 675
Biggs, W. R. and Swinehart, J. H. (1976). "Metal Ions in Biological Systems", Vol. 6, p. 142 (Sigel, H., ed), M. Dekker, New York
Bishai, H. M. (1962). *Hydrobiologia*, **19**, 357
Black, W. A. P. and Mitchell, R. L. (1952). *J. Mar. Biol. Ass. U.K.* **30**, 575
Blanchard, R. L. (1967). "Radioecological Concentration Processes", p. 281 (Åberg, B. and Hungate, F. P., eds), Pergamon Press Ltd., Oxford, New York
Block, C. and Dams, R. (1973). *Analyt. Chim. Acta*, **68**, 11
Block, C. and Dams, R. (1974). *Analyt. Chim. Acta*, **71**, 53
Block, C. and Dams, R. (1975). *Envir. Sci. Technol.* **9**, 146
Blondin, G. A., Kessler, R. J. and Green, D. E. (1977). *Proc. Nat. Acad. Sci. Wash.* **74**, 3667
Blumberg, W. E. and Peisach, J. (1965). *J. Biol. Chem.* **240**, 870
Boeri, E. (1963). "Comprehensive Biochemistry", Vol. 8, p. 38 (Florkin, M. and Stotz, E. H., eds), Elsevier, Amsterdam, London, New York
Bolin, B. (1970). *Sci. Amer.* **223**, no. 3, 125
Bokhoven, C. and Theeuwen, H. H. J. (1956). *Koninkl. Ned. Akad. Wetenschap. Proc. Ser.* **B59**, 78
Bolter, E., Turekian, K. K. and Schutz, D. F. (1964). *Geochim. Cosmochim. Acta*, **28**, 1459
Borg, D. C. and Cotzias, G. C. (1958). *Nature, Lond.* **182**, 1677
Borle, A. B. (1975). "Calcium Transport in Contraction and Secretion", p. 77 (Carafoli, E., Clementi, F., Drabikowski, W. and Margrett, A., eds), N. Holland, Amsterdam
Borle, A. B. and Anderson, J. H. (1976). *Sympos. Soc. Exp. Biol.* **30**, 141
Borowitzka, M. A. (1977). *Oceanog. Mar. Biol. Annu. Rev.* **15**, 189
Bortels, H. (1930). *Arch. Mikrobiol.* **1**, 333
Böstrom, K., Peterson, M. N. A., Joensuu, O. I. and Fisher, D. E. (1969). *J. Geophys. Res.* **74**, 3261
Boussingault, J. B. (1851). "Die Landwirtschaft in ihren Beziehungen zur Chemie, Physik und Meteorologie", C. Graeger
Bowden, F. L. (1975). "Techniques and Topics in Bio-Inorganic Chemistry", p. 207 (McAuliffe, C. A., ed), Macmillan, London
Bowen, H. J. M. (1956a). *J. Mar. Biol. Ass. U.K.* **35**, 451
Bowen, H. J. M. (1956b). *J. Nucl. Energy*, **2**, 255
Bowen, H. J. M. (1960). *Biochem. J.* **77**, 79
Bowen, H. J. M. (1966). "Trace Elements in Biochemistry", Academic Press, London, New York
Bowen, H. J. M. (1972). *Analyst*, **97**, 1003
Bowen, H. J. M. (1974). *J. Radioanal. Chem.* **19**, 215
Bowen, H. J. M. (1975a). *Atomic Energy Rev.* **13**, 451
Bowen, H. J. M. (1975b). *Education in Chem.* **12**, 72
Bowen, H. J. M. (1977a). "International Conference on Heavy Metals in the Environment", Vol. 1, p. 1 (Hutchinson, T. C., ed), University of Toronto, Canada
Bowen, H. J. M. (1977b). "The Chemical Environment", p. 1 (Lenihan, J. M. A. and Fletcher, W. W., eds), Blackie, Glasgow
Bowen, H. J. M. and Cawse, P. A. (1962). *Radiat. Bot.* **1**, 215
Bowen, H. J. M. and Dymond, J. A. (1955). *Proc. Roy. Soc.* **B144**, 355

Bowen, R. (1966). "Paleotemperature Analysis", Elsevier, Amsterdam, London, New York

Bowen, V. T. and Sutton, D. (1951). *J. Mar. Res.* **10**, 153

Bowling, D. J. F. (1968). *J. Exp. Bot.* **19**, 381

Boyd, E. M. (1975). "Toxicity of Pure Foods", Chem. Rubber Co., Cleveland, Ohio

Boyden, C. R. (1975). *Mar. Pollut. Bull.* **6**, 180

Boyle, E. A., Collier, R., Dengler, A. T., Edmond, J. M., Ng, A. C. and Stallard, R. F. (1974). *Geochim. Cosmochim. Acta*, **38**, 1719

Boyle, E. A., Edmond, J. M. and Sholkevitz, E. R. (1977). *Geochim. Cosmochim. Acta*, **41**, 1313

Boyle, E. A., Sclater, F. R. and Edmond, J. M. (1977). *Earth Planet Sci. Lett.* **37**, 38

Boyle, R. W. (1968). *Geol. Surv. Canad. Bull.* **160**

Boyle, R. W. and Lynch, J. J. (1968). *Econ. Geol.* **63**, 421

Braman, R. S. and Forebach, C. C. (1973). *Science*, **182**, 1247

Braman, R. S. and Tompkins, M. A. (1979). *Analyt. Chem.* **51**, 12

Brar, S. S., Nelson, D. M., Kline, J. R. and Gustafson, P. F. (1970). *J. Geophys. Res.* **75**, 2939

Brätter, P., Möller, P. and Rösick, U. (1971). "Activation Analysis in Geochemistry and Cosmochemistry", p. 461 (Brunfelt, A. O. and Steinnes, E., eds), Universitetsforlaget, Oslo

Brätter, P., Gawlik, D., Lausch, J. and Rösick, U. (1977). *J. Radioanal. Chem.* **37**, 393

Breeze, V. G. (1973). *J. Appl. Ecol.* **10**, 513

Bressan, D. J., Carr, R. A., Hannan, P. J. and Wilkniss, P. E. (1974). *J. Radioanal. Chem.* **19**, 373

Brewer, P. G. (1975). "Chemical Oceanography", p. 415 (Riley, J. P. and Skirrow, G., eds), Academic Press, London, New York

Briggs, F. N. and Solaro, R. J. (1976). "Metal Ions in Biological Systems", Vol. 6, p. 323 (Sigel, H., ed), M. Dekker, New York

Brock, T. D., Brock, K. M., Belly, R. T. and Weiss, R. L. (1972). *Arch. Mikrobiol.* **84**, 54

Broecker, W. S. (1974). "Chemical Oceanography", Harcourt Brace Jovanovich, New York

Broecker, W. S., Goddard, J. and Sarmiento, J. L. (1976). *Earth Planet Sci. Lett.* **32**, 220

Brooks, R. R. (1972) "Geobotany and Biochemistry in Mineral Exploration", Harper and Row, London, New York

Brooks, R. R., Lee, J., Reeves, R. D. and Jaffre, T. (1977a). *J. Geochem. Explor.* **7**, 49

Brooks, R. R., McCleave, J. A., and Malaisse, F. (1977b). *Proc. Roy. Soc.* **B197**, 231

Brooks, R. R. and Wither, E. D. (1977). *J. Geochem. Explor.* **7**, 295

Brostrom, C. O., Brostrom, M. A. and Wolff, D. J. (1977). *J. Biol. Chem.* **252**, 5677

Brown, E. B., Aisen, P., Fielding, J. and Crichton, R. R. (1977). "Proteins of Iron Metabolism", Grune and Stratton, New York

Brown, R. and Vossen, P. G. T. (1970). *Analyt. Chem.* **42**, 1820

Brownell, P. F. and Wood, J. G. (1957). *Nature, Lond.* **179**, 635

Browning, E. M. (1969). "Toxicity of Industrial Metals", 2nd Ed., Butterworth, London

Broyer, T. C., Carlton, A. B., Johnson, C. M. and Stout, P. R. (1954). *Plant Physiol.* **29**, 526

Broyer, T. C., Lee, D. C. and Asher, C. J. (1966). *Plant Physiol.* **41**, 1425
Bruce, R. S. and Carr, T. E. F. (1961). *Reactor Sci. Technol.* **14**, 9 and 145
Bruland, K. W., Franks, R. P., Knauer, G. A. and Martin, J. H. (1979). *Analyt. Chim. Acta* **105**, 233
Brunfelt, A. O., Roelandts, I. and Steinnes, E. (1974). *Analyt. Chim. Acta*, **99**, 277
Bryan, G. W. (1973). *J. Mar. Biol. Ass. U.K.* **53**, 145
Bryan, G. W. (1976). "Marine Pollution", p. 185 (Johnston, R., ed), Academic Press, London, New York
Bryan, G. W. and Hummerstone, L. G. (1973). *J. Mar. Biol. Ass. U.K.* **53**, 705
Bryan, W. G., Potts, G. W. and Forster, G. R. (1977). *J. Mar. Biol. Ass. U.K.* **57**, 379
Buchauer, M. J. (1973). *Envir. Sci. Technol.* **7**, 131
Buffoni, F. and Blaschko, H. (1964). *Proc. Roy. Soc.* **B161**, 153
Bunker, C. M. and Bush, C. A. (1968). *U.S. Geol. Surv. Prof. Paper*, **600B**, 71
Burgermeister, W. and Winkler-Oswatitsch, R. (1977). *Topics Curr. Chem.* **69**, 91
Burk, R. F. (1976). "Trace Elements in Human Health and Disease", p. 105 (Prasad, A. S., ed), Academic Press, London, New York
Burkholder, P. R., Pfister, R. M. and Leitz, F. M. (1966). *Appl. Microbiol.* **14**, 649
Burns, R. C. and Hardy, R. W. F. (1975). "Nitrogen Fixation in Bacteria and Higher Plants", Springer Verlag, Berlin, Heidelberg, New York
Burreson, B. J., Moore, R. E. and Roller, P. P. (1976). *J. Agric. Food Chem.* **24**, 856
Burrows, W. D. (1977). *CRC Crit. Rev. Envir. Control* **7**, 167
Burton, J. D. (1975). "Chemical Oceanography", Vol. 3, p. 91 (Riley, J. P. and Skirrow, G., eds), Academic Press, London, New York
Burton, J. D. and Leatherland, T. M. (1971) *Nature, Lond.* **231**, 440
Burton, J. D. and Liss, P. S. (1976). "Estuarine Chemistry", Academic Press, London, New York
Burton, J. D. and Massie, K. S. (1971). *J. Mar. Biol. Ass. U.K.* **51**, 679
Butcher, R. W. (1946). *J. Inst. Sewage Purif.* **2**, 92
Bütschli, O. (1908). *Deuts. Südpol. Exped. (1901–3)*, **9**, 239
Byrne, A. R. and Kosta, L. (1978). *Sci. Total Envir.* **10**, 17
Byrne, A. R., Ravnik, V. and Kosta, L. (1976). *Sci. Total Envir.* **6**, 65
Cadle, R. D. (1973). "Chemistry of the Lower Atmosphere", p. 69 (Rasool, S. I., ed), Plenum Press, New York, London
Cadle, R. D., Wartburg, A. F., Pollock, W. H., Gandrud, B. W. and Shedlovsky, J. P. (1973). *Chemosphere*, **2**, 231
Calder, J. A. and Parker, P. L. (1973). *Geochim. Cosmochim. Acta*, **37**, 133
Cali, J. P. (1976). National Bureau of Standards certificates of analysis of SRM 1570, 1571, 1573, 1575
Campbell, C. A., Paul, E. A., Rennie, D. A. and McCallum, K. J. (1967). *Soil Sci.* **104,** 217
Canalas, R. E., Alexander, E. C. and Manuel, O. K. (1968). *J. Geophys. Res.* **73**, 3331
Cannon, H. L. (1960). *Science*, **132**, 159
Cannon, H. L. (1963). *Soil Sci.* **96**, 196
Cannon, H. L., Shacklette, H. T. and Bastron, H. (1968). *U.S. Geol. Surv. Bull.* **1278-A**
Cantor, L. M. (1967). "A World Geography of Irrigation", Oliver and Boyd, Edinburgh

Carithers, R. P., Yoch, D. C. and Arnon, D. I. (1977). *J. Biol. Chem.* **252,** 7453
Carlbom, L., Nielsen, N. and Skjöldebrand, R. (1956). *Nature, Lond.* **177**, 988
Carlisle, D. B. (1958). *Nature, Lond.* **181**, 933
Carlisle, D. B. (1968). *Proc. Roy. Soc.* **B171**, 31
Carlisle, E. M. (1972). *Fed. Proc.* **31**, 700
Carlisle, E. M. (1974). *Fed. Proc.* **33**, 704
Carpenter, R. (1969). *Geochim. Cosmochim. Acta*, **33**, 1153
Carroll, K. G., Spinelli, F. R. and Goyer, R. A. (1970). *Nature, Lond.* **227**, 1056
Carroll, M. D. and Loneregan, J. F. (1968). *Austral. J. Agric. Res.* **19**, 859
Casagrande, D. J. and Erchall, L. D. (1977). *Geochim. Cosmochim. Acta*, **41**, 1391
Cawse, P. A. (1974). U.K. Atomic Energy Authority Report AERE-R 7669. H.M. Stationery Office, London
Cawse, P. A. (1976). U.K. Atomic Energy Authority Report AERE-R 8398. H.M. Stationery Office, London
Cawse, P. A. (1977). U.K. Atomic Energy Authority Report AERE-R 8869. H.M. Stationery Office, London
Caygill, C. J. P., Lucy, J. A. and Diplock, A. T. (1971). *Biochem. J.* **125**, 407
CEC (1978). "Commission of European Communities: Criteria (Dose/Effect Relationships) for Cadmium", Pergamon Press Ltd., Oxford, New York
Cecilioni, V. A. (1972). *Fluoride*, **5**, 172
Cerbón, J. (1970). *J. Bact.* **102**, 87
Cerwenka, E. A. and Cooper, W. C. (1961). *Arch. Envir. Health*, **3**, 189
Chambers, L. A., Trudinger, P. A., Smith, J. W. and Burns, M. S. (1976). *Geochim. Cosmochim. Acta*, **40**, 1191
Chameides, W. L. and Stedman, D. H. (1976). *Envir. Sci. Technol.* **10**, 150
Chang, C. W. and Thompson, C. R. (1965). *Int. J. Air Water Pollut.* **9**, 685
Chang, H. W. and Neumann, E. (1976). *Proc. Nat. Acad. Sci. Wash.* **73**, 3364
Chapman, H. D. (ed) (1966). "Diagnostic Criteria for Plants and Soils", Univ. California; Div. Agric. Sci.
Chappell, W. R. (1975). "Heavy Metals in the Aquatic Environment", p. 167 (Krenkel, P. A., ed), Pergamon Press Ltd., Oxford, New York
Chappell, W. R. and Petersen, K. K. (eds) (1976). "Molybdenum in the Environment", M. Dekker, New York
Chatin, A. (1852). *C.R. Acad. Sci. Paris*, **34**, 14 and 51
Chattopadhyay, A. and Jervis, R. E. (1974). *Analyt. Chem.* **46**, 1630
Chen, Y., Senesi, N. and Schnitzer, M. (1978). *Geoderma*, **20**, 87
Cheng, H. H., Bremner, J. P. and Edwards, A. P. (1964). *Science*, **146**, 1574
Cherdyntsev, V. V. and Mikhailov, V. F. (1963). *Geochem. USSR*, **1**
Cherry, R. D. and Shannon, L. V. (1974). *Atomic Energy Rev.* **12**, 3
Chester, R. and Aston, S. R. (1976). "Chemical Oceanography", Vol. 6, p. 281 (Riley, J. P. and Skirrow, G., eds), Academic Press, London, New York
Chester, R., Gardner, D., Riley, J. P. and Stoner, J. (1973). *Mar. Pollut. Bull.* **4**, 28
Chester, R. and Stoner, J. H. (1973). *Nature, Lond.* **246**, 138
Chichilo, P. and Whittaker, C. W. (1961). *Agron. J.* **53**, 139
Chipman, W. A. (1966). "Radioactivity and Human Diet", p. 419 (Russell, R. S., ed), Pergamon Press Ltd., Oxford, New York
Chipman, W. A. and Thommeret, J. (1970). *Bull. Inst. Océanog. Monaco*, **69**, no. 1402
Chlebowski, J. F. and Coleman, J. E. (1976). "Metal Ions in Biological Systems", Vol. 6, p. 1 (Sigel, H., ed), M. Dekker, New York

Chow, T. J. and Goldberg, E. D. (1960). *Geochim. Cosmochim Acta*, **20**, 192
Chow, T. J. and Patterson, C. C. (1966). *Earth Planet. Sci. Lett.* **1**, 397
Christensen, H. E. (1972). U.S. Dept. Commerce Report NTIS PB-213 613
Chuang, L. S. and Emery, J. F. (1978). *J. Radioanal. Chem.* **45**, 169
Clark, M. C. and Swaine, D. J. (1962). "Trace Elements in Coal", CSIRO Tech. Comm. 45, Chatswood, N.S.W., Australia
Clemente, G. F. (1976). *J. Radioanal. Chem.* **32**, 25
Cleveland, J. M. and Rees, T. F. (1976). *Envir. Sci. Technol.* **10**, 802
Clifton, R. J., Farrow, M. and Hamilton, E. I. (1971). *Annls. Occup. Hyg.* **14**, 303
Cloud, P. E., Friedman, I., Sisler, F. D. and Dibeler, V. H. (1958). *Science*, **127**, 1394
Colbourn, P., Alloway, B. J. and Thornton, I. (1975). *Sci. Total Envir.* **4**, 359
Cole, J. A. (1976). *Adv. Microbiol. Physiol.* **14**, 1
Cole, M. M. (1973). *J. Appl. Ecol.* **10**, 269
Collander, R. (1941). *Plant Physiol.* **16**, 691
Colman, P. M., Freeman, H. C., Guss, J. M., Murata, M., Norris, V. A., Ramshaw, J. A. M. and Venkatappa, M. P. (1978). *Nature, Lond.* **272,** 319
Colman, P. M., Jansonius, J. N. and Matthews, B. W. (1972). *J. Molec. Biol.* **70**, 701
Comar, C. L. and Bronner, F. (eds) (1962). "Mineral metabolism", Academic Press, London, New York
Cone, J. E., del Rio, R. M., Davis, J. N. and Stadtman, T. C. (1976). *Proc. Nat. Acad. Sci. Wash.* **73**, 2659
Connor, J. J. and Shacklette, H. T. (1975). *U.S. Geol. Surv. Prof. Paper* **574-F**
Cooke, G. W. (1975). "Fertilizing for Maximum Yield", Crosby Lockwood, St. Albans
Cormier, M. J., Hori, K. and Anderson, J. M. (1974). *Biochim. Biophys. Acta*, **346**, 137
Cornelis, R., Speecke, A. and Hoste, J. (1975). *Analyt. Chim. Acta.* **78**, 317
Corner, E. D. S. and Davies, A. G. (1971). *Adv. Mar. Biol.* **9**, 101
Cotzias, G. C., Papavasiliou, P. S., Hughes, E. R., Tang, L. and Borg, D. C. (1968). *J. Clin. Invest.* **47**, 992
Cowen, J. P., Hodge, V. F. and Folsom, T. R. (1977). *Analyt. Chem.* **49**, 495
Cowgill, U. M. (1972). *Develop. Appl. Spectrosc.* **10**, 331
Cowgill, U. M. (1973). *Sci. Total Envir.* **2**, 259
Cox, D. P. and Alexander, M. (1973). *Bull. Envir. Contam. Toxicol.* **9**, 84
Crafts, A. S. (1961). "Translocation in Plants", Holt, Rinehart and Winston, New York
Craig, H. (1953). *Geochim. Cosmochim. Acta*, **3**, 53
Craigie, J. S. and Gruenig, D. E. (1967). *Science*, **157**, 1058
Cranston, R. E. and Murray, J. W. (1978). *Analyt. Chim. Acta*, **99**, 275
Crecelius, E. A. (1975). *Limnol. Oceanog.* **20**, 441
Crecelius, E. A., Johnson, C. J. and Hofer, G. C. (1974). *Water Air Soil Pollut.* **3**, 337
Creek, R. D., Parker, H. E., Hauge, S. M., Andrews, E. N. and Carrick, C. W. (1954). *Poultry Sci.* **33**, 1052
Crocket, J. H., Macdougall, J. D. and Harriss, R. C. (1973). *Geochim. Cosmochim. Acta*, **37**, 2547
Cross, J. D., Dale, I. M., Smith, H. and Smith, L. B. (1978a). *Radiochem. Radioanal. Lett.* **35**, 281
Cross, J. D., Raie, R. M., Smith, H. and Smith, L. B. (1978b). *Radiochem. Radioanal. Lett.* **35**, 291

Cross, J. D. and Smith, H. (1978). *Forensic Sci.* **11**, 147
Crutzen, P. J. (1970). *Quart. J. Roy. Meteorol. Soc.* **96**, 320
Crutzen, P. J. (1974). *Ambio*, **3**, 201
Csongor, E. (1973). *Acta Phys.* **34**, 249
Cullen, S. and Gross, E. (1951). *Science*, **113**, 580
Cutter, G. A. (1978). *Analyt. Chim. Acta*, **98**, 59
Dahl, J. L. and Hokin, L. E. (1974). *Annu. Rev. Biochem.* **43**, 327
Dale, I. M., Duncan, H. J. and McDonald, C. (1973). *Radiochem. Radioanal. Lett.* **15**, 77
Dall'Aglio, M. (1968). "Origin and Distribution of the Elements", p. 1065 (Ahrens, L. H., ed), Pergamon Press Ltd., Oxford, New York
Daly, R. A. (1968). "Igneous Rocks and the Depths of the Earth", Harper and Row, New York, London
Dams, R., Billiet, J., Block, C., Demuunck, M. and Janssens, M. (1975). *Atmos. Envir.* **9**, 1099
Dams, R., Robbins, J. A., Rahn, K. A. and Winchester, J. W. (1970). *Analyt. Chem.* **42**, 861
Dantas, C. C. and Ruf, H. (1975). *Radiochim. Acta* **22**, 192
Darley, W. M. (1974a). "Algal Physiology and Biochemistry", p. 636 (Stewart, W. D. P., ed), Blackwell, Oxford
Darley, W. M. (1974b). *Bot. Monog.* **10**, 655
David, O. J. (1974). *Envir, Health Perspec.* **7**, 17
Davies, B. E. (1976). *Geoderma*, **16**, 183
Davies, I. J. T. (1972). "The Clinical Significance of the Essential Biological Metals", Heinemann, London
Davies, R. E., Reid, B. L., Kurnich, A. A. and Couch, J. R. (1960). *J. Nutrit.* **70**, 193
de Bortoli, M. C. and Gaglione, P. (1969). *Health Phys.* **17**, 701
de Bortoli, M. C. and Gaglione, P. (1972). *Health Phys.* **22**, 43
Deevey, E. S. (1970). *Sci. Amer.* **223**, no. 3, 149
Deevey, E. S., Stuiver, M. and Nakai, N. (1963). "Radioecology", p. 471 (Schultz, V. and Klement, A. W., eds), Chapman and Hall, London
De Filippi, L. J. and Hultquist, D. E. (1978). *J. Biol. Chem.* **253**, 2940 and 2946
Degens, E. T. (1965). "Geochemistry of Sediments", Prentice Hall, New Jersey
Degens, E. T., Guillard, R. R. L. and Hellebust, J. A. (1968). *Deep Sea Res.* **15**, 1
Degens, E. T. and Ross, D. A. (eds) (1969). "Hot Springs and Recent Heavy Metal Deposition in the Red Sea", Springer-Verlag, Berlin, Heidelberg, New York
De Groot, A. J. and Allersma, E. (1975). "Heavy Metals in the Aquatic Environment", p. 85 (Krenkel, P. A., ed), Pergamon Press Ltd., Oxford, New York
de Haan, F. A. M. and Zwerman, P. J. (1976). "Soil Chemistry", p. 192 (Bolt, G. H. and Bruggenwert, M. G. M., eds), Elsevier, Amsterdam, London, New York
Deleens, E. and Garnier-Dardart, J. (1977). *Planta*, **135**, 241
Delwiche, C. C. (1970). *Sci. Amer.* **223**, no. 3, 137
Delwiche, C. C. and Steyn, P. L. (1970). *Envir. Sci. Technol.* **4**, 929
Denaeyer-De Smet, S. (1973). *Bull. Soc. Roy. Bot. Belg.* **106**, 149
De Niro, M. J. and Epstein, S. (1978). *Geochim. Cosmochim. Acta*, **42**, 495
de Ong, E. R. (1956). "Chemistry and Uses of Pesticides", Reinhold, New York
Deuser, W. G., Degens, E. T. and Guillard, R. R. L. (1968). *Geochim. Cosmochim. Acta*, **32**, 657

Deuser, W. G., Degens, E. T., Harvey, G. R. and Rubin, M. (1973). *Science*, **181**, 51
D'Hondt, P., Lievens, P., Versieck, J. and Hoste, J. (1977). *Radiochem. Radioanal. Lett.* **31**, 231
Dickert, J. W. and Rozacky, E. A. (1969). *Arch. Biochem. Biophys.* **134**, 473
D'Itri, F. M. (1972). "The Environmental Mercury Problem", Chem. Rubber Co., Cleveland, Ohio
Dixon, J. B. and Weed, S. O. (eds) (1977). "Minerals in Soil Environments", Soil Sci. Soc. Amer., Madison, Wisc.
Dixon, N. E., Gazzola, C., Blakeley, R. L. and Zerner, B. (1975). *J. Amer. Chem. Soc.* **97**, 4131
Doisy, R. J., Streeten, D. H. P., Freiberg, J. M. and Schneider, A. J. (1976). "Trace Elements in Human Health and Disease", p. 79 (Prasad, A. S., ed), Academic Press, London, New York
Dougherty, H. W., Sadowski, S. J. and Baker, E. E. (1978). *J. Biol. Chem.* **253**, 5220
Dounce, A. L. and Beyer, G. T. (1948). *J. Biol. Chem.* **174**, 859
Drabikowski, W. (ed) (1974). "Calcium-Binding Proteins", Elsevier, Amsterdam, London, New York
Dreisigacker, E. and Roether, W. (1978). *Earth Planet. Sci. Lett.* **38**, 301
Droop, M. R. (1969). "Methods in Microbiology", 3B, p. 269, Academic Press, London, New York
Duce, R. A., Roy, B. J., Hoffman, G. L. and Walsh, P. R. (1976). *Geophys. Res. Lett.* **3**, 339
Dunitz, J. D., Hawley, D. M., Mikloš, D., White, D. N. J., Berlin, Y., Manišic, R. and Prelog, V. (1971). *Helv. Chim. Acta*, **54**, 1709
Dunster, H. J., Garner, R. J., Howells, H. and Wix, L. F. U. (1964). *Health Phys.* **10**, 353
Durbin, P. W. (1960). *Health Phys.* **2**, 225
Durum, W. H. and Haffty, J. (1963). *Geochem. Cosmochim Acta*, **27**, 1
Duursma, E. K. (1972). *Oceanog. Mar. Biol. Annu. Rev.* **10**, 137
Dymond, J. and Hogan, L. (1973). *Earth Planet. Sci. Lett.* **20**, 131
Eagle, H. (1956). *Arch. Biochem. Biophys.* **61**, 356
Edgington, D. N., Gordon, S. A., Thommes, M. M. and Almodovar, L. R. (1970). *Limnol. Oceanog.* **15**, 945
Egorov, V. V., Zhigalovskaya, T. N., Malakhov, S. G. and Shilina, A. I. (1971). *Trudy. Inst. Eksp. Meteorol.* **14**, 73
Ehhalt, D., Münnich, K. O., Roether, W., Schölch, J. and Stich, W. (1964). 3rd Internat. Conf., Peaceful uses of atomic energy, 14, 45, U.N.
Ehrlich, H. L. (1971). "Soil Biochemistry", Vol. 2, p. 361 (McLaren, A. D. and Skujinš, J., eds), M. Dekker, New York
Eichmann, R. and Schidlowski, M. (1975). *Geochim. Cosmochim. Acta*, **39**, 585
Eisenbud, M. (1973). "Environmental Radioactivity", Academic Press, London, New York
Eisenbud, M. and Petrow, H. G. (1964). *Science*, **144**, 288
Ekman, L. (1967). "Radioecological Concentration Processes", p. 547 (Åberg, B. and Hungate, F. P., eds), Pergamon Press Ltd., Oxford, New York
Ellfolk, N. and Sievers, G. (1971). *Acta Chem. Scand.* **25**, 3532
Elliott, H. L., Dryburgh, F., Fell, G. S., Sabet, S. and Macdougall, A. I. (1978). *Br. Med. J.* 1101
Ellis, R. E. (1965). *Phys. Med. Biol.* **10**, 153

Elofson, R. M., Schulz, K. F. and Hitchon, B. (1977). *Geochim. Cosmochim. Acta*, **41**, 567
Emery, T. (1971). *Biochem.* **10**, 1483
Emiliani, C. (1978). *Earth Planet. Sci. Lett.* **37**, 349
Engel, A. E. J., Nagy, B., Nagy, L. A., Engel, C. G., Kemp, G. O. W. and Drew, C. M. (1968). *Science*, **161**, 1005
Eppley, R. W. (1962). "Physiology and Biochemistry of Algae", p. 255 (Lewin, R. A., ed), Academic Press, London, New York
Epstein, S., Thompson, P. and Yapp, C. J. (1977). *Science*, **198**, 1209
Epstein, S. and Yapp, C. J. (1976). *Earth Planet. Sci. Lett.* **30**, 252
Erämetsä, O., Haarala, A. R. and Yliruokanen, I. (1973). *Suomi Kemistilehti*, **B46**, 234
Erämetsä, O. and Haukka, M. (1970). *Suomi Kemistilehti*, **B43**, 183
Erämetsä, O., Sihvonen, M. L. and Forssén, A. (1968). *Ann Med. Exp. Fenn.* **46**, 179
Erämetsä, O. and Yliruokanen, I. (1971). *Suomi Kemistilehti*, **B44**, 372
Eriksson, E. (1960). *Tellus*, **12**, 63
Erlenmayer, H., Schoenauer, W. and Süllmann, H. (1936). *Helv. Chim. Acta*, **19**, 1376
Ernst, E. and Juricskay, S. (1971). *Acta Biochim. Biophys.* **6**, 165
Ernst, W. (1974). "Schwermetall Vegetation der Erde", G. Fischer Verlag, Stuttgart
Essing, H. G., Schaller, K. H., Szadkowski, D. and Lehnert, G. (1969). *Arch. Hyg. Bakteriol.* **153**, 490
Eyster, C. (1967). U.S. Air Force Aerospace Med. Rept. SAM-TR-67-40
Ezdakova, L. A. (1964). *Bot. Zhur.* **49**, 1798
Fagerström, T. and Jernelöv, A. (1972). *Water Res.* **6**, 1193
Falchuk, K. H., Mazus, B., Ulpino, L. and Vallee, B. L. (1976). *Biochem.* **15**, 4468
Falk, J. E. (1964). "Porphyrins and Metalloporphyrins", Elsevier, Amsterdam, London, New York
Faller, L. D., Baroudy, B. M., Johnson, A. M. and Ewall, R. X. (1977). *Biochem.* **16**, 3864
Faulkner, D. J. and Andersen, R. J. (1974). "The Sea", Vol. 5, p. 679 (Goldberg, E. D., ed), John Wiley and Sons Ltd., London, New York
Faulkner, D. J., Stallard, M. O., Fayos, J. and Clardy, J. (1973). *J. Amer. Chem. Soc.* **95**, 3413
Faure, G. (1977). "Principles of Isotope Geology", John Wiley and Sons Ltd., London, New York
Feeley, H. W. (1977). U.S. Dept. Agric. Health and Safety Lab. Rept HASL 328
Fenton, D. E. (1977). *Chem. Soc. Rev.* **6**, 325
Ferguson, R. L. and Rublee, P. (1976). *Limnol. Oceanog.* **21**, 141
Fergusson, G. J. (1965). Proc. 6th Internat. Conf., Radiocarbon and Tritium dating, U.S. Dept. Commerce, Washington, DC
Ferry, B. W., Baddeley, M. S. and Hawksworth, D. L. (eds) (1973). "Air Pollution and Lichens", Athlone Press, London
Fiabane, A. M. and Williams, D. R. (1977). "The Principles of Bioinorganic Chemistry", Chemical Society, London
Finkel, A. J., Miller, C. E. and Katz, J. J. (1963). "Noble Gas Compounds", p. 309 (Hyman, H. H., ed), Chicago Univ. Press
Fisenne, I. M. and Perry, P. M. (1978). *Radiochem. Radioanal. Lett.* **33**, 259
Fisher, G. L. (1975). *Sci. Total Envir.* **4**, 373
Flanagan, F. J. (1973). *Geochim. Cosmochim. Acta*, **37**, 1189

Fleischer, M., Sarofim, A. F., Fassett, D. W., Hammond, P., Shacklette, H. T., Nisbet, I. C. T. and Epstein, S. (1974). *Envir. Health Perspect.* **7**, 253.
Fleming, G. A. (1963). *J. Sci. Food Agric.* **14**, 203
Flesch, P. (1968). *J. Soc. Cosmet. Chem.* **19**, 675
Florence, T. M. and Batley, G. E. (1977). *Talanta*, **24**, 151
Florkin, M. and Jeuniaux, C. (1964). "The Physiology of Insecta", Vol. 3, p. 110 (Rockstein, M., ed), Academic Press, London, New York
Fogg, G. E. (1975). "Chemical Oceanography", Vol. 2, p. 385 (Riley, J. P. and Skirrow, G., eds), Academic Press, London, New York
Folsom, T. R., Young, D. R., Johnson, J. N. and Pillai, K. C. (1963). *Nature, Lond.* **200**, 327
Forman, H. T. and Kennedy, J. (1978). *Arch. Biochem. Biophys.* **191**, 23
Fortmann, L. C., Gay, D. D., Wirtz, K. O. (1977). "Trace Substances in Environmental Health", Vol. 11, p. 117 (Hemphill, D. D., ed), Columbia Univ., Missouri
Foster, R. F. (1963). "Radioecology", p. 569 (Schultz, V. and Klement, A. W., eds), Reinhold, New York
Fowler, E. B. (ed) (1965). "Radioactive Fall out, Soils, Plants, Foods, Man", Elsevier, Amsterdam, London, New York
Fox, M. R. S. (1976). "Trace Elements in Human Health and Disease", p. 401 (Prasad, A. S., ed), Academic Press, London, New York
Francis, C. W., Chesters, G. and Haskin, L. A. (1970). *Envir. Sci. Technol.* **4**, 586
Franke, W. and Mönch, J. (1948). *Biochem. Zeits.* **319**, 174
Frei, E. and Aebi, H. (1958). *Helv. Chim. Acta*, **41**, 241
Friberg, L., Piscator, M. and Nordberg, G. (1971). "Cadmium in the Environment", Chem. Rubber Co., Cleveland, Ohio
Fried, M. and Shapiro, R. E. (1961). *Annu. Rev. Plant. Physiol.* **12**, 91
Friedman, A. M. (1976). "Actinides in the Environment", Amer. Chem. Soc., Washington
Friedman, M. (1973). "The Chemistry and Biochemistry of the Sulphydryl Group in Aminoacids, Peptides and Proteins", Pergamon Press Ltd., Oxford, New York
Friend, J. P. (1973). "Chemistry of the Lower Atmosphere", p. 177 (Rasool, S. I., ed), Plenum Press, New York, London
Fries, L. (1975). *Z. Pflanzenphysiol.* **76**, 366
Frissel, M. J. (ed) (1978). "Cycling of Mineral Nutrients in Agricultural Ecosystems", Elsevier, Amsterdam, London, New York
Frost, D. V. (1967). *Fed. Proc.* **26**, 194
Fuex, A. N. (1977). *J. Geochem. Explor.* **7**, 155
Fuge, R. and James, K. H. (1974). *Mar. Pollut. Bull.* **5**, 9
Fujinaga, T., Kusaka, Y., Koyama, M., Tsuji, H., Mitsuji, T., Iwai, S., Okuda, J., Takamatsu, T. and Ozaki, T. (1973). *J. Radioanal. Chem.* **13**, 301
Fujita, M., Iwamoto, J., Kondo, M., Yabe, A. and Akaishi, J. (1969). *Health Phys.* **17**, 41
Fukai, R. and Meinke, W. W. (1959). *Nature, Lond.* **84**, 815
Fukai, R. and Meinke, W. W. (1962). *Limnol. Oceanog.* **7**, 186
Furr, A. K., Lawrence, A. W., Tong, S. S. C., Grandalfo, M. C., Hofstader, R. A., Bache, C. A., Gutermann, W. H. and Lisk, D. H. (1976). *Envir. Sci. Technol.* **10**, 683
Fushimi, H., Harrison, C. R. and Ravin, H. A. (1971). *J. Biochem.* **69**, 1041
Gabrielli, L. F. and Marletta, G. P. (1969). *Tec. Ital.* **34**, 427

Galbally, I. (1968). *Nature, Lond.* **218**, 456

Games, L. M., Hayes, J. M. and Gunsalus, R. P. (1978). *Geochim. Cosmochim. Acta* **14**, 1295

Gardner, A. W. and Hall-Patch, P. K. (1962). *Vet Record*, **74**, 113

Gardner, D. and Riley, J. P. (1974). *J. Cons. Int. Explor. Mer.* **35**, 202

Garrels, R. M. and Mackenzie, F. T. (1971). "Evolution of Sedimentary Rocks", W. W. Norton, New York

Garten, C. T., Gardner, R. H. and Dahlman, R. C. (1978). *Health Phys.* **34**, 611 and 705.

Gauch, H. G. (1972). "Inorganic Plant Nutrition", Dowden, Hutchinson and Ross, Philadephia

Gensicke, F. (1966). *Z. Naturforsch.* **B21**, 567

Gerretsen, F. C. and de Hoop, H. (1954). *Plant Soil*, **5**, 349

Ghiretti, F., Salvato, B., Carlucci, S. and De Pieri, R. (1972). *Experientia*, **28**, 232

Gibbs, R, J, (1967). *Bull. Geol. Soc. Amer.* **78**, 1203

Gibbs, R. J. (1972). *Geochim. Cosmochim. Acta*, **36**, 1061

Gieseking, J. E. (ed) (1975). "Soil Components", Springer Verlag, Berlin, Heidelberg, New York

Giggenbach, W. F. (1975). *Bull. Volcanol.* **39**, 15 and 132

Gilbert, I. G. F. and Radley, J. M. (1964). *Biochim. Biophys. Acta*, **79**, 575

Gillespie, F. L., Shimmins, J. and Lenihan, J. M. A. (1970). *Radiochem. Radioanal. Lett.* **4**, 43

Girling, C. A. (1978). Thesis, University of London

Gissel-Nielsen, G. (1971). *Agric. Food Chem.* **19**, 564

Gladney, E. S., Owens, J. W. and Starner, J. W. (1976). *Analyt. Chem.* **48**, 973

Glasby, G. P. (1973). *Mar. Chem.* **1**, 105

Glasby, G. P. (ed) (1977). "Marine Manganese Nodules", Elsevier, Amsterdam, London, New York

Gogotova, G. I., Galimov, E. M. and Ivanov, M. V. (1973). *Dokl. Akad. Nauk. SSSR*, **208**, 1224

Gohda, S. (1975). *Bull. Chem. Soc. Japan*, **48**, 1213

Goldberg, E. D., Koide, M., Schmitt, R. A. and Smith, R. H. (1963). *J. Geophys. Res.* **68**, 4209

Goldhaber, M. B. and Kaplan, I. R. (1974). "The Sea", Vol. 5, p. 569 (Hill, M. N., Maxwell, A. E. and Goldberg, E. D., eds), John Wiley and Sons Ltd., London, New York

Goldwater, L. J. and Stopford, W. (1977). "The Chemical Environment", p. 38 (Lenihan, J. M. A. and Fletcher, W. W., eds), Blackie, Glasgow

Golley, F. B., McGinnis, J. T., Clements, R. G., Child, G. I. and Duever, M. J. (1969). *Bioscience*, **19**, 693 and 697

Gong, H., Rose, A. W. and Suhr, N. H. (1977). *Geochim. Cosmochim. Acta*, **41**, 1687

Goodall, D. W. and Gregory, F. C. (1947). *Imp. Bur. Hortic. Plantation Crops Tech. Comm.* **17**

Gopal-Ayengar, A. R. and Mistry, K. B. (1962). "Radioisotopes in Soil-Plant Nutrition Studies", 3, I.A.E.A., Vienna

Gordon, H. T. (1959). *Annls. N.Y. Acad. Sci.* **77**, 290

Gorham, E. and Swaine, D. J. (1965). *Limnol. Oceanog.* **10**, 268

Gorshkov, G. V., Zyabkin, V. A. and Tsvetkov, O. S. (1964). *Atom. Energ. USSR*, **17**, 492

Gorski, L., Heinonen, J. and Suschny, O. (1974). Report IAEA/RL/25, Vienna

Gosling, A. W., Jenne, E. A. and Chao, T. T. (1971). *Econ. Geol.* **66**, 309
Govett, J. S. and Govett, M. H. (eds) (1976). "World Mineral Supplies". Elsevier, Amsterdam, London, New York
Grancini, G., Stievano, M. B., Girardi, F., Guzzi, G. and Pietra, R. (1976). *J. Radioanal. Chem.* **34**, 65
Granstein, W. C., Cromack, K. and Sollins, P. (1977). *Science*, **198**, 1252
Gravelle, M. and Lelubre, M. (1957). *Bull. Soc. Géol. Fr.* **7**, 435
Gray, D., McKown, D. M., Kay, M., Eichov, M. and Vogt, J. R. (1972). *IEEE Trans. Nucl. Sci. NS*, **19**, 194
Gregor, B. (1965). *Geochim. Cosmochim. Acta*, **29**, 1007
Greenberg, D. M. (ed) (1975). "Metabolism of Sulphur Compounds", 3rd Ed., Academic Press, London, New York
Griffin, J. D. and Goldberg, E. D. (1975). *Limnol. Oceanog.* **20**, 456
Griffith, E. J., Beeton, A., Spencer, J. M. and Mitchell, D. T. (1973). "Environmental Phosphorus Handbook", John Wiley and Sons, London, New York
Griffith, S. M. and Schnitzer, M. (1976). "Environmental Biogeochemistry", Vol. 1, p. 117 (Nriagu, J. O., ed), Ann Arbor Science, Ann Arbor, Mich.
Griffiths, A. J., Hughes, D. and Thomas, D. (1975). "Minerals and the Environment", p. 387 (Jones, M. J., ed), Inst. Mining Metallurgy, London
Gris, E. (1844). *C.R. Acad. Sci. Paris*, **19**, 1118
Gross, M. G. (1967). "Estuaries", p. 273 (Lauff, G. H., ed), Amer. Ass. Adv. Sci., Washington, DC
Gryzhankova, L. N., Sayenko, G. N., Karyakin, A. V. and Laktionova, N. V. (1973). *Oceanology*, **13**, 206
Gulbrandsen, R. A. (1966). *Geochim. Cosmochim. Acta*, **30**, 769
Hadzistelios, I. and Papadopolou, C. (1977). *J. Radioanal. Chem.* **36**, 427
Halcrow, W., Mackay, D. W. and Thornton, I. (1973). *J. Mar. Biol. Ass. U.K.* **53**, 721
Hall, D. O., Cammach, R. and Rao, K. K. (1974). "Iron in Biochemistry and Medicine", p. 279 (Jacobs, A. and Worwood, M., eds), Academic Press, London, New York
Hall, R. J. (1977). "Trace Substances in Environmental Health", Vol. 11, p. 156 (Hemphill, D. D., ed), Univ. Missouri, Columbia
Hallbauer, D. K. (1978). *Gold Bull.* **11**, 18
Hallberg, L., Harwerth, H. G. and Vanotti, A. (1970). "Iron Deficiency", Academic Press, London, New York
Haller, W. A., Rancitelli, L. A. and Cooper, J. A. (1968). *J. Agric. Food Chem.* **16**, 1036
Hallsworth, E. G., Wilson, S. B. and Adams, W. A. (1965). *Nature, Lond.* **205**, 307
Halstead, R. L. and McKercher, R. B. (1975). "Soil Biochemistry", Vol. 4, p. 31 (Paul, E. A. and McLaren, A. D., eds), Academic Press, London, New York
Hamaguchi, H., Ohta, N., Onuma, N. and Kawasaki, K. (1960). *Nippon Kagaku Zasshi*, **81**, 920
Hambidge, K. M. and Nichols, B. L. (eds) (1978). "Zinc and Copper in Clinical Medicine", Spectrum, Jamaica, N.Y.
Hamilton, E. I. (1974). *Sci. Total Envir.* **3**, 8
Hamilton, E. I. and Minski, M. J. (1973). *Sci. Total Envir.* **1**, 375
Hamilton, E. I., Minski, M. J. and Cleary, J. J. (1973). *Sci. Total Envir.* **1**, 341
Hamner, C. L. (1942). *Bot. Gaz.* **104**, 161

Hampson, B. L. and Tennant, D. (1973). *Analyst*, **98**, 873
Hanawalt, R. B. and Whittaker, R. H. (1977). *Soil. Sci.* **123**, 25
Hanna, W. J. and Grant, C. L. (1962). *Bull. Torrey Bot. Club*, **89**, 293
Hanson, W. C., Watson, D. G. and Perkins, R. W. (1967). "Radioecological Concentration Processes", p. 233 (Åberg, B. and Hungate, F. P., eds), Pergamon Press Ltd., Oxford, New York
Hardy, E. P., Meyer, M. W., Allen, J. S. and Alexander, L. T. (1968). *Nature, Lond.* **219**, 584
Hardy, K. A., Akselson, R., Nelson, J. W. and Winchester, J. W. (1976). *Envir. Sci. Technol.* **10**, 176
Hardy, R. W. F. (ed) (1977). "A Treatise on Dinitrogen Fixation", Vol. 3, John Wiley and Sons, London, New York
Harris, E. D., Gonnerman, W. A., Savage, J. E. and O'Dell, B. L. (1974). *Biochim. Biophys. Acta*, **341**, 332
Hart, E. B., Steenbock, H., Waddell, J. and Elvehjem, C. A. (1928). *J. Biol. Chem.* **77**, 797
Hartman, R. E. and Hayes, R. L. (1969). *J. Pharm. Exp. Therap.* **168**, 193
Hashimoto, T. (1971). *Analyt. Chim. Acta*, **56**, 347
Haskin, L. A., Frey, F. A., Schmitt, R. A. and Smith, R. H. (1966). *Phys. Chem. Earth*, **7**, 167
Haskin, L. A. and Gehl, M. A. (1962). *J. Geophys. Res.* **67**, 2537
Haskin, L. A., Haskin, M. A., Frey, F. A. and Wildermann, T. R. (1968). "Origin and Distribution of the Elements", p. 889 (Ahrens, L. H., ed), Pergamon Press Ltd., Oxford, New York
HASP (1960). High altitude sampling program Rept. DASA 532, Defense Atomic Support Agency, Washington DC
Hattula, T. and Johanson, M. (1978). *Radiochem. Radioanal. Lett.* **32**, 35
Hauck, R. D. and Bremner, J. M. (1976). *Adv. Agron.* **28**, 219
Hayes, D. W. and MacMurdo, K. W. (1977). *Health Phys.* **32**, 215
Hazard, J. T., Yokota, M. and Drysdale, J. W. (1977). *Blood* **49**, 139
Headlee, A. J. W. and Hunter, R. G. (1953). *Indust. Eng. Chem.* **45**, 548
Heide, F. and Ködderitzsch, H. (1964). *Naturwiss.* **51**, 104
Heindryckx, R. and Dams, R. (1974). *J. Radioanal. Chem.* **19**, 339
Heinen, W. (1965). *Arch. Biochem. Biophys.* **110**, 137
Heinrichs, H. (1977) *Naturwiss.* **64**, 479
Helmke, P. A. and Haskin, L. A. (1973). *Geochim. Cosmochim. Acta*, **37**, 1513
Helsen, J. N., Shaw, D. M. and Crocket, J. H. (1978). *Geochim. Cosmochim. Acta* **42**, 1067
Hem, J. D. (1968). *Adv. Chem.* **73**, 98
Hemler, M., Lands, W. E. M. and Smith, W. L. (1976). *J. Biol. Chem.* **251**, 5575
Henkin, R. I. (1978). "Zinc and Copper in Clinical Medicine", p. 35 (Hambidge, K. M. and Nichols, B. L., eds), Spectrum, Jamaica, N.Y.
Herron, M. M., Langway, C. C., Weiss, H. V. and Cragin, J. H. (1977). *Geochim. Cosmochim. Acta*, **41**, 915
Herzinger, F. (1940). *Bodenk. PflErnähr.* **16**, 141
Hewitt, E. J. (1966). "Soil and Water Culture Methods Used in the Study of Plant Nutrition", 2nd Ed., Royal Commonwealth Agric. Bureau, Farnham
Hewitt, E. J.. and Smith, T. A. (1975). "Plant Mineral Nutrition", English Univ. Press, Hodder and Stoughton, London

Hidy, G. M. (1973). "Chemistry of the Lower Atmosphere", p. 121 (Rasool, S. I., ed), Plenum Press, New York, London
Higgins, E. S., Richert, D. and Westerfield, W. W. (1956). *J. Nutrit.* **59**, 539
Hill, A. E. and Hill, B. S. (1973). *Int. Rev. Cytol.* **35**, 299
Hind, G., Nakatani, H. Y. and Izawa, S. (1969). *Biochim. Biophys. Acta*, **172**, 277
Hinkle, M. E. and Grenshaw, G. L. (1975). *U.S. Geol. Surv. Bull.* **1408**, 51
Hinneri, S. (1975). *Ann. Bot. Fenn.* **12**, 135
Hirose, A., Kobori, K. and Ishii, D. (1978). *Analyt. Chim. Acta*, **97**, 303
Hirsch-Holb, H., Kolb, H. J. and Greenberg, J. M. (1971). *J. Biol. Chem.* **246**, 395
Hobson, G. D. and Tiratsoo, E. N. (1975). "Introduction to Petroleum Geology", Scientific Press, Beaconsfield
Hodge, V. F., Hoffman, F. L. and Folsom, T. R. (1974). *Health Phys.* **27**, 29
Hoering, T. C. (1956). *Nat. Acad. Sci. Publ.* **400**, 39
Hoering, T. C. and Ford, H. T. (1960). *J. Amer. Chem. Soc.* **82**, 376
Hoffman, D. C., Lawrence, F. O., McWherter, J. L. and Rourke, F. M. (1972). *Nature, Lond.* **234**, 132
Hoffmann, J. L. (1977). *Arch. Biochem. Biophys.* **179**, 136
Hofstader, R. A., Milner, O. I. and Rummels, J. H. (eds) (1976). *Adv. Chem.* **156**, 1
Högberg, H. E., Thomson, R. H. and King, T. J. (1976). *J. Chem. Soc. Perkin*, **I**, 1696
Holden, A. V. (1972). "Mercury Contamination in Man and his Environment", p. 143, Tech. Rept. 137, I.A.E.A., Vienna
Holdsworth, E. S. and Arshad, J. H. (1977). *Arch. Biochem. Biophys.* **183**, 361
Holeman, J. N. (1968). *Water Resources Res.* **4**, 737
Höll, W. and Hampp, R. (1975). *Residue Rev.* **54**, 79
Holland, H. D. (1978). "The Chemistry of the Atmosphere and Oceans", John Wiley and Sons Ltd., London, New York
Hollins, J. G. (1969). *Health Phys.* **17**, 497
Holm, E. and Persson, B. R. R. (1976). "Transuranium Nuclides in the Environment", p. 435, I.A.E.A., Vienna
Holm, E. and Persson, B. R. R. (1977). *Health Phys.* **33**, 471
Holm, E. and Persson, B. R. R. (1978). *Nature, Lond.* **273**, 289
Holser, W. T. and Kaplan, I. R. (1966). *Chem. Geol.* **1**, 93
Honda, S. I. and Robertson, R. N. (1956). *Austral. J. Biol. Sci.* **9**, 305
Hood, D. W. and McRoy, C. P. (1971). "Impingement of Man on the Oceans", p. 667 (Hood, D. W., ed), John Wiley and Sons Ltd., London, New York
Hopkins, L. L. and Mohr, H. E. (1971). "Newer Trace Elements in Nutrition", p. 195 (Mertz, W. and Cornatzer, W. E., eds), M. Dekker, New York
Höpner, T. and Trautwein, A. (1972). *Z. Naturforsch.* **27B**, 1075
Horowitz, C. T., Schock, H. H. and Horowitz-Kisimova, L. A. (1974). *Plant Soil*, **40**, 397
Horsfall, J. G. (1956). "Principles of Fungicidal Action", Chronica Botanica, Waltham, Mass.
Hosohara, K., Kozuma, H., Kawasaki, K. and Tsuruta, T. (1961). *J. Chem. Soc. Japan*, **82**, 1107
House, A. L. (1961). *Annu. Rev. Entomol.* **6**, 13
Houtman, J. P. W. (1973). *Delft Progr. Rept., Chem. & Phys.* **1**, 5
Howard, J. H. (1977). *Geochim. Cosmochim. Acta.* **41**, 1665
Howlett, C. and Taylor, A. (1978). *Analyst*, **103**, 916
Hubbell, O. W. and Glenn, J. L. (1973). *U.S. Geol. Surv. Prof. Paper*, **433L**

Hubert, G., Rieder, N., Schmitt, G. and Send, W. (1975). *Z. Naturforsch.* **30C**, 422
Hudson, T. G. F. (1964). "Vanadium, Toxicity and Biological Significance", Elsevier, Amsterdam, London, New York
Huebner, G., Neef, H., Fischer, G. and Schellenberger, A. (1975). *Zeits. Chem.* **15**, 221
Huey, C. L., Brinckmann, F. E., Grim, S. O. and Iverson, W. P. (1974). Proc. Int. Conf., Transport of persistent chemicals in aquatic ecosystems II, 73
Huey, C. L., Brinckmann, F. E., Iverson, W. P. and Grim, S. O. (1975). Abstract C-214, Int. Conf., Heavy metals in the environment, University of Toronto, Canada
Huff, J. W., Bossardt, D. K., Miller, O. P. and Barnes, R. H. (1956). *Proc. Soc. Exp. Biol. Med.* **92**, 216
Hunter, D. R., Haworth, R. A. and Southard, J. H. (1976). *J. Biol. Chem.* **251**, 5069
Huntress, W. T. (1977). *Chem. Soc. Rev.* **6**, 295
Huntzicker, J. J., Friedlander, S. K. and Davidson, C. I. (1975). *Envir. Sci. Technol.* **9**, 448
Hutchinson, G. E. (1943). *Quart. Rev. Biol.* **18**, 242
Hutchinson, G. E. (1970). *Sci. Amer.* **223**, no. 3, 45
Hutner, S. H. (1972). *Annu. Rev. Microbiol.* **26**, 313
Hwang, J. Y. (1972). *Analyt. Chem.* **44**, 20A
Ichikawa, R. and Ohno, S. (1974). *Bull. Jap. Soc. Sci. Fish*, **40**, 501
Ichioka, P. S. and Arnon, D. I. (1955). *Physiol. Plant.* **8**, 552
ICRP (1964). Publ. 6, International Commission for Radiological Protection, Pergamon Press, Oxford, New York
ICRP (1966). Publ. 9, International Commission for Radiological Protection, Pergamon Press, Oxford, New York
ICRP (1972). Publ. 19, International Commission for Radiological Protection, Pergamon Press, Oxford, New York
Imamura, Y. (1967). *Osaka City Med. J.* **3**, 167
Ingraham, J. L. and Emerson, R. (1954). *Amer. J. Bot.* **41**, 146
Ishibashi, M., Fujinaga, T., Yamamoto, T., Fujita, T. and Watanabe, K. (1965). *J. Chem. Soc. Japan*, **86**, 728
Itoh, S. and Schwartz, I. L. (1956). *Nature, Lond.* **178**, 494
Iverson, W. P., Huey, C., Brinckman, F. E., Jewett, K. J. and Blair, W. (1975). "Heavy Metals in the Aquatic Environment", p. 193 (Krenkel, P. A., ed), Pergamon Press Ltd., Oxford, New York
Iyengar, G., Kollmer, W. E. and Bowen, H. J. M. (1978). "Elemental Composition of Human Tissues and Body Fluids", Springer Verlag, Berlin, Heidelberg, New York
Jaakola, T. (1969). *Annals. Acad. Sci. Fenn. AII*, **150**, 1
Jackson, A. H. (1976). "Chemistry and Biochemistry of Plant Pigments", p. 1 (Goodwin, T. W., ed), Academic Press, London, New York
Jackson, J. F. and Chapman, K. S. R. (1975). "Trace Elements in Soil-Plant-Animal Systems", p. 213 (Nicholas, D. J. D. and Egan, A. R., eds), Academic Press, London, New York
Jackson, T. A. (1975). *Soil Sci.* **119**, 56
Jacobs, A. and Worwood, M. (eds) (1974). "Iron in Biochemistry and Medicine", Academic Press, London, New York
Jacobs, M. B. (1967). "The Analytical Toxicology of Industrial Inorganic Poisons", Interscience, New York, London

Jacques, L. W., Brown, E. B., Barrett, J. M., Brey, W. S. and Weltner, W. (1977). *J. Biol. Chem.* **252**, 4533
James, P. W. (1973). "Air Pollution and Lichens", p. 143 (Ferry, B. W., Baddeley, M. S. and Hawksworth, D. L., eds), Athlone Press, London
Janakidevi, K., Dewey, V. C. and Kidder, G. W. (1965). *J. Biol. Chem.* **240**, 1754
Jansson, S. L. (1963). *Soil. Sci.* **95**, 31
Jeffrey, P. D., Shaw, D. C. and Treacy, G. B. (1976). *Biochem.* **15**, 5527
Jenkins, C. E. (1969). *Health Phys.* **17**, 507
Jenkinson, D. S. and Rayner, J. H. (1977). *Soil Sci.* **123**, 298
Jenne, E. A. (1968). *Adv. Chem.* **73**, 337
Jenne, E. A. (1977). "Molybdenum in the Environment", Vol. 2, p. 555 (Chappell, W. R. and Petersen, K. K., eds), M. Dekker, New York
Jennett, J. C., Bolter, E., Gale, N., Tranter, W. and Hardie, M. (1975). "Minerals and the Environment", p. 13 (Jones, M. J., ed), Inst. Mining Metallurgy, London
Jennings, K. D. (1977). *Mar. Biol.* **44**, 223
Jernelov, A. (1975). "Internat. Conf., Heavy Metals in the Environment", Vol. 2, p. 845 (Hutchinson, T. C., ed), University of Toronto, Canada
Jervis, R. E., Tiefenbach, B. and Chattopadhyay, A. (1974). *Canad. J. Chem.* **52**, 3008
Jeuniaux, C. (1971). "Chemical Zoology", Vol. 6, p. 63 (Florkin, M. and Scheer, B. T., eds), Academic Press, London, New York
Jocelyn, P. C. (1972). "Biochemistry of the SH Group", Academic Press, London, New York
Joensuu, O. I. (1971). *Science*, **172**, 1027
John, W., Kaifer, R., Rahn, K. and Wesolowski, J. J. (1973). *Atmos. Envir.* **7**, 107
Johnson, C. M., Stout, P. R., Broyer, T. C. and Carlton, A. B. (1957). *Plant Soil*, **8**, 337
Johnson, F. N. and Johnson, S. (eds) (1978). "Lithium in Medical Practice", MTP Press, Leicester
Johnson, N. M., Likens, G. E. and Bormann, F. H. (1968). *Geochim. Cosmochim Acta*, **32**, 531
Johnson, R. N. and Volcani, B. E. (1978). *Biochem. J.* **172**, 557
Jones, C. M., Harries, J. M. and Martin, A. E. (1971). *J. Sci. Food Agric.* **22**, 602
Jones, J. B. (1972). "Micronutrients in Agriculture", p. 319 (Mortvedt, J. J., Giordano, P. M.. and Lindsay, W. L., eds), Soil Sci. Soc. Amer., Madison, Wisc.
Jones, J. R. E. (1939). *J. Exp. Biol.* **16**, 425
Jones, J. R. E. (1940). *J. Animal Ecol.* **9**, 188
Jones, J. R. E. (1964). "Fish and River Pollution", Butterworth, London
Jordan, C. F. and Kline, J. R. (1976). *Health Phys.* **30**, 199
Judson, S. and Ritter, D. F. (1964). *J. Geophys. Res.* **49**, 3395
Jugo, S. (1977). *Envir. Res.* **13**, 401
Junge, C. E. (1963). "Air Chemistry and Radioactivity", Academic Press, London, New York
Junge, C. E. (1972). *Quart. J. Roy. Meteorol. Soc.* **98**, 711
Kaakinen, J. W., Jordan, R. M., Lawasami, M. H. and West, R. E. (1975). *Envir. Sci. Technol.* **9**, 862
Kadota, H. and Ishida, Y. (1972). *Annu. Rev. Microbiol.* **26**, 127
Kameda, K. (1962). *J. Radiat. Res.* **3**, 89
Kaplan, I. R. and Rittenberg, S. C. (1962). *Nature, Lond.* **194**, 1098

Karvanek, M. and Janicek, G. (1972). *Sb. Vys. Sk. Chem. Technol. Praze, Potraviny*, **E35**, 75
Kashket, S. and Bunick, F. J. (1978). *Arch. Oral Biol.* **23**, 993
Kato, T., Sato, N. and Suzuki, N. (1976). *Talanta*, **23**, 517
Katz, J. J. and Crespi, H. L. (1970). "Isotope Effects in Chemical Reactions", p. 286 (Collins, C. J. and Bowman, N. S., eds), D. Van Nostrand, Florence, Ky
Katz, J. J., Uphams, R. A., Crespi, H. L. and Blake, M. I. (1975). "Isotopes and Chemical Principles", p. 184 (Rock, P. A., ed), Amer. Chem. Soc., Washington, DC
Katz, M. (1975). "Heavy Metals in the Aquatic Environment", p. 25 (Krenkel, P. A., ed), Pergamon Press, Oxford, New York
Kauranen, P. and Miettinen, J. K. (1967). "Radioecological Concentration Processes", p. 275 (Åberg, B. and Hungate, F. P., eds), Pergamon Press Ltd., Oxford, New York
Keele, B. M., McCord, J. M. and Fridovich, I. (1970). *J. Biol. Chem.* **245**, 6176
Keeling, C. D. (1961). *Geochim. Cosmochim. Acta*, **24**, 299
Keeling, C. D. (1973). "Chemistry of the Lower Atmosphere", p. 251 (Rasool, S. I., ed), Plenum Press, New York, London
Kehoe, R. A. (1965). *Arch. Envir. Health*, **11**, 736
Keilin, D. and Mann, T. (1940). *Biochem. J.* **34**, 1163
Keith, M. L. and Weber, J. N. (1964). *Geochim. Cosmochim Acta*, **28**, 1787
Keller, W. D. and Smith, G. E. (1967). *Geol. Soc. Amer. Spec. Paper*, **90**, 48
Kellner, K. (1955). *Biol. Zentr.* **74**, 662
Kellogg, C. E. (1975) "Agricultural Development; Soil, Food, People and Work", Soil Sci. Soc. Amer., Madison, Wisc.
Kellogg, W. W., Cadle, R. D., Allen, E. R., Lazrus, A. L. and Martell, E. A. (1972). *Science*, **175**, 587
Kemp, A. G. L., Thode, H. G. (1968). *Geochim. Cosmochim. Acta*, **32**, 71
Kemmerer, A. R., Elvehjem, C. A. and Hart, E. B. (1931). *J. Biol. Chem.* **92**, 623
Kern, H. and Naef-Roth, S. (1971). *Phytopath. Z.* **71**, 231
Kessler, E. (1974). "Algal Physiology and Biochemistry", p. 456 (Stewart, W. D. P., ed), Blackwell, Oxford
Kester, D. R. (1975). "Chemical Oceanography", Vol. 1, p. 497 (Riley, J. P. and Skirrow, G., eds), Academic Press, London, New York
Kharkar, D. P., Thomson, J., Turekian, K. K. and Forster, W. O. (1976). *Limnol. Oceanog.* **21**, 294
Kharkar, D. P., Turekian, K. K. and Bertine, K. K. (1968). *Geochim. Cosmochim, Acta*, **32**, 285
Kimura, A. and Nishimura, H. (1970). *J. Antibiot.* **23**, 461
Kimura, H., Mittal, C. K. and Murad, F. (1976). *J. Biol. Chem.* **251**, 7769
King, R. C. (1957). *Amer. Nat.* **41**, 319
Kipling, M. D. (1977). "The Chemical Environment", p. 93 (Lenihan, J. M. A. and Fletcher, W. W., eds), Blackie, Glasgow
Kiss, S., Drăgan-Balarda, M. and Rădulescu, D. (1976). *Adv. Agron.* **27**, 25
Kittrick, J. A. (1976). *Soil Sci. Soc. Amer. J.* **40**, 314
Kizloff, L. M. and Lute, M. (1977). *J. Biol. Chem.* **252**, 7715
Kjaer, A. (1963). "Chemical Plant Taxonomy", p. 453 (Swain, T., ed), Academic Press, London, New York
Klechkovsky, V. M., Polikarpov, G. G. and Aleksakhin, R. M. (eds) (1973). "Radioecology", John Wiley and Sons, London, New York

Klein, D. H. (1975). *Water, Air, Soil Pollut.* **4**, 3
Klein, D. H., Andren, A. W., Carter, J. A., Emery, J. F., Feldman, C., Fulkerson, W., Lyon, W. S., Ogle, J. C. and Talmi, Y. (1975). *Envir. Sci. Technol.* **9**, 973
Kline, J. and Brar, S. (1969). *Soil. Sci. Soc. Amer. Proc.* **33**, 234
Klinman, J. P. (1975). *J. Biol. Chem.* **250**, 2569
Knauer, G. A. and Martin, J. H. (1972). *Limnol. Oceanog.* **17**, 868
Kneifel, H. and Bayer, E. (1973). *Angew. Chem.* **12**, 508
Kneip, T. J. and Lauer, G. J. (1973). *Progr. Analyt. Chem.* **5**, 43
Knizhnikov, V. A. and Barkhudarov, R. M. (1975). *Atomic Energy Rev.* **13**, 171
Koch, R. C. and Roesmer, J. (1962). *J. Food Sci.* **27**, 309
Koda, Y. (1977). *J. Radioanal. Chem.* **36**, 35
Kojima, Y., Berger, C., Vallee, B. L. and Kagi, J. H. R. (1976). *Proc. Nat. Acad. Sci. Wash.* **73**, 3413
Kokubu, N. and Hidaka, T. (1965). *Nature, Lond.* **205**, 1028
Koljonen, T. (1973). *Geol. Soc. Finland Bull.* **54**, 119
Kono, Y., Takahashi, M. and Asada, K. (1976). *Arch. Biochem. Biophys.* **174**, 454
Koons, R. D. and Helmke, P. A. (1978). *Soil Sci. Soc. Amer. J.* **42**, 237
Koontz, W. A. and Shiman, R. (1976). *J. Biol. Chem.* **251**, 368
Korbonskaya, Y. A. and Amanova, N. M. (1975). *Mikol. Fitopatol.* **9**, no. 6, 527
Kosman, D. J., Ettinger, M. J., Bereman, R. D. and Giordano, R. S. (1977). *Biochem.* **16**, 1597
Kosta, L., Ravnik, V., Byrne, A. R., Stirn, J., Dermelj, M. and Stegner, P. (1978). *J. Radioanal. Chem.* **44**, 317
Koutras, D. A. (1968). "Uses of Activation Analysis in Studies of Mineral Element Metabolism in Man", Rept. IAEA-122, I.A.E.A., Vienna
Kovalsky, V. V., Vorotnitskaya, I. E. and Lekarev, V. S. (1967). "Radioecological Concentration Processes", p. 329 (Åberg, B. and Hungate, F. P., eds), Pergamon Press Ltd., Oxford, New York
Kovalsky, V. V., Vorotnitskaya, I. E., Lekarev, V. S. and Nikitina, E. V. (1968). *Trudy. Biogeochem. Lab.* **12**, 5
Kraus, A. and Simon, H. (1975). *Z. Physiol. Chem.* **356,** 1477
Krauskopf, K. B. (1959). "Silica in Sediments", p. 4 (Ireland, H. A., ed), Soc. Econ. Paleontol. Mineral. Spec. Pub. 7, Tulsa, Okl.
Kreger, D. R. and Boeré, H. (1969). *Acta Bot. Neerl.* **18**, 143
Kretsinger, R. H. (1976). *Annu. Rev. Biochem.* **45**, 239
Krogmann, D. W. (1973). "The Biochemistry of Green Plants", Prentice Hall, New Jersey
Kroopnick, P., Weiss, R. F. and Craig, H. (1972). *Earth Planet. Sci. Lett.* **16**, 103
Kubota, J. (1977). "Molybdenum in the Environment", Vol. 2, p. 555 (Chappell, W. R. and Peterson, K. K., eds), M. Dekker, New York
Kubota, J. and Allaway, W. H. (1972). "Micronutrients in Agriculture", p. 525 (Mortvedt, J. J. *et al.*, eds), Soil Sci. Soc. Amer., Madison, Wisc.
Kuehner, E. C., Alvarez, R., Paulson, P. J. and Murphy, T. J. (1972). *Analyt. Chem.* **44**, 2050
Kumpulainen, J.. and Koivistoinen, P. (1977). *Residue Rev.* **68**, 37
Kurup, C. K. R. and Brodie, A. F. (1967). *Biochem. Biophys. Res. Comm.* **28**, 862
Kustin, H. and McLeod, G. C. (1977). *Topics Curr. Chem.* **69**, 1
Kuykendall, W. E., Fite, L. E. and Wainerdi, R. E. (1974). *J. Radioanal. Chem.* **19**, 351
Låg, J. and Steinnes, E. (1976). *Geoderma*, **16**, 317

Lagerwerff, J. W. (1972). "Micronutrients in Agriculture", p. 593 (Mortvedt, J. J. *et al.*, eds), Soil Sci. Soc. Amer., Madison, Wisc.
Lagoutte, B. and Duranton, J. (1975). *F.E.B.S. Lett.* **51**, 21
Lamb, H. H. (1972). "Climate, Present, Past and Future", Methuen, London
Landström, O., Samsahl, K. and Wenner, C. G. (1967). Aktiebolaget Atomenergi AE-296, Stockholm
Landström, O. and Wenner, C. G. (1965). Aktiebolaget Atomenergi AE-204, Stockholm
Langham, W. H. (1972). *Health Phys.* **22**, 943
Langmyhr, F. J., Solberg, R. and Thomassen, Y. (1977). *Analyt. Chim. Acta*, **92**, 105
Larkum, A. W. D. (1968). *Nature, Lond.* **218**, 447
Laskowski, W. (1956). *Genetics, Princeton*, **41**, 98
Laurens, S. and Commanay, L. (1969). *C.R. Soc. Biol.* **163**, 2754
Lavelle, F. and Michelson, A. M. (1975). *Biochimie*, **57**, 375
Lawrey, J. D. (1977). *Bull. Torrey Bot. Club*, **104**, 368
Lazrus, A. L., Lorange, E. and Lodge, J. P. (1970). *Envir. Sci. Technol.* **4**, 55
Lee, C., Kim, N. B. and Lee, I. C. (1976). *J. Korean Nucl. Soc.* **8**, 195
Lee, C. O. and Armstrong, W. M. (1972). *Science*, **175**, 1261
Lee, S. D. (ed) (1977). "Biochemical Effects of Environmental Pollutants", Ann Arbor Science, Ann Arbor, Mich.
Leh, F. K. V. and Lak, R. V. C. (1974). "Environment and Pollutions", C. C. Thomas, Springfield, Ill.
Lehnert, G., Stadelmann, G., Schaller, K. H. and Szadkowski, D. (1969). *Arch. Hyg. Bakteriol.* **153**, 403
Lehr, J. J., Wybenga, J. M. and Rosanow, M. (1958). *Plant Physiol.* **33**, 421
Lemberg, R. and Barrett, J. (1973). "Cytochromes", Academic Press, London, New York
Lerman, A., Mackenzie, F. T. and Garrels, R. M. (1975). *Mem. Geol. Soc. Amer.* **142**, 205
LeRoy, L. W. and Koksoy, M. (1962). *Econ. Geol.* **57**, 107
Lester, G. (1958). *J. Bact.* **75**, 426
Levitzki, A. and Steer, M. L. (1974). *Eur. J. Biochem.* **41**, 171
Lewin, J. and Reimann, B. E. F. (1969). *Annu. Rev. Plant Physiol.* **20**, 289
Lewis, B. G. (1976). "Environmental Biogeochemistry", Vol. 1, p. 389 (Nriagu, J. O., ed), Ann Arbor Science, Ann Arbor, Mich.
Lewis, R. G. (1977). *Residue Rev.* **68**, 123
Liebig, G. F., Vanselow, A. P. and Chapman, H. D. (1943). *Soil Sci.* **56**, 173
Liebig, J. (1840). "Die Organische Chemie in ihren Anwendung auf Agricultur und Physiologie", Leipzig
Lieser, K. H. and Neitzert, V. (1976). *J. Radioanal. Chem.* **31**, 397
Lieth, H. and Whittaker, R. H. (eds) (1975). "The Primary Productivity of the Biosphere", Springer Verlag, Berlin, Heidelberg, New York
Likens, G. E. and Boormann, F. H. (1974). *Science*, **184**, 1176
Linday, E. M. and Syrett, P. J. (1960). *J. Biochem. Microbiol.* **2**, 25
Linke, W. F. (ed) (1958). "Solubilities of Inorganic and Metal-Organic Compounds", Amer. Chem. Soc., Washington, DC
Linnell, J. C., Hoffbrand, A. V., Hussein, H. A. A., Wise, I. J. and Matthews, D. M. (1974). *Clin. Sci. Molec. Med.* **46**, 163
Linnmann, L., Andersson, A., Nilsson, K. O., Lind, B., Kjellstrom, T. and Friberg, L. (1973). *Arch. Envir. Health*, **27**, 45

Linstedt, K. D. and Kruger, P. (1970). *Analyt. Chem.* **42**, 113
Linton, R. W., Williams, P., Evans, C. A. and Natusch, D. F. S. (1977). *Analyt. Chem.* **49**, 1514
Little, P. and Martin, M. H. (1972). *Envir. Pollut.* **3**, 241
Livingston, H. D., Schneider, D. L. and Bowen, V. T. (1975). *Earth Planet. Sci. Lett.* **25**, 361
Livingston, H. D. and Thompson, G. (1971). *Limnol. Oceanog.* **16**, 786
Livingstone, D. A. (1963). *U.S. Geol. Surv. Prof. Paper*, **440-G**
Lockwood, A. P. M. (1963). "Animal Body Fluids and their Regulation", Heinemann, London
Logan, J. A., Prather, M. J., Wofsy, S. C. and McElroy, M. B. (1978). *Phil. Trans. Roy. Soc.* **A290**, 187
London, J. and Kelley, J. (1974). *Science*, **184**, 987
Loneregan, J. F., Snowball, K. and Simmons, W. J. (1968). *Austral. J. Agric. Res.* **19**, 845
Long, C. (1961). "Biochemists Handbook", Spon., London
Lounamaa, J. (1956). *Suomal. Eläin-ja Kasvit. Seur. Van. Julk.* **29**, no. 4
Love, S. K. (1951). *Indust. Eng. Chem.* **43**, 1541
Lovelock, J. E., Maggs, R. J. and Rasmussen, R. A. (1972). *Nature, Lond.* **237**, 452
Lovelock, J. E., Maggs, R. J. and Wade, R. J. (1973). *Nature, Lond.* **241**, 194
Lovenberg, W. (ed) (1977), "Iron-Sulphur Proteins", Academic Press, London, New York
Lovering, T. S. and Engel, C. (1967). *U.S. Geol. Surv. Prof. Paper*, **594B**
Lowenstam, H. A. (1974). "The Sea", Vol. 5, p. 715 (Goldberg, E. D., ed), John Wiley and Sons, London, New York
Lowman, F. G., Stevenson, R. A., Escalera, R. M. and Ufret, S. L. (1967). "Radioecological Concentration Processes", p. 735 (Åberg, B. and Hungate, F. P., eds), Pergamon Press Ltd., Oxford, New York
Lowry, M. W., Huggins, W. C. and Forrest, L. A. (1936). *Georgia Exp. Sta. Bull.* **193**, 3
Lunde, G. (1970). *J. Sci. Food Agric.* **21**, 416
Lunde, G. (1973). *Acta Chem. Scand.* **27**, 1586
Lupton, J. E. (1976). *Earth Planet. Sci. Lett.* **32**, 371
Lüttge, U. (1975). "Ion Transport in Cells and Tissues", p. 336 (Baker, D. A. and Hall, J. L., eds), Elsevier, Amsterdam, London, New York
Lyon, G. L. (1974). "Natural Gases in Marine Sediments", p. 141 (Kaplan, I. R., ed), Plenum Press, New York, London
Lyon, G. L., Peterson, P. J. and Brooks, R. R. (1969). *N.Z. J. Sci.* **12**, 541
Lyr, H. (1977). "Plant Disease", Vol. 1, p. 239 (Horsfall, J. G. and Cowling, E. B., eds), Academic Press, London, New York
Macfarlane, M. G. and Spencer, A. G. (1953). *Biochem. J.* **54**, 569
MacLennan, D. H. and Wong, P. T. S. (1971). *Proc. Nat. Acad. Sci. Wash.* **68**, 1231
MacRobbie, E. A. C. (1962). *J. Gen. Physiol.* **45**, 861
MacRobbie, E. A. C. (1966). *Austral. J. Biol. Sci.* **19**, 371
MacRobbie, E. A. C. (1971). *Biol. Rev.* **46**, 429
MacRobbie, E. A. C. (1974). "Algal Physiology and Biochemistry", p. 676 (Stewart, W. D. P., ed), Blackwell, Oxford
Madsen, P. and Jensen, K. O. (1978). Personal communication Risø, Denmark)
Maenhaut, W. and Zoller, W. H. (1977). *J. Radioanal. Chem.* **37**, 637
Magno, P. J., Groulx, P. R. and Apidianakis, J. C. (1970). *Health Phys.* **18**, 383

Magno, P. J., Kauffmann, P. E. and Shleien, B. (1967). *Health Phys.* **13**, 1325
Magnusson, G. (1963). *Acta Pharm. Toxicol.* **20** (Suppl.), 3
Mah, R. A., Ward, D. M., Baresi, L. and Glass, T. L. (1977). *Annu. Rev. Microbiol.* **31**, 309
Mahler, H. R. (1961). "Mineral Metabolism", Vol. IB, p. 743 (Comar, C. L. and Bronner, F., eds), Academic Press, London, New York
Maines, M. D. and Kappas, A. (1976). *Proc. Nat. Acad. Sci. Wash.* **73**, 4428
Malaisse, F., Gregoire, J., Brooks, R. R., Morrison, R. S. and Reeves, R. D. (1978). *Science*, **199**, 887
Malkin, R. and Malmstrom, B. G. (1970). *Adv. Enzymol.* **33**, 177
Malone, C., Koeppe, D. E. and Miller, R. J. (1974). *Plant Physiol.* **53**, 388
Mamuro, T., Matsuda, Y., Mizohata, A., Takeuchi, T. and Fujita, A. (1970). *Annu. Rep. Radiat. Center, Osaka Pref.* **11**, 1
Marei, A. N., Barkhudarov, R. M., Novikova, N. J., Petakhova, E. V., Dubova, L. D. and Briganina, V. M. (1972). *Health Phys.* **22**, 9
Marowsky, G. and Wedepohl, K. H. (1971). *Geochim. Cosmochim. Acta*, **35**, 1255
Marschner, H. (1975). *Progr. Bot.* **37**, 98
Marshall, C. E. (1964). "The Physical Chemistry and Mineralogy of Soils", John Wiley and Sons, London, New York
Martin, H. (1972). "Insecticide and Fungicide Handbook", Blackwell, Oxford
Martin, J. H. (1970). *Limnol. Oceanog.* **15**, 576
Martin, J. H. and Knauer, G. A. (1973). *Geochim. Cosmochim. Acta*, **37**, 1639
Martonosi, A. (1972). *Curr. Topics Membranes Transport*, **3**, 83
Marumo, S. (1959). *Bull. Agric. Chem. Soc. Japan*, **23**, 428
Marumo, S., Hattori, H. and Abe, H. (1971). *Analyt. Biochem.* **40**, 488
Mascarenho, J. P. and Machlis, L. (1964). *Plant Physiol.* **39**, 70
Mason, B. (1966). "Principles of Geochemistry", John Wiley and Sons, London, New York
Masuda, E. (1935). *J. Pharm. Soc. Japan*, **55**, 625
Matschke, J., Mass, I., Geissler, C. and Steinbrenner, K. (1972). *Zentralbl. Bakt. Parasit. Infek. Hyg.* **127**, 166
Matsumoto, H., Hirasawa, E., Morimura, S. and Takahashi, E. (1976). *Plant Cell Physiol. Tokyo*, **17**, 627
Matsumoto, H., Morimura, S. and Takahashi, E. (1977). *Plant Cell Physiol. Tokyo*, **18**, 987
Matsumura, S., Kokubu, N., Watanabe, S. and Sameshima, Y. (1955). *Mem. Fac. Sci. Kyushu Univ. Ser.* **C2**, 81
Matthews, A. D. and Riley, J. P. (1969). *Analyt. Chim. Acta*, **48**, 25
Matthews, A. D. and Riley, J. P. (1970), *Analyt. Chim. Acta*, **51**, 455
Mauchline, J. and Templeton, W. L. (1964). *Oceanog. Mar. Biol. Annu. Rev.* **2**, 229
May, P. F., Till, A. R. and Cumming, M. J. (1973). *J. Appl. Ecol.* **10**, 607
May, S., Piccot, D. and Pinte, G. (1975). *J. Radioanal. Chem.* **27**, 353
Maynard, J. B. (1976). *Geochim. Cosmochim. Acta*, **40**, 1523
Mazé, P. (1914). *Annl. Inst. Pasteur*, **28**, 21
McBride, B. C. and Wolfe, R. S. (1971). *Biochem.* **10**, 4312
McCance, R. A. and Widdowson, E. P. (1960). "The Composition of Foods", H.M. Stationery Office, London
McClachan, J. and Craigie. J. S. (1967). *Nature, Lond.* **214**, 604
McConnell, G., Ferguson, D. M. and Pearson, C. R. (1975). *Endeavour*, **34**, 13

McCready, R. G. L., Kaplan, I. R. and Din, G. A. (1974). *Geochim. Cosmochim. Acta*, **38**, 1239
McCready, R. G. L., Laishley, E. J. and Krouse, H. R. (1975). *Canad. J. Microbiol.* **21**, 235
McDonald, R. B. and Merriman, W. R. (1977). "The Modern Inorganic Chemicals Industry", p. 168 (Thompson, R., ed), Chemical Society, London
McEwan, T. (1964). *Nature, Lond.* **201**, 827
McFarlane, J. C. (1976). *Envir. Exp. Biol.* **16**, 9
McKeague, J. A. and Kloosterman, B. (1974). *Canad. J. Soil Sci.* **54**, 503
McLean, F. T. and Gilbert, B. E. (1927). *Soil Sci.* **24**, 163
McLendon, H. R., Stewart, O. M., Boni, A. L., Corey, J. C., McLeod, K. W. and Pinder, J. E. (1976). "Transuranium Nuclides in the Environment", p. 347, I.A.E.A., Vienna
Means, J. L., Crerar, D. A., Boreszik, M. P. and Duguid, J. O. (1978). *Geochim. Cosmochim. Acta* **42**, 1763
Medina, E. and Troughton, J. H. (1974). *Plant Sci. Lett.* **2**, 357
Meehan, W. R. and Smythe, L. E. (1967). *Envir. Sci. Technol.* **1**, 839
Megumi, K. and Mamuro, T. (1972). *J. Geophys. Res.* **77**, 3052
Megumi, K. and Mamuro, T. (1973). *J. Geophys. Res.* **78**, 1804
Megumi, K. and Mamuro, T. (1974). *J. Geophys. Res.* **79**, 3357
Megumi, K. and Mamuro, T. (1977). *J. Geophys. Res.* **82**, 353
Meisch, H. U., Reinle, W. and Richter, B. A. (1978). *Z. Naturforsch.* **33C**, 616
Meisch, H. U., Schmitt, J. A. and Reinle, W. (1977). *Z. Naturforsch.* **32C**, 172
Mekhtiyeva, V. L. and Pankina, R. G. (1968). *Geokhimiya*, **6**, 739
Melander, L. (1960). "Isotope Effects on Reaction Rates", Roland Press, New York
Meredith, P. A., Moore, M. R. and Goldberg, A. (1977). *Enzyme* **22**, 22
Mertz, W. (1969). *Physiol. Rev.* **49**, 163
Mertz, W. (1974). "Geochemistry and the Environment", Vol. 1, p. 29, U.S. Nat. Acad. Sci., Washington, DC
Mertz, W. and Cornatzer, W. E. (eds) (1971). "Newer Trace Elements in Nutrition", M. Dekker, New York
Metzler, D. E. (1977). "Biochemistry", Academic Press, London, New York
Meyer, B. (1977). "Sulfur, Energy and Environment", Elsevier, Amsterdam, London, New York
Miettinen, J. K. (1975). "Ecological Toxicology Research", p. 215 (McIntyre, A. D. and Mills, C. F., eds), Plenum Press, New York, London
Miettinen, J. K., Jokelainen, A., Roine, P., Liden, K., Naversten, Y., Bengtsson, G., Hasanen, E. and McCall, R. C. (1963). *Annl. Acad. Sci. Fenn.* **A2**, 120
Millbank, J. W. and Kershaw, K. A. (1973). "The Lichens", p. 289 (Ahmadjian, V. and Hale, M. E., eds), Academic Press, London, New York
Miller, L. P. (1973). "Phytochemistry", Vol. 3, 1, D. Van Nostrand, Florence, Ky
Miller, M. S., Friedlander, S. K. and Hidy, G. M. (1972). "Aerosols and Atmospheric Chemistry", p. 301 (Hidy, G. M., ed), Academic Press, London, New York
Miller, R. J. and Koeppe, D. E. (1970). "Trace Substances in Environmental Health", Vol. 6, p. 181 (Hemphill, D. D., ed), Columbia University, Missouri
Millman, A. P. (1957). *Geochim. Cosmochim. Acta*, **12**, 85
Milne, P. R. and Wells, J. R. E. (1970). *J. Biol. Chem.* **245,** 1566
Minagawa, T. (1958). *Exp. Cell Res.* **14**, 333
Mining Annual Review (1977). *Mining J.*

Minski, M. J., Girling, C. A. and Peterson, P. J. (1977). *Radiochem. Radioanal. Lett.* **30**, 179
Misra, H. P. and Fridovich, I. (1977). *J. Biol. Chem.* **252**, 6421
Mitchell, R. L. (1964). "Chemistry of the Soil", p. 320 (Bear, F. E., ed), 2nd Ed., Chapman and Hall, London
Miyake, Y. and Wada, E. (1971). *Rec. Oceanog. Wks. Japan*, **11**, 1
Miyake, Y. and Tsunogai, S. (1963). *J. Geophys. Res.* **68**, 3989
Mizutani, Y. and Rafter, T. A. (1969). *N.Z. J. Sci.* **12**, 60
Mlodecki, H., Lasota, W. and Matys, J. (1973). *Bromatol. Chem. Toksykol.* **6**, 41
MOAFF (1971). Survey of mercury in food, H.M. Stationery Office, London
MOAFF (1972). Survey of lead in food, H.M. Stationery Office, London
MOAFF (1973). Survey of cadmium in food, H.M. Stationery Office, London
Molloy, L. F. and Richards, E. L. (1971). *J. Sci. Food Agric.* **22**, 397
Moore, C. and Boström, K. (1978). *Chem Geol.* **23**, 1
Moore, C. V. and Dubach, R. (1962). "Mineral Metabolism", Vol. 2B, p. 288 (Comar, C. L. and Bronner, F. E., eds), Academic Press, London, New York
Moore, D. P. (1972). "Micronutrients in Agriculture", p. 171 (Mortvedt, J. J. *et al.*, eds), Soil Sci. Soc. Amer., Madison, Wisc.
Moore, H. (1977). *Atmos. Envir.* **11**, 1239
Moore, H. E., Martell, E. A. and Poet, S. E. (1976). *Envir. Sci. Technol.* **10**, 586
Moore, L. F. and Traquair, J. A. (1976). *Planta*, **128**, 179
Moraghan, J. T. and Buresh, R. J. (1977). *Soil Sci. Soc. Amer. Proc.* **41**, 47
Morgan, E. H. (1974). "Iron in Biochemistry and Medicine", p. 29 (Jacobs, A. and Worwood, M., eds), Academic Press, London, New York
Morisawa, M. (1968). "Streams, their Dynamics and Morphology", McGraw Hill, New York
Morisawa, M. and Mohri, H. (1972). *Exp. Cell Res.* **70**, 311
Morrison, M. and Schonbaum, G. R. (1976. *Annu. Rev. Biochem.* **45**, 861
Mortenson, L. E. and Chen, J. S. (1974). "Microbial Iron Metabolism", p. 231 (Neilands, J. B., ed), Academic Press, London, New York
Mosbach, R. (1973). *Biochim. Biophys. Acta*, **73**, 204
Moskalev, Y. I. (1972). *Health Phys.* **22**, 723
Moss, W. D. and Campbell, E. E. (1972). *Health Phys.* **22**, 933
Moura, J. J. G. and Xavier, A. V. (1978). "New Trends in Bio-Inorganic Chemistry", p. 79 (Williams, R. J. P. and Da Silva, J. R. R. F., eds), Academic Press, London, New York
Müller, P. J. (1977). *Geochim. Cosmochim. Acta*, **41**, 765
Mullette, K. J. (1975). *Plant Soil*, **42**, 495
Mullin, J. B. and Riley, J. P. (1956). *J. Mar. Res.* **15**, 103
Murozumi, M., Chow, T. J. and Patterson, C. C. (1969). *Geochim. Cosmochim. Acta*, **33**, 1247
Murr, L. E., Torma, A. E. and Brierly, J. A. (eds) (1978). "Metallurgical Applications of Bacterial Leaching", Academic Press, London, New York
Murthy, G. K., Rhea, U. and Peeler, J. T. (1971). *Envir. Sci. Technol.* **5**, 434
Muth, O. H. (ed) (1967). "Selenium in Biomedicine", Avi Co., Westport, Conn.
Muth, O. H. and Oldfield, J. E. (1970). "Symposium on Sulphur in Nutrition", Avi Co., Westport, Conn.
Myers, A. J. and Hamilton, J. C. (1961). *U.S. Geol. Surv. Prof. Paper*, **424B**, 286
Myron, D. R., Givard, S. H., Hopkins, L. L. and Nielsen, F. H. (1975). *Fed. Proc.* **34**, 923

Nadkarni, R. A. (1975). *Radiochem. Radioanal. Lett.* **21**, 161
Nadkarni, R. A. (1977). *Radiochem. Radioanal. Lett.* **30**, 329
Nagai, S. (1965). *J. Bact.* **90**, 220
Nagaya, Y. and Nakamura, K. (1974). *Nippon Kaiyo Gakkai-Shi*, **30**, 179
Naidenov, M. and Travesi, A. (1977). *Soil. Sci.* **124**, 152
Nakajima, T., Maitra, S. K. and Ballou, C. E. (1976). *J. Biol. Chem.* **251**, 174
Nakatani, H. Y., Barber, J. and Minski, M. J. (1979). *Biochim. Biophys. Acta* **545**, 24
Nakao, M. (1960). *Osaka Shiritsu Daigaku Igaku Zasshi*, **9**, 541
Navrot, J. and Singer, A. (1976). *Soil Sci.* **121**, 337
Neidleman, S. L. (1975). *C.R.C. Crit. Rev. Microbiol.* **3**, 333
Neilands, J. B. (ed) (1974). "Microbial Iron Metabolism", Academic Press, London, New York
Neuberger, M. and Fourcy, A. (1969). *Int. J. Appl. Radiat. Isot.* **20**, 641
Neufeld, A. H. (1936). *Canad. J. Res.* **14B**, 160
Newton, D., Rundo, J. and Harrison, G. E. (1977). *Health Phys.* **33**, 45
Nicholas, D. J. D. (1963a). *Biol. Rev.* **38**, 530
Nicholas, D. J. D. (1963b). "Plant Physiology", Vol. 3, p. 363 (Steward, F. C., ed), Academic Press, London, New York
Nicholas, D. J. D., Fisher, D. J., Redmond, W. J. and Wright, M. A. (1960). *J. Gen. Microbiol.* **22**, 191
Nicholls, G. D. (1968). "Coal and Coal-Bearing Strata", p. 269 (Murchison, D. and Westoll, T. S., eds), Oliver and Boyd, Edinburgh
Nicholls, G. D., Curl, H. and Bowen, V. T. (1959). *Limnol. Oceanog.* **4**, 472
Nicholls, P. and Elliott, W. B. (1974). "Iron in Biochemistry and Medicine", p. 221 (Jacobs, A. and Worwood, M., eds), Academic Press, London, New York
Nicol, J. A. C. (1960). "The Biology of Marine Animals", Pitman, London, New York
Nielsen, F. H. (1971). "Newer Trace Elements in Nutrition", p. 215 (Mertz, W. and Cornatzer, W. E., eds), M. Dekker, New York
Nielsen, F. H., Givand, S. H. and Myron, D. R. (1975). *Fed. Proc.* **34**, 923
Nielsen, F. H. and Sandstead, H. H. (1974). *Amer. J. Clin. Nutrit.* **27**, 515
Nielsen, F. H. and Sauberlich, H. E. (1970). *Proc. Soc. Exp. Biol. Med.* **134**, 845
Nilsson, R. and Kearns, D. R. (1973). *Photochem. Photobiol.* **17**, 65
Nissen, P. (1974). *Annu. Rev. Plant Physiol.* **25**, 53
Nissenbaum, A. and Kaplan, I. R. (1972). *Limnol. Oceanog.* **17**, 570
Nissenbaum, A. and Swaine, D. J. (1976). *Geochim. Cosmochim. Acta.* **40**, 809
Nobel, P. S. (1975). "Ion Transport in Plant Cells and Tissues", p. 101 (Baker, D. A. and Hall, J. L., eds), North Holland, Amsterdam
Nomoto, S., Decsy, M. L., Murphy, J. R. and Sunderman, W. F. (1973). *Biochem. Med.* **8**, 171
Nordin, B. E. C. (ed) (1976). "Calcium, Phosphate and Magnesium Metabolism", Churchill Livingstone, Edinburgh
Norris, J. R. and Jensen, H. L. (1957). *Nature, Lond.* **180**, 1493
Norrish, K. (1975). "Trace Elements in Soil-Plant-Animal Systems", p. 55 (Nicholas, D. J. D. and Egan, A. R., eds), Academic Press, London, New York
Northrop, D. B. (1975). *Biochem.* **14**, 2644
Northrop, D. B. (1977). "Isotope Effects on Enzyme Catalysed Reactions", p. 122 (Cleland, W. W., O'Leary, M. H. and Northrop, D. B., eds), University Park Press, Baltimore, Md.

Noshkin, V. E. (1972). *Health Phys.* **22**, 537
Nozaki, M. and Ishimura, Y. (1974). "Microbial Iron Metabolism", p. 417 (Neilands, J. B., ed), Academic Press, London, New York
Nriagu, J. O. (ed) (1976). "Environmental Biogeochemistry". Ann Arbor Science, Ann Arbor, Mich.
Nydal, R. and Lövseth, K. (1970). *J. Geophys. Res.* **75**, 2271
Nye, P. H. and Tinker, P. B. (1977). "Solute Movement in the Soil-Root System", Blackwell, Oxford
Nyhan, J. W., Miera, F. R. and Neher, R. E. (1976). *J. Envir. Quality*, **5**, 431
Oakes, T. W., Shank, K. E., Easterly, C. E. and Quintana, L. R. (1977). "Trace Substances in Environmental Health", Vol. 11, p. 123 (Hemphill, D. D., ed), Univ. Missouri, Columbia
Ochiai, E. I. (1977). "Bioinorganic Chemistry", Allyn and Bacon, London, Rockleigh N.J.
O'Dell, B. L., Newberne, F. M. and Savage, J. E. (1958). *J. Nutrit.* **65**, 503
Odum, E. P. (1975). "Ecology", 2nd Ed., Holt, Rinehart and Winston, New York
Oelrichs, P. B. and McEwan, T. (1961). *Nature, Lond.* **190**, 808
Oglesby, L. C. (1969). "Chemical Zoology", Vol. 4, p. 211 (Florkin, M. and Scheer, B. T., eds), Academic Press, London, New York
Oh, S. H., Ganther, H. E. and Hoekstra, W. G. (1974). *Biochem.* **13**, 1825
Ohnishi, T., Salerno, J. C., Winter, D. B., Lim, J., Yu, C. A., Yu, L. and King, T. E. (1976). *J. Biol. Chem.* **251**, 2094
Ohno, S. and Yatazawa, M. (1970). *Radioisot. Tokyo*, **19**, 565
Oka, T. and Simpson, F. J. (1971). *Biochem. Biophys. Res. Comm.* **43**, 1
O'Kelley, J. C. (1974). "Algal Physiology and Biochemistry", p. 610 (Stewart, W. D. P., ed), Blackwell, Oxford
O'Leary, M. H. (1977). "Isotope Effects on Enzyme Catalysed Reactions", p. 233 (Cleland, W. W. *et al.*, eds), University Park Press, Baltimore, Md.
Olson, K. W. and Skogerboe, R. K. (1975). *Envir. Sci. Technol.* **9**, 227
Omdahl, J. L., Hunsaker, L. A.. and Aschenbrenner, V. A. (1977). *Arch. Biochem. Biophys.* **184**, 172
Omura, T. (1961). *J. Biochem. Tokyo*, **50**, 264 and 389
Ondov, J. M., Zoller, W. H., Olmez, I., Aras, N. K., Gordon, G. E., Rancitelli, L. A., Abel, K. H., Filby, R. H., Shah, K. R. and Ragaini, R. C. (1975). *Analyt. Chem.* **47**, 1102
Orii, Y., Manabe, M. and Yoneda, M. (1977). *J. Biochem. Tokyo*, **81**, 505
Osmund, J. K. and Cowart, J. B. (1976). *Atom. Energy Rev.* **14**, 621
Österberg, R. (1976). "An Introduction to Bioinorganic Chemistry", p. 13 (Williams, D. R., ed), C. C. Thomas, Springfield, Ill.
Owens, M. and Wood, G. (1968). *Water Res.* **2**, 151
Page, A. L., Bingham, F. T. and Nelson, C. (1972). *J. Envir. Quality*, **1**, 288
Page, A. L. and Gange, T. J. (1970). *Envir. Sci. Technol.* **4**, 140
Paliwal, K. V. (1974). *Proc. Indian Nat. Sci. Acad.* **A40**, 331
Palmer, H. E. and Beasley, T. M. (1967). "Radioecological Concentration Processes", p. 259 (Åberg, B. and Hungate, F. P., eds), Pergamon Press Ltd., Oxford, New York
Papadopoulos, S., Macfarlane, S. and Harden, R. McG. (1967). *J. Endocrinol.* **38**, 381
Pardee, A. B. (1966). *J. Biol. Chem.* **241**, 5886
Parfenov, Y. D. (1974). *Atom. Energy Rev.* **12**, 75

Parwel, A., Ryhage, R. and Wickman, F. E. (1957). *Geochim. Cosmochim. Acta*, **11**, 165

Pate, J. S. (1975). "Encyclopaedia of Plant Physiology", Vol. 1, p. 451 (Zimmermann, M. H. and Milburn, J. A., eds), Springer Verlag, Berlin, Heidelberg, New York

Patten, B. C. and Witkamp, M. (1967). *Ecology*, **48**, 813

Patterson, C. C. (1965). *Arch. Envir. Health*, **11**, 344

Patterson, C. C. (1971). "Impingement of Man on the Oceans", p. 245 (Hood, D. W., ed), John Wiley and Sons, London, New York

Patterson, C. C. (1972). "Methods of Chemical and Metallurgical Investigation of Ancient Coinage", p. 149 (Hall, E. T. and Metcalf, D. M., eds), Royal Numismatic Society, London

Patterson, C. C. and Settle, D. (1977). *Mar. Biol.* **39**, 289

Patterson, J. B. E. (1971). *Min. of Agric. Fish & Food Tech. Bull.* **21**, 193: "Trace Elements in Soils and Crops", H.M. Stationery Office, London

Patterson, S. H. and Roberson, C. E. (1961). *U.S. Geol. Surv. Prof. Paper*, **424C**, 195

Paul, E. A. (1976). "Environmental Biogeochemistry", Vol. 1, p. 225 (Nriagu, J. O., ed), Ann Arbor Science, Ann Arbor, Mich.

Paul, K. G. and Stigbrand, T. (1970). *Biochim. Biophys. Acta*, **221**, 255

Peck, H. P., Bramlett, R. and Der Vartanian, D. V. (1972). *Z. Naturforsch.* **27B**, 1084

Peden, J. D. (1973). *Mar. Pollut. Bull.* **4**, 7

Pedersen, M. (1969). *Physiol. Plant.* **22**, 680

Pedersen, M., Saenger, P. and Fries, L. (1974). *Phytochem.* **13**, 2273

Peisach, J., Aisen, P. and Blumberg, W. E. (eds) (1966). "The Biochemistry of Copper", Academic Press, London, New York

Pentreath, R. C. (1977). *Oceanog. Mar. Biol. Annu. Rev.* **15**, 365

Perhac, R. M. (1972). *J. Hydrol.* **15**, 177

Perlman, D. and O'Brien, E. (1954). *J. Bact.* **68**, 167

Perrin, R. M. S. (1965). "Experimental Pedology", p. 73 (Hallsworth, E. G. and Crawford, D. V., eds), Butterworth, London

Persson, H. (1948). *Revue Bryol. Lichenol.* **17**, 1

Peters, R. A. and Shorthouse, M. (1964). *Nature, Lond.* **202**, 21

Peters, R. A. and Shorthouse, M. (1967). *Nature, Lond.* **216**, 80

Peterson, P. J. (1971). *Sci. Progr.* 57, 505

Peterson, P. J., Burton, M. A. S., Gregson, M., Nye, S. M. and Porter, E. K. (1976). "Trace Substances in Environmental Health", Vol. 10, p. 123 (Hemphill, D. W., ed), Columbia University, Missouri

Petrow, H. G., Cover, A., Schiessle, W. and Parsons, E. (1964). *Analyt. Chem.* **36**, 1600

Petrow, H. G. and Strehlow, C. D. (1967). *Analyt. Chem.* **39**, 265

Pettijohn, E. J. (1977). "Sedimentary Rocks", 3rd Ed., Harper and Row, New York, London

Petzelt, C. and Auel, D. (1977). *Proc. Nat. Acad. Sci. Wash.* **74**, 1610

Pigford, T. H. (1974). *Annu. Rev. Nucl. Sci.* **24**, 515

Pike, E. R., Graham, L. C. and Fogden, M. W. (1975). *J. Ass. Publ. Anal.* **13**, 19 and 48

Pilkey, O. H. and Harriss, R. C. (1966). *Limnol. Oceanog.* **11**, 381

Pillay, K. K. S. and Thomas, C. C. (1971). *J. Radioanal. Chem.* **7**, 107

Piscator, M. (1976). "Trace Elements in Health and Disease", p. 431 (Prasad, A. S., ed), Academic Press, London, New York

Pistorius, E. K., Axelrod, B. and Palmer, G. (1976). *J. Biol. Chem.* **251**, 7144

Planas, J. (1967). *Nature, Lond.* **215**, 289

Polar, E. (1975). *Planta,* **123**, 97

Pollack, S. and Lasky, F. D. (1977). "Proteins of Iron Metabolism", p. 393 (Brown, E. B. *et al.*, eds), Grune and Stratton, New York

Pollard, A. S., Parr, A. J. and Loughman, B. C. (1977). *J. Exp. Bot.* **28**, 831

Ponomarenko, A. N., Trufanov, G. V. and Golubev, S. N. (1974). *Soviet J. Ecol.* **5**, 279

Ponsford, D. R. A. (1955). *Bull. Geol. Surv. Great Britain,* **10**, 24

Porter, E. K. and Peterson, P. J. (1975). *Sci. Total Envir.* **4**, 365

Porter, H. (1971). *Biochim. Biophys. Acta,* **229**, 143

Porter, J. R. (1946). "Bacterial Chemistry and Physiology", John Wiley and Sons, London, New York

Portnoy, B. and Molokhia, M. (1974). *Brit. J. Dermatol.* **91**, 701

Potter, L., Kidd, D. and Standiford, D. (1975). *Envir. Sci. Technol.* **9**, 41

Prasad, A. S. (ed) (1976). "Trace Elements in Human Health and Disease", Academic Press, London, New York

Pratt, J. M. (1972). "Inorganic Chemistry of Vitamin B_{12}", Academic Press, London, New York

Pratt, J. M. (1975). "Techniques and Topics in Bioinorganic Chemistry", p. 109 (McAuliffe, C. A., ed), Macmillan, London

Prelog, V. (1971). *Pure Appl. Chem.* **25**, 197

Presant, E. W. and Tupper, W. M. (1965). *Canad. J. Soil Sci.* **45**, 305

Pressman, B. C. (1976). *Annu. Rev. Biochem.* **45**, 501

Preston, A. (1973). *Nature, Lond.* **242**, 95

Preston, A. (1974). "The Sea", Vol. 5, p. 817 (Goldberg, E. D., ed), John Wiley and Sons, London, New York

Preston, A., Jefferies, D. F., Dutton, J. W. R., Harvey, B. R. and Steele, A. K. (1972). *Envir. Pollut.* **3**, 69

Price, C. A., Clark, H. E. and Funkhouser, E. A. (1972). "Micronutrients in Agriculture", p. 231 (Mortvedt, J. J. *et al.*, eds), Soil Sci. Soc. Amer., Madison, Wisc.

Price, N. B. (1976). "Chemical Oceanography", Vol. 6, p. 1 (Riley, J. P. and Skirrow, G., eds), Academic Press, London, New York

Puchelt, H. (1967). *Sitzber. Heidelb. Akad. Wiss. Math. Nat. Kl.* **4**, Abh

Purves, D. (1966). *Nature, Lond.* **210**, 1077

Purves, D. (1977). "Trace Element Contamination of the Environment", Elsevier, Amsterdam, London, New York

Putrament, A., Baranowska, H., Ejchart, A. and Jachymezyk, W. (1977). *Molec. Gen. Genet.* **151**, 69

Pytkowicz, R. M. (1973). *Schweiz. J. Hydrol.* **35**, 8

Quinche, J. P. (1976). *Rev. Suisse Agric.* **8**, 143

Rabinowitz, M. B., Wetherill, G. W. and Kopple, J. D. (1973). *Science,* **182**, 725

Rafter, T. A. and O'Brien, B. J. (1970). "Radiocarbon Variations and Absolute Chronology", p. 355 (Olsson, I. U., ed), John Wiley and Sons, London, New York

Rafter, T. A. and Stout, J. D. (1970). "Radiocarbon Variations and Absolute

Chronology", p. 401 (Olsson, I. U., ed), John Wiley and Sons, London, New York
Rahn, K. A. and Winchester, J. W. (1971). Univ. Michigan Tech. Rep. ORA 089030, Ann Arbor, Michigan
Rahola, T., Aaran, R. K. and Miettinen, J. K. (1972). "The Assessment of Radioactive Organs and Body Burdens", p. 553, I.A.E.A., Vienna
Rahola, T., Aaran, R. K. and Miettinen, J. K. (1973). *Proc. Int. Radiat. Prot. Ass., 2nd Eur. Congr. Radiat. Prot.*, 213 (Bujdoso, E., ed), Akad. Kiado, Budapest
Raleigh, G. J. (1939). *Plant Physiol.* **14**, 883
Randhava, N. S. and Broadbent, F. E. (1965). *Soil Sci.* **99**, 295
Rankama, K. (1954). "Isotope Geology", Pergamon Press Ltd., Oxford, New York
Rankama, K. (1963). "Progress in Isotope Geology", Interscience, London, New York
Rankama, K. and Sahama, T. (1950). "Geochemistry", Chicago University Press
Rantu, R. and Sporn, A. (1970). *Nahrung*, **14**, 25
Rasmussen, R. A. and Went, F. W. (1965). *Proc. Nat. Acad. Sci. Wash.* **53**, 215
Rastorfer, J. R. (1974). *Ohio J. Sci.* **74**, 55
Ravindranath, S. D. and Fridovich, I. (1975). *J. Biol. Chem.* **250**, 6107
Rawlings, J., Shaw, V. K., Chisnell, J. R., Brill, W. J., Zimmermann, R., Münch, E. and Orme-Johnson, W. H. (1978). *J. Biol. Chem.* **253**, 1001
Ray, B. J. and Johnson, D. L. (1972). *Analyt. Chim. Acta*, **62**, 196
Reeder, S. W., Hitchon, B. and Levinson, A. A. (1972). *Geochim. Cosmochim. Acta*, **36**, 825
Reeves, A. L. (1965). *Arch. Envir. Health*, **11**, 209
Regnard, C. and Schlösung, T. (1897). *C.R. Acad. Sci. Paris*, **124**, 302
Reiner, E. (1971). "Mechanisms of Toxicity", p. 111 (Aldridge, W. N., ed), Macmillan, London
Renger, G. (1977). *Topics Curr. Chem.* **69**, 39
Rennie, D. A. and Paul, E. A. (1975). "Isotope Ratios and Pollutant Source Behaviour", I.A.E.A., Vienna
Rice, T. R. and Wolfe, D. A. (1971). "Impingement of Man on the Oceans", p. 325 (Hood, D. W., ed), John Wiley and Sons, London, New York
Richards, F. A. (1971). "Impingement of Man on the Oceans", p. 201 (Hood, D. W., ed), John Wiley and Sons, London, New York
Rick, G. R., Petrova, O. N., Misyulk, L. A. and Platonova, L. V. (1961). *Biophys. USSR*, **6**, 834
Riemer, D. N. and Toth, S. J. (1968). *New Jersey Agric. Exp. Bull.* **820**
Riley, J. P. and Chester, R. (1971). "Introduction to Marine Chemistry", Academic Press, London, New York
Riley, J. P. and Roth, I. (1971). *J. Mar. Biol. Ass. U.K.* **51**, 63
Riley, J. P. and Segar, D. A. (1970). *J. Mar. Biol. Ass. U.K.* **50**, 721
Riley, J. P. and Skirrow, G. (1975). "Chemical Oceanography", Academic Press, London, New York
Ringwood, A. E. (1966). *Geochim. Cosmochim. Acta*, **30**, 41
Rissanen, K. and Miettinen, J. K. (1972). "Mercury Contamination in Man and His Environment", p. 5, Tech. Rep. 137, I.A.E.A., Vienna
Ritchie, J. C., McHenry, J. R. and Gill, A. C. (1974). *Ecology*, **55**, 887
Roberts, K. R. and Marzluf, G. A. (1971). *Arch. Biochem. Biophys.* **142**, 651
Robertson, D. E. (1970). *Geochim. Cosmochim. Acta*, **34**, 553
Robinson, E. and Robbins, R. C. (1970a). *Annls. N.Y. Acad. Sci.* **174**, 89

Robinson, E. and Robbins, R. C. (1970b). "Global Effects of Environmental Pollution", p. 50 (Singer, S. F., ed), D. Reidel, Dordrecht, Holland

Roche, J., Fontaine, M. and Leloup, J. (1963). "Comparative Biochemistry", Vol. 5, p. 493 (Florkin, M. and Mason, M. S., eds), Academic Press, London, New York

Roche, J. and Yagi, M. (1952). *C.R. Soc. Biol.* **146**, 642

Rodin, L. E. and Bazilevich, N. I. (1967). "Production and Mineral Cycling in Terrestrial Vegetation", Oliver and Boyd, Edinburgh

Rodin, L. E., Bazilevich, N. I. and Rozov, N. N. (1974). "Productivity of World Ecosystems" (Reichle, D. E. *et al.*, ed), Nat. Acad. Sci., Washington, DC

Roginski, E. and Mertz, W. (1967). *J. Nutrit.* **93**, 240

Romney, E. M. and Childress, J. D. (1963). "Radioecology", p. 521 (Schultz, V. and Klement, A. W., eds), Reinhold, New York

Romney, E. M., Childress, J. D. and Alexander, G. V. (1962). *Science*, **135**, 786

Rosenfeld, T. and Beath, O. A. (1964). "Selenium in Geobotany, Biochemistry, Toxicity and Nutrition", Academic Press, London, New York

Roskoski, R. (1969). *Arch. Biochem. Biophys.* **130**, 561

Rösler, H. J., Schrön, W. and Voland, B. (1968). "Origin and Distribution of the Elements", p. 1127 (Ahrens, L. H., ed), Pergamon Press Ltd., Oxford, New York

Roth, E. (1956). *C.R. Acad. Sci. Paris*, **242**, 3097

Routson, R. C. and Cataldo, D. A. (1978). *Health Phys.* **34**, 685

Rowe, J. J. and Steinnes, E. (1977). *Talanta*, **24**, 438

Roy, A. B. and Trudinger, P. A. (1970). "The Biochemistry of Inorganic Compounds of Sulphur", Cambridge University Press

Ruch, R. R., Cahill, R. A., Frost, J. K., Camp, L. R. and Gluskoter, H. J. (1977). *J. Radioanal. Chem.* **38**, 415; *see also* Illinois State Geol. Surv. Circ. 499 (1977)

Rundo, J. (1967). "Strontium Metabolism", p. 131 (Lenihan, J. M. A., Loutit, J. F. and Martin, J. H., eds), Academic Press, London, New York

Russell, R. S. (ed) (1966). "Radioactivity and Human Diet", Pergamon Press Ltd., Oxford, New York

Russell, W. A., Papanastassiou, D. A. and Tombrello, T. A. (1978). *Geochim. Cosmochim. Acta* **42**, 1075

Ruzicka, F. J. and Beinert, H. (1977). *J. Biol. Chem.* **252**, 8440

Ryden, L. and Björk, I. (1976). *Biochem.* **15**, 3411

Rygh, O. (1949). *Bull. Soc. Chim. Biol.* **31**, 1052 and 1403

Saeki, M., Kunii, K., Seki, T., Sugiyama, K., Suzuki, T. and Shishido, S. (1977). *Envir. Res.* **13**, 256

Saenko, G. D., Koryakova, M. V., Makienko, V. F. and Dobrosmyslova, I. G. (1976). *Mar. Biol.* **34**, 169

Sagers, R. D. (1974). "Microbial Iron Metabolism", p. 445 (Neilands, J. B., ed), Academic Press, London, New York

Salbu, B., Steinnes, E. and Pappas, A. C. (1975). *Analyt. Chem.* **47**, 1011

Salmon, L., Atkins, D. H. F., Fisher, E. M. R., Healy, C. and Law, D. V. (1977). U.K. Atomic Energy Authority Rep. AERE-R 8680, H.M. Stationery Office, London

Sanchez, P. A. (1976). "Properties and Management of Soils in the Tropics", John Wiley and Sons, London, New York

Sandalls, F. J. and Hatton, D. B. (1977). *Atmos. Envir.* **11**, 321

Sandalls, F. J. and Penkett, S. A. (1977), *Atmos. Envir.* **11**, 197

Sandholm, M., Oksanen, H. E. and Personen, L. (1973). *Limnol. Oceanog.* **18**, 496

Sang, J. H. (1956). *J. Exp. Biol.* **33**, 45

San Pietro, A. (ed) (1971). "Methods in Enzymology", Vol. 23, pp. 29 and 110, Academic Press, London, New York

Sarosiek, J. and Klys, B. (1962). *Sudety Acta Soc. Bot. Polon.* **31**, 737

Sawada, M. (1965). *Bull. Tokyo Univ. Forest.* **59**, 33

Scadden, E. M. (1969). *Geochim. Cosmochim. Acta,* **33**, 633

Scharpenseel, H. W., Tamers, M. A. and Pietig, F. (1968). *Z. Pflanzenernähr. Bodenk.* **119**, 44

Scharrer, K. (1955). "Biochemie der Spurenelemente", Parey, Berlin

Scharrer, K. H. and Linser, H. (eds) (1969). "Handbuch der Pflanzenernährung und Dungung", Springer Verlag, Berlin, Heidelberg, New York

Schatzmann, H. J. (1975). *Curr. Topics Membranes Transport,* **6**, 125

Schelenz, R. and Diehl, J. F. (1973). *Z. Lebensm. Unters. Forsch.* **151**, 369

Schelhorn, H., Pfrepper, G. and Geisler, M. (1976). *J. Radioanal. Chem.* **32**, 187

Schell, W. R. (1970). "Radiocarbon Variations and Absolute Chronology", p. 447 (Olsson, I. U., ed), John Wiley and Sons, London, New York

Schlettwein-Gsell, D. and Mommsen-Straub, S. (1970a). *Int. J. Vit. Nutrit. Rev.* **40**, 659

Schlettwein-Gsell, D. and Mommsen-Straub, S. (1970b). *Int. J. Vit. Nutrit. Rev.* **40**, 673

Schlettwein-Gsell, D. and Mommsen-Straub, S. (1971a). *Int. J. Vit. Nutrit. Rev.* **41**, 268

Schlettwein-Gsell, D. and Mommsen-Straub, S. (1971b). *Int. J. Vit. Nutrit. Rev.* **41**, 429

Schlettwein-Gsell, D. and Mommsen-Straub, S. (1971c). *Int. J. Vit. Nutrit. Rev.* **41**, 554

Schmalz, R. F. (1965). *Science,* **149**, 993

Schmitt, J. A., Meisch, H. U. and Reinle, W. (1977). *Z. Naturforsch.* **32C**, 712

Schmitt, J. A., Meisch, H. U. and Reinle, W. (1978). *Z. Naturforsch.* **33C**, 608

Schmitt, N., Brown, G., Devlin, E. L., Larsen, A. A., McCausland, E. D. and Saville, J. M. (1971). *Arch. Envir. Health,* **23**, 185

Schneider, P. W. (1976). "Metal Ions in Biological Systems", p. 197 (Sigel, H., ed), M. Dekker, New York

Schnitzer, M. and Skinner, S. I. M. (1963). *Soil Sci.* **96**, 86 and 181

Schnitzer, M. and Skinner, S. I. M. (1965). *Soil. Sci.* **99**, 278

Schoenborn, B. P., Watson, H. C. and Kendrew, J. C. (1965). *Nature, Lond.* **207**, 28

Schowen, R. L. (1977). "Isotope Effects on Enzyme Catalysed Reactions", p. 64 (Cleland, W. W. *et al.*, eds), University Park Press, Baltimore, Md.

Schraer, H. (ed) (1970). "Biological Calcification", North Holland, Amsterdam

Schramel, P., Samsahl, K. and Pavlu, J. (1974). *J. Radioanal. Chem.* **19**, 329

Schreiber, B. (1967). "Radioecological Concentration Processes", p. 763 (Åberg, B. and Hungate, F. P., eds), Pergamon Press, Oxford, New York

Schreiner, H. R. (1962). *Proc. 22nd Internat. Congr. Physiol., Leiden,* **1**, 535

Schrempf, M., Salter, R. L. and Galston, A. W. (1976). *Plant Physiol.* **58**, 190

Schroeder, H. A. and Balassa, J. J. (1966a). *J. Chron. Dis.* **19**, 85

Schroeder, H. A. and Balassa, J. J. (1966b). *J. Chron. Dis.* **19**, 573

Schroeder, H. A. and Balassa, J. J. (1967). *J. Chron Dis.* **20**, 211

Schroeder, H. A., Balassa, J. J. and Tipton, I. H. (1962). *J. Chron. Dis.* **15**, 51
Schroeder, H. A., Balassa, J. J. and Tipton, I. H. (1963). *J. Chron. Dis.* **16**, 1047
Schroeder, H. A., Balassa, J. J. and Tipton, I. H. (1964). *J. Chron. Dis.* **17**, 483
Schroeder, H. A., Balassa, J. J. and Tipton, I. H. (1966). *J. Chron. Dis.* **19**, 545
Schroeder, H. A. and Darrow, D. K. (1973). "Chemical Analysis of the Environment", p. 81 (Ahuja, S. *et al.*, eds), Plenum Press, New York, London
Schroeder, H. A., Frost, D. V. and Balassa, J. J. (1970). *J. Chron. Dis.* **23**, 227
Schroeder, H. A. and Nason, A. P. (1971). *Clin. Chem.* **17**, 461
Schroeder, H. A., Nason, A. P., Tipton, I. H. and Balassa, J. J. (1967). *J. Chron. Dis.* **20**, 179
Schroeder, H. A., Tipton, I. H. and Nason, A. P. (1972). *J. Chron. Dis.* **25**, 491
Schröter, K., Läuchli, A. and Sievers, A. (1975). *Planta*, **122**, 213
Schulz, R. K., Tompkins, G. A., Leventhal, L. and Babcock, K. L. (1976). *J. Envir. Quality*, **5**, 406
Schutte, K. H. (1964). "The Biology of the Trace Elements", Crosby Lockwood, St. Albans
Schutyser, P., Govaerts, A., Dams, R. and Hoste, J. (1977). *J. Radioanal. Chem.* **37**, 651
Schutz, D. F. and Turekian, K. K. (1965). *Geochim. Cosmochim. Acta*, **29**, 259
Schütz, K., Junge, C., Beck, R. and Albrecht, B. (1970). *J. Geophys. Res.* **75**, 2230
Schwarz, K. (1971). "Newer Trace Elements in Nutrition", p. 314 (Mertz, W. and Cornatzer, W. E., eds), M. Dekker, New York
Schwarz, K. (1973). *Proc. Nat. Acad. Sci. Wash.* **70**, 1608
Schwarz, K. and Foltz, M. (1957). *J. Amer. Chem. Soc.* **79**, 3293
Schwarz, K. and Mertz, W. (1959). *Arch. Biochem. Biophys.* **85**, 292
Schwarz, K. and Milne, D. B. (1971). *Science*, **174**, 426
Schwarz, K. and Milne, D. B. (1972a). *Bioinorg. Chem.* **1**, 331
Schwarz, K. and Milne, D. B. (1972b). *Nature, Lond.* **239**, 333
Schwarz, K., Milne, D. B. and Vinyard, E. (1970). *Biochem. Biophys. Res. Comm.* **40**, 22
Sclater, F. R., Boyle, E. and Edmond, J. M. (1976). *Earth Planet. Sci. Lett.* **31**, 119
Scrutton, M. C., Utter, M. F. and Mildvan, A. S. (1966). *J. Biol. Chem.* **241**, 3480
Scurfield, G. (1960). *For. Abstr.* **21**, 1
Segalen, P. (1973). "L'aluminium dans les Sols", ORSTOM, Paris
Segar, D. A., Collins, J. D. and Riley, J. P. (1971). *J. Mar. Biol. Ass. U.K.* **51**, 131
Seltzer, J. L., Welgus, H. G., Jeffrey, J. J. and Eisen, A. Z. (1976). *Arch. Biochem. Biophys.* **173**, 355
Sengupta, S. R. and Pal, B. (1971). *Indian J. Nutr. Diet*, **8**, 66
Shacklette, H. T. (1965). *U.S. Geol. Surv. Bull.* **1198D**
Shacklette, H. T. (1967). *U.S. Geol. Surv. Bull.* **1198G**
Shacklette, H. T. (1970). *U.S. Geol. Surv. Prof. Paper*, **713**, 35
Shacklette, H. T. (1972). *U.S. Geol. Surv. Bull.* **1314G**
Shacklette, H. T., Boerngen, J. G., Cahill, J. P. and Rahill, R. L. (1973). *U.S. Geol. Surv. Circ.* 673
Shacklette, H. T. and Cuthbert, M. E. (1967). *U.S. Geol. Surv. Spec. Paper*, **90**, 31
Shah, K. R., Filby, R. H. and Haller, W. A. (1970). *J. Radioanal. Chem.* **6**, 185 and 413
Shah, V. K. and Brill, W. J. (1977). *Proc. Nat. Acad. Sci. Wash.* **74**, 3249
Shaikh, A. U. and Tallman, D. E. (1978). *Analyt. Chim. Acta*, **98**, 251
Sharp, J. H. (1973). *Limnol. Oceanog.* **18**, 441

Shaw, D. M., Dostal, J. and Keays, R. R. (1976). *Geochim. Cosmochim. Acta*, **40**, 73
Shawe, D. R. (1976). *U.S. Geol. Surv. Prof. Paper*, **937**
Sheppard, J. C. and Funk, W. H. (1975). *Envir. Sci. Technol.* **9**, 638
Shidlovsky, M. (1965). *Nature, Lond.* **205**, 895
Shiegl, W. E. and Vogel, J. C. (1970). *Earth Planet. Sci. Lett.* **7**, 307
Shils, M. E. (1964). *Amer. J. Clin. Nutrit.* **15**, 133
Shimmins, J., Hilditch, T., Harden, R. McG. and Alexander, W. D. (1965). *J. Clin. Endoc. Metab.* **28**, 575
Shirahata, K., Deguchi, T., Hayashi, T., Matsubara, T. and Suzuki, T. (1970). *J. Antibiot.* **23**, 546
Shrift, A. (1972). "Phytochemical Ecology", p. 145 (Harborne, J. B., ed), Academic Press, London, New York
Shrift, A. (1973). "Organic Selenium Compounds", p. 763 (Klayman, D. L. and Günther, W. H. H., eds), John Wiley and Sons, London, New York
Shukla, P. N., Kothari, B. K. and Goel, P. S. (1978). *Analyt. Chim. Acta* **96**, 259
Shum, A. C. and Murphy, J. C. (1972). *J. Bact.* **110**, 447
Sibley, D. F. and Wilblad, J. T. (1977). *Geochim. Cosmochim. Acta*, **41**, 545
Siebert, G., Lang, K. and Lang, H. (1951). *Biochem. Z.* **321**, 543
Siebert, G. and Langendorf, H. (1970). *Naturwiss*, **57**, 119
Siegel, B. Z., Siegel, S. M. and Thoraninsson, F. (1973). *Nature, Lond.* **241**, 526
Siegel, G. J. and Fogt, S. K. (1976). *Arch. Biochem. Biophys.* **174**, 744
Siegel, L. M., Faeder, E. J. and Kamin, H. (1972). *Z. Naturforsch.* **27B**, 1087
Siegel, S. M. (1962). "The Plant Cell Wall", Pergamon Press Ltd., Oxford, New York
Siegel, S. M. and Siegel, B. Z. (1975). *Envir. Sci. Technol.* **9**, 473
Silker, W. B. (1964). *Limnol. Oceanog.* **9**, 540
Silver, S. and Clark, D. (1971). *J. Biol. Chem.* **246**, 569
Silver, S., Johnseine, P. and King, K. (1970). *J. Bact.* **104**, 1299
Silverman, M. P. and Ehrlich, H. L. (1964). *Adv. Appl. Microbiol.* **6**, 153
Simkiss, K. (1967). "Calcium in Reproductive Physiology", Chapman and Hall, London
Simkiss, K. (1975). "Bone and Biomineralization", E. Arnold, London
Simkiss, K. (1976). *Sympos. Soc. Exp. Biol.* **30**, 423
Simonson, R. W. (1970). "Nutrient Mobility in Soils", p. 21 (Engelstad, O. P., ed), Soil Sci. Soc. Amer., Madison, Wisc.
Singer, S. F. (ed) (1970). "Global Effects of Environmental Pollution", D. Riedel, Holland
Singh, H. B., Salas, L., Shigeishi, H. and Crawford, A. (1977). *Atmos. Envir.* **11**, 819
Singh, J. F., Babcock, D. F. and Lardy, H. A. (1978). *Biochem. J.* **172**, 549
Skirrow, G. (1975). "Chemical Oceanography", Vol. 2, p. 1 (Riley, J. P. and Skirrow, G., eds), Academic Press, London, New York
Slayman, C. L. and Slayman, C. W. (1968). *J. Gen. Physiol.* **52**, 424
Slowey, J. F. and Hood, D. W. (1971). *Geochim. Cosmochim. Acta*, **35**, 121
Slykhouse, T. O. and Fee, J. A. (1976). *J. Biol. Chem.* **251**, 5472
Smales, A. A. and Salmon, L. (1955). *Analyst*, **80**, 37
Smith, B. N. and Epstein, S. (1971). *Plant Physiol.* **47**, 380
Smith, C. L., De Laeter, J. R. and Rosman, K. J. R. (1977). *Geochim. Cosmochim. Acta*, **41**, 676
Smith, I. K. (1976). *Plant Physiol.* **58**, 358

Smith, I. C. (1978). "Trace Metals in the Environment", Vol. 1 (Thallium), John Wiley and Sons, London, New York

Smith, I. C. and Carson, B. L. (1978). "Trace Metals in the Environment", Vol. 2 (Silver), Vol. 3 (Zirconium), Vol. 4 (Palladium and Osmium), Vol. 5 (Indium), John Wiley and Sons, London, New York

Smith, J. C. (1969). "Trace Substances in Environmental Health", Vol. 2, p. 223 (Hemphill, D., ed), Columbia University, Missouri

Smith, J. D. and Burton, J. D. (1972). *Geochim. Cosmochim. Acta*, **36**, 621

Smith, K. A. and Mercer, E. R. (1970). *J. Radioanal. Chem.* **5**, 303

Smith, L. E. H., Mohr, L. H. and Raftery, M. A. (1973). *J. Amer. Chem. Soc.* **95**, 7497

Smith, P. F. (1962). *Annu. Rev. Plant Physiol.* **13**, 81

Smith, R. M. and Gawthorne, J. M. (1975). "Trace Elements in Soil-Plant-Animal Systems", p. 243 (Nicholas, D. J. D. and Egan, A. R., eds), Academic Press, London, New York

Snyder, W. S. (ed) (1975). "Report of Task Group on Reference Man", Pergamon Press Ltd, Oxford, New York

Solberg, Y. J. (1967). *Annls. Bot. Fenn.* **4**, 29

Soldano, B. A., Bien, P. and Kwan, P. (1975). *Atmos. Envir.* **9**, 941

Solomon, E. I., Hare, J. W. and Gray, H. B. (1976). *Proc. Nat. Acad. Sci. Wash.* **73**, 1389

Somers, E. (1959). *Nature, Lond.* **184**, 475

Sommer, A. L. (1931). *Plant Physiol.* **6**, 339

Sone, N. (1971). *J. Biochem. Tokyo*, **69**, 753

Sorenson, J. R. J. (1976). "Environmental Biogeochemistry", Vol. 2, p. 427 (Nriagu, J. O., ed), Ann Arbor Science, Ann Arbor, Mich.

Sorenson, J. R. J., Campbell, I. R., Tepper, L. B. and Lingg, R. D. (1974). *Envir. Health Perspect.* **8**, 3

Soroiu, M. M. and Cerci, M. (1967). *Geochim. Cosmochim. Acta*, **31**, 1271

Spanswick, R. M. and Williams, E. J. (1964). *J. Exp. Bot.* **15**, 193

Spector, W. G. (1953). *Proc. Roy. Soc.* **B141**, 268

Spector, W. S. (1956a). "Handbook of Biological Data", W. B. Saunders, Philadelphia

Spector, W. S. (1956b). "Handbook of Toxicology", Vol. 1, W. B. Saunders, Philadelphia

Spedding, D. J. (1974). "Air Pollution", Clarendon Press, Oxford

Spiers, F. W. (1968). "Radioisotopes in the Human Body", Academic Press, London, New York

Spotte, S. H. (1970). "Fish and Invertebrate Culture", John Wiley and Sons, London, New York

Sprague, J. P., Elsom, P. F. and Saunders, R. L. (1965). *Int. J. Air Water Pollut.* **9**, 531

Sprung, J. L. (1976). *Adv. Envir. Sci. Technol.* **7**, 263

Sreekumaran, C., Pillai, K. C. and Folsom, T. R. (1968). *Geochim. Cosmochim. Acta*, **32**, 1229

Stahl, W. (1976). "Nuclear Techniques in Geochemistry and Geophysics", p. 213, IAEA, Vienna

Stärk, H., Süss, A. and Trojan, K. (1971). *Landwirts. Forschung.* **24**, 129

Steenkamp, D. J., Singer, T. P. and Beinert, H. (1978). *Biochem. J.* **169**, 361

Steevens, D. R., Walsh, L. M. and Keeney, D. R. (1972). *J. Envir. Quality*, **1**, 317

Steinnes, E. (1977a). *Talanta*, **24**, 121

Steinnes, E. (1977b). *Kjeller Rep.* **154**, Oslo, Norway

Steinnes, E. and Mukherjee, A. D. (1973). *J. Radioanal. Chem.* **14**, 129

Steinnes, E. and Rowe, J. J. (1976). *Analyt. Chim. Acta*, **87**, 451

Stella, R., Genova, N. and Di Casa, M. (1977). *Radiochem. Radioanal. Lett.* **30**, 65

Stenflo, J. (1976). *J. Biol. Chem.* **251**, 355

Stern, A. C. (ed) (1968). "Air Pollution", Academic Press, London, New York

Stevens, C. M., Krout, L., Walling, D., Venters, A., Engelkemier, A. and Ross, L. E. (1972). *Earth Planet. Sci. Lett.* **16**, 147

Stevenson, F. J. (1965). "Soil Nitrogen", Vol. 1 (Bartholomew, W. V. and Clark, F. E., eds), Amer. Sci. Agron., Madison, Wisc.

Stevenson, F. J. (1977). *Soil Sci.* **123**, 10

Stevenson, R. A. and Ufret, S. L. (1966). *Limnol. Oceanog.* **11**, 11

Stewart, B. A. (1970). "Nutrient Mobility in Soils", p. 47 (Engelstad, O. P., ed), Soil Sci. Soc. Amer., Madison, Wisc.

Stickland, L. H. (1949). *Biochem. J.* **44**, 190

Stijve, T. and Besson, R. (1976). *Chemosphere*, **5**, 151

Stiles, W. (1958). "Encyclopaedia of Plant Physiology", Vol. 4, p. 558 (Ruhland, W., ed), Springer Verlag, Berlin, Heidelberg, New York

Stirpe, F. and Novello, F. (1971). "Mechanisms of Toxicity", p. 147 (Aldridge, W. N., ed), Macmillan, London

Strahler, A. N. (1972). "Planet Earth", Harper and Row, New York, London

Strohal, P. and Pinter, J. (1973). *Limnol. Oceanog.* **18**, 250

Struempler, A. W. (1975). *Envir. Sci. Technol.* **9**, 1164

Stock, A. and Cucuel, F. (1934). *Naturwiss.* **22**, 390

Stuiver, M. (1967). *Geochim. Cosmochim. Acta*, **31**, 2151

Stumm, W. (ed) (1977). "Global Chemical Cycles and their Alteration by Man", Dahlem Conference, Berlin

Stumm, W. and Brauner, P. A. (1975). "Chemical Oceanography", p. 173 (Riley, J. P. and Skirrow, G., eds), Academic Press, London, New York

Stumm, W. and Morgan, J. W. (1970). "Aquatic Chemistry", John Wiley and Sons, London, New York

Suelter, G. H. (1970). *Science*, **168**, 789

Sugimura, Y., Suzuki, Y. and Miyake, Y. (1976). *J. Oceanog. Soc. Japan*, **32**, 235

Sugiwara, K. (1968). "Origin and Distribution of the Elements", p. 1017 (Ahrens, L. H., ed), Pergamon Press Ltd., Oxford, New York

Sullivan, B., Bonaventura, J., Bonaventura, C. and Godette, G. (1976). *J. Biol. Chem.* **251**, 7644

Summers, A. O. and Jacoby, G. A. (1977). *J. Bact.* **129**, 276

Summers, A. O. and Sugarman, L. I. (1974). *J. Bact.* **119**, 242

Suzuki, S. and Kondo, Y. (1971). *Gen. Comp. Endocrinol.* **17**, 402

Swain, C. G., Stivers, E. C., Reuwer, J. F. and Schaad, L. J. (1958). *J. Amer. Chem. Soc.* **80**, 5885

Swaine, D. J. (1955). Commonwealth Bur. Soil Sci. Tech. Comm. 48, Commonwealth Agric. Bur., Farnham

Swaine, D. J. (1962). Commonwealth Bur. Soil Sci. Tech. Comm. 52, Commonwealth Agric. Bur., Farnham

Szabo, B. J. (1967). *Geochim. Cosmochim. Acta*, **31**, 1321

Takashima, Y., Momoshima, N. and Hamidian, M. R. (1978b). *Radiochem. Radioanal. Lett.* **36**, 69

Takashima, Y., Momoshima, N., Nakayama, Y., Yamashita, H. and Nakashima, H. (1978a). *Radiochem. Radioanal. Lett.* **36**, 61
Tal, M. (1968). *Biochim. Biophys. Acta*, **169**, 564
Tammes, P. M. L. and Van Die, J. (1964). *Acta Bot. Neerl.* **13**, 76
Tanner, J. T. and Friedman, M. H. (1977). *J. Radioanal. Chem.* **37**, 529
Task group on metal accumulation (1973). *Envir. Physiol. Biochem.* **3**, 65
Taves, D. R. (1968). *Nature, Lond.* **217**, 1050
Taves, D. R. (1978). *Nature, Lond.* **272**, 361
Taylor, H. P. (1978). *Earth Planet. Sci. Lett.* **38**, 177
Taylor, S. R. (1964). *Geochim. Cosmochim. Acta*, **28**, 1273
Tazieff, H. (1970). *Bull. Volcanol.* **34**, 421
Tellam, R., Winzor, D. J. and Nichol, L. W. (1978). *Biochem. J.* **173**, 185
Tepper, L. B., Hardy, H. L. and Chamberlin, R. I. (1961). "Toxicity of Beryllium Compounds", Elsevier, Amsterdam, London, New York
Theis, T. L. and Singer, P. C. (1973). "Trace Metals and Metal-Organic Interactions in Natural Waters", p. 303 (Singer, P. C. ed), Ann Arbor Science, Ann Arbor, Mich.
Thiers, R. E. and Vallee, B. L. (1957). *J. Biol. Chem.* **226**, 911
Thode, H. G., Kleerekoper, H. and McElcheran, D.. (1951). *Research*, **4**, 581
Thode, H. G., Monster, J. and Dunford, H. B. (1961). *Geochim. Cosmochim. Acta*, **25**, 159
Thode, H. G., Wanless, R. K. and Wallouch, R. (1954). *Geochim. Cosmochim. Acta*, **5**, 286
Thomas, B., Roughan, J. A. and Watters, E. D. (1974). *J. Sci. Food Agric.* **25**, 771
Thomas, W. A. and Jacobs, D. G. (1969). *Soil Sci.* **108**, 305
Thompson, J. N. and Scott, M. L. (1969). *J. Nutrit.* **97**, 305
Thomson, J. F. (1960). *Arch. Biochem. Biophys.* **90**, 1
Thornton, I., Watling, H. and Darracott, A. (1975). *Sci. Total Envir.* **4**, 325
Thornton, I. and Webb, J. S. (1975). "Trace Substances in Environmental Health", Vol. 9, p. 77 (Hemphill, D., ed), Columbia University, Missouri
Tiffin, L. O. (1972). "Micronutrients in Agriculture", p. 199 (Mortvedt, J. J. *et al.*, eds), Soil. Sci. Soc. Amer., Madison, Wisc.
Timofeeva-Resovskaya, E. A., Timofeev-Resovsky, N. V. and Gileva, E. A. (1961). *Dokl. Akad. Nauk. SSSR*, **140**, 1437
Tipton, K. F. (1965). *Biochim. Biophys. Acta*, **110**, 414
Tjell, J. C. and Hovmand, M. F. (1978). *Acta Agric. Scand.* **28**, 81
Todd, J. R. (1962). *Vet. Record*, **74**, 116
Townsley, P. M. and Neilands, J. B. (1957). *J. Biol. Chem.* **224**, 695
Tracy, O. (1969). *J. Med. Microbiol.* **2**, 309
Tsivoglou, E. C., Bartsch, A. F., Rushing, D. E. and Holaday, D. A. (1958). *Sewage Indust. Wastes*, **30**, 1012
Tsubaki, T. and Irukayama, K. (eds) (1977). "Minamata Disease", Elsevier, Amsterdam, London, New York
Tuominen, Y. and Jaakola, T. (1973). "The Lichens", p. 185 (Ahmadjian, V. and Hale, M. E., eds), Academic Press, London, New York
Turekian, K. K. (1969). "Handbook of Geochemistry", p. 296 (Wedepohl, K. H., ed), Springer Verlag, Berlin, Heidelberg, New York
Turekian, K. K. (1971). "Impingement of Man on the Oceans", p. 9 (Hood, D. W., ed), John Wiley and Sons, London, New York
Turekian, K. K., Katz, A. and Chan, L. (1973). *Limnol. Oceanog.* **18**, 240

Turekian, K. K. and Wedepohl, K. H. (1961). *Bull. Geol. Soc. Amer.* **72**, 175
Turel, Z. R., Merchant, J. R. and Haldar, B. C. (1978). *J. Radioanal. Chem.* **45**, 203
Turner, M. A. and Rust, R. H. (1971). *Soil. Sci. Soc. Amer. Proc.* **35**, 755
Turner, R. C., Radley, J. M. and Mayneord, W. V. (1958). *Health Phys.* **1**, 268
Tweedle, M. F., Wilson, L. J., Garcia-Iniguez, L., Babcock, G. T. and Palmer, G. (1978). *J. Biol. Chem.* **253**, 8065
Tyler, G. (1974). *Plant Soil*, **41**, 303
Tyler, G. (1976). *Soil Biol. Biochem.* **8**, 327
Tyutina, N. A., Aleskovsky, V. B. and Vasilev, P. I. (1959). *Geochemistry, USSR*, no. 6 668
Udodov, P. A. and Parilov, Y. S. (1961). *Geokhimiya*, no. 8, 703
Underwood, E. J. (1976). "Trace Elements in Human Health and Disease", Vol. 2, p. 269 (Prasad, A. S., ed), Academic Press, London, New York
Underwood, E. J. (1977). "Trace Elements in Human and Animal Nutrition", 4th Ed., Academic Press, London, New York
Underwood, E. J. and Filmer, J. F. (1935). *Austral. Vet. J.* **11**, 84
U.N. (1977). World Energy Supplies, 1971–5, series J, no. 20, United Nations, New York
Unesco (1971). "Discharge of Selected Rivers of the World"
Ure, A. M. and Bacon, J. R. (1978). *Analyst*, **103**, 807
Urey, H. C., Lowenstam, H. A., Epstein, S. and McKinney, C. R. (1951). *Bull. Geol. Soc. Amer.* **62**, 399
U.S. Dept. Agric. (1975). Handbook 436, Soil Taxonomy, U.S. Govt., Washington, DC
U.S. Geol. Survey (1976). Water Supply Paper 2160
Valente, I. (1978). Ph.D. Thesis, Reading University
Valkovic, V. (1977). "Trace Elements in Human Hair", Garland S.T.P.M.
Vallee, B. L. (1976). *Trends Biochem. Sci.* **1**, 88
Van As, D., Fourie, H. O. and Loggaar, C. M. V. (1975). *S. Afr. J. Sci.* **71**, 151
Van der Klugt, N., Poelstra, P. and Zwemmer, E. (1977). *J. Radioanal. Chem.* **35**, 109
Van der Sloot, H. A. and Das, H. A. (1974). *Analyt. Chim. Acta*, **73**, 235
Van der Sloot, H. A., Massie, R. and Das, H. A. (1975). *J. Radioanal. Chem.* **25**, 99
Van Goor, B. J. and Wiersma, D. (1976). *Physiol. Plant.* **36**, 213
Van Hook, R. I., Harris, W. F. and Henderson, G. S. (1977). *Ambio*, **6**, 281
Van Stosch, H. A. (1963). Proc. 4th Internat. Seaweed Sympos., p. 142
Van der Weijen, C. H. (1975). Reactor Centrum Nederland Rep. RCN 221, Petten, Holland
Vaskovsky, V. E., Korotchenko, O. D., Kosheleva, L. P. and Levin, V. S. (1972). *Comp. Biochem. Physiol.* **41B**, 777
Vassilaki, M., Salmon, L. and Gibson, J. A. B. (1966). *Geochim. Cosmochim. Acta*, **30**, 601
Veal, D. J. (1966). *Analyt. Chem.* **38**, 1080
Veeh, H. H. and Turekian, K. K. (1968). *Limnol. Oceanog.* **13**, 304
Vega, J. M. and Kamin, M. (1977). *J. Biol. Chem.* **252**, 896
Venugopal, B. and Luckey, T. D. (1978). "Metal Toxicity in Mammals", Vol. 2, Plenum Press, New York, London
Verkhovskaya, I. N., Vavilov, P. P. and Maslov, V. I. (1967). "Radioecological

Concentration Processes", p. 313 (Åberg, B. and Hungate, F. P., eds), Pergamon Press Ltd., Oxford, New York
Vinogradov, A. P. (1953). "The Elementary Chemical Composition of Marine Organisms", Sears Foundation, New Haven, Conn.
Vinogradov, A. P. (1959). "Geochemistry of Rare and Dispersed Elements in Soils", Chapman and Hall, London
Vinogradov, A. P., Kutyurin, V. M. and Zadorozhnyi, I. K. (1959). *Geochem. USSR*, no. 4, 221
Vinogradova, Z. A. and Kovalsky, V. V. (1962). *Dokl. Akad. Nauk. SSSR*, **147**, 1458
Visser, S. A. and Villeneuve, J. P. (1975). *Verh. Int. Verein. Limnol.* **19**, 1416
Volchok, H. L. and Bogen, D. C. (1973). *Progr. Analyt. Chem.* **5**, 43
Volini, M., Van Sweringen, B. and Chen, F. S. (1978). *Arch. Biochem. Biophys.* **191**, 205
Voronkov, M. G., Zelchan, G. I. and Lukevits, E. J. (1971). "Silicon and Life", Zinatne Press, Riga
Wacker, W. E. C. and Vallee, B. L. (1959). *J. Biol. Chem.* **234**, 3257
Wada, E. and Hattori, A. (1976). *Geochim. Cosmochim. Acta*, **40**, 249
Wagener, K. (1975). Phys. Chem. Sci. Rep. 1, 433; The nature of sea water (Goldberg, E. D., ed)
Wager, L. M. and Brown, G. M. (1968). "Layered Igneous Rocks", Oliver and Boyd, Edinburgh
Wahlgren, M. A., Alberts, J. J., Nelson, D. M. and Orlandini, K. A. (1976). "Transuranium Nuclides in the Environment", p. 9, I.A.E.A., Vienna
Wakatsuki, T., Furukawa, H. and Kyuma, H. (1977). *Geochim. Cosmochim. Acta*, **41**, 891
Waksman, S. A. (1952). "Soil Microbiology", John Wiley and Sons, London, New York
Walch, R. M. and Huffman, E. W. D. (1973). *Plant Physiol.* **52**, 183
Waldron, H. A. and Stöfen, D. (1974). "Sub-Clinical Lead Poisoning", Academic Press, London, New York
Walker, E. C., Cuttitta, F. and Senftle, F. E. (1958). *Geochim. Cosmochim. Acta*, **15**, 183
Walker, G. (1977). *Mar. Biol.* **39**, 343
Walker, G., Rainbow, P. S., Foster, P. and Holland, D. L. (1975). *Mar. Biol.* **33**, 161
Walker, J. B. (1953). *Arch. Biochem. Biophys.* **46**, 1
Wallace, J. M. and Hobbs, P. V. (1977). "Atmospheric Science", Academic Press, London, New York
Wangen, L. E. and Gladney, E. S. (1978). *Analyt. Chim. Acta*, **96**, 271
Ward, F. N. and Lakin, H. W. (1954). *Analyt. Chem.* **26**, 1168
Ward, N. I., Brooks, R. R. and Roberts, E. (1977). *Bryologist* **80**, 304
Warren, H. V. and Delavault, R. E. (1960). *Bull. Geol. Soc. Amer.* **61**, 123
Warrington, K. (1923). *Annls. Bot.* **27**, 629
Waslien, C. I. (1976). "Trace Elements in Human Health and Disease", p. 347 (Prasad, A. S., ed), Academic Press, London, New York
Wasserman, R. H., Corradino, R. A., Carafoli, E., Kretsinger, R. H., MacLennan, D. H. and Siegel, F. L. (eds) (1977). "Calcium-Binding Proteins and Calcium Function", North-Holland, Amsterdam
Watling, H. R. and Watling, R. J. (1976). *Mar. Pollut. Bull.* **7**, 91

Weaver, J. N. (1973). *Analyt. Chem.* **45**, 1950

Webb, J. L. (1965). "Enzymes and Metabolic Inhibitors", Vol. 2, p. 729, Academic Press, London, New York

Webb, J. (1975). "Techniques and Topics in Bio-Inorganic Chemistry", p. 271 (McAuliffe, C. A., ed), Macmillan, London

Weber, J. N. (1968). *Geochim. Cosmochim. Acta*, **32**, 33

Weber, J. N., Deines, P., Weber, P. H. and Baker, P. A. (1976). *Geochim. Cosmochim. Acta*, **40**, 31

Wedepohl, K. H. (1968). "Origin and Distribution of the Elements", p. 999 (Ahrens, L. H., ed), Pergamon Press Ltd., Oxford, New York

Wedepohl, K. H. (1969). "Handbook of Geochemistry", pp. 227 and 250 (Wedepohl, K. H., ed), Springer Verlag, Berlin, Heidelberg, New York

Wedepohl, K. H. (1969–1974). "Handbook of Geochemistry", Springer Verlag, Berlin, Heidelberg, New York

Weinberg, E. D. (1971). *J. Infect. Dis.* **124,** 401

Weinstock, B. and Chang, T. Y. (1976). "Environmental Biogeochemistry", Vol. 1, p. 39 (Nriagu, J. O., ed), Ann Arbor Science, Ann Arbor, Mich.

Weisiger, R. A. and Fridovich, I. (1973). *J. Biol. Chem.* **248**, 3582

Weiss, H. V. and Crozier, T. E. (1972). *Analyt. Chim. Acta*, **58**, 231

Wellman, R. P., Cook, F. D. and Krouse, H. R. (1968). *Science*, **161**, 269

Wester, P. O. (1965). *Biochim. Biophys. Acta*, **109**, 268

Wester, P. O. (1971). *Acta Med. Scand.* **190**, 155

Wester, P. O. (1973). *Acta Med. Scand.* **194**, 505

Wester, P. O. (1974). *Atherosclerosis* **20**, 207

Whanger, P. D., Pedersen, N. D. and Weswig, P. H. (1973). *Biochem. Biophys. Res. Comm.* **53**, 1031

Whicker, F. W., Farris, G. C. and Dahl, A. H. (1967). "Radioecological Concentration Processes", p. 621 (Åberg, B. and Hungate, F. P., eds), Pergamon Press Ltd., Oxford, New York

Whittaker, R. H. (1975). "Communities and Ecosystems", 2nd Ed., Macmillan, London

Whitton, B. A. (ed) (1975). "River Ecology", Blackwell, Oxford

Wiessner, W. (1962). "Physiology and Biochemistry of Algae", p. 267 (Lewin, R. A., ed), Academic Press, London, New York

Wigglesworth, V. B. (1964). "The Life of Insects", Weidenfeld and Nicolson, London

Wilgus, H. S., Norris, L. C. and Heuser, G. F. (1936). *Science*, **84**, 252

Williams, J. (ed) (1978). "Carbon Dioxide, Climate and Society", Pergamon Press Ltd., London, New York

Williams, P. J. Le B. (1975). "Chemical Oceanography", Vol. 2, p. 301 (Riley, J. P. and Skirrow, G., eds), Academic Press, London, New York

Williams, R. J. P. and Da Silva, J. R. R. F. (eds) (1978). "New Trends in Bio-Inorganic Chemistry", Academic Press, London, New York

Williams, S. R. (1977). "Nutrition and Diet Therapy", C. V. Mosby, St. Louis, Mo.

Williamson, M. B. and Gulick, A. (1944). *J. Cell. Comp. Physiol.* **23**, 77

Williamson, P. and Evans, P. R. (1972). *Bull. Envir. Contam. Toxicol.* **8**, 280

Williamson, P. and Evans, P. R. (1973). *Pedobiol.* **13**, 16

Wilson, A. T. (1978). *Nature, Lond.* **273**, 40

Wilson, A. T. and Grinsted, M. J. (1975). *Nature, Lond.* **257**, 387

Wilson, S. B. and Nicholas, D. J. D. (1967). *Phytochem.* **6**, 1057

Winge, D. R., Premakumar, R., Wiley, R. D. and Rajagopalan, K. V. (1975). *Arch. Biochem. Biophys.* **170**, 253
Winton, E. F., Tardiff, R. G. and McCabe, L. J. (1971). *J. Amer. Water Works Ass.* **63**, 95
Wither, E. D. and Brooks, R. R. (1977). *J. Geochem. Explor.* **8**, 579
Witherspoon, J. D. (1964). *Ecol. Monog.* **34**, 403
Wolfenden, E. B. (1965). *Geochim. Cosmochim. Acta*, **29**, 1051
Wolff, J. and Corelli, J. (1969). *Eur. J. Biochem.* **9**, 371
Wong, G. T. F. and Brewer, P. G. (1976). *Analyt. Chem. Acta*, **81**, 81
Wong, G. T. F. and Brewer, P. G.. (1977). *Geochim. Cosmochim. Acta*, **41**, 151
Wong, P. T. S., Chan, Y. K. and Luxon, P. L. (1975). *Nature, Lond.* **253**, 263
Wong, W. W. and Sackett, W. M. (1978). *Geochim. Cosmochim. Acta* **42**, 1809
Wood, J. M. (1974). *Science*, **183**, 1049
Wood, J. M., Kennedy, F. S. and Roser, C. G. (1968). *Nature, Lond.* **220**, 173
Wood, J. M., Segall, H. J., Ridley, W. P., Cheh, A., Chudyk, W. and Thayer, J. S. (1977). "International Conference on Heavy Metals in the Environment", Vol. 1, p. 49 (Hutchinson, T. C., ed), University of Toronto, Canada
Woolard, F. X., Moore, R. E. and Roller, P. P. (1976). *Tetrahedron*, **32**, 2843
Woolfolk, C. A. and Whiteley, H. R. (1962). *J. Bact.* **84**, 647
Woolley, J. T. (1957). *Plant Physiol.* **32**, 317
Wozniak, W. (1975). *Rocz. Akad. Roln. Poznan.* **79**, 129
Wrenn, M. E. and Cohen, N. (1967). *Health Phys.* **13**, 1075
Wu, M. and Wacksman, J. T. (1971). *J. Bact.* **105**, 1222
Wyn Jones, R. G. and Lunt, O. R. (1967). *Bot. Rev.* **33**, 407
Yamada-Ankei, T., Iwasaki, H. and Mori, T. (1977). *J. Biochem. Tokyo*, **81**, 835
Yamagata, N. (1962). *Nature, Lond.* **196**, 83
Yamagata, N., Kurioka, W. and Shimiza, T. (1963). *J. Radiat. Res.* **4**, 8
Yamagata, N. and Shigematsu, I. (1970). *Bull. Inst. Publ. Health*, **9**, 1
Yamagata, N., Yamagata, T. and Matsuda, S. (1959). *Bull. Chem. Soc. Japan*, **32**, 407
Yamamoto, T., Fujita, T. and Ishibashi, M. (1970). *Rec. Oceanog. Works, Japan*, **10**, 125
Yamamoto, T., Fujita, T. and Shigematsu, T. (1969). *Rec. Oceanog. Works, Japan*, **10**, 29
Yamamoto, T., Fujita., Shigematsu, T. and Ishibashi, M. (1968). *Rec. Oceanog. Works, Japan*, **9**, 209
Yamamoto, T., Yamaoka, T., Fujita, T. and Isoda, C. (1973). *Rec. Oceanog. Works, Japan*, **12**, 20
Yamanaka, T. and Okunuki, K. (1974). "Microbial Iron Metabolism", p. 349 (Neilands, J. B., ed), Academic Press, London, New York
Yapp, C. J. and Epstein, S. (1977). *Earth Planet. Sci. Lett.* **34**, 333
Yen, T. F. (ed) (1975). "The Role of Trace Metals in Petroleum", Ann Arbor Science, Ann Arbor, Mich.
Yliruokanen, I. (1975a). *Bull. Geol. Soc. Finland*, **47**, 71
Yliruokanen, I. (1975b). *Annl. Acad. Sci. Fenn.* **A176**, 1
Yoch, D. C., Arnon, D. I. and Sweeney, W. V. (1975). *J. Biol. Chem.* **250**, 8330
Young, E. G., Begg, R. W. and Pentz, E. I. (1944). *Arch. Biochem.* **5**, 121
Young, E. G. and Langille, W. M. (1958). *Canad. J. Bot.* **36**, 301
Young, R. S. (1935). *Mem. Cornell Univ. Agric. Exp. Sta.* **174**

Young, R. S. (1979). "Cobalt in Biology and Biochemistry", Academic Press, London, New York

Zaghloul, R., Obeid, M. and Staerk, H. (1973). *Radiochem. Radioanal. Lett.* **15**, 363

Ziegler, H. (1975). "Transport in Plants", Vol. 1, p. 85 (Zimmermann, M. H. and Milburn, J. A., eds), Springer Verlag, Berlin, Heidelberg, New York

Ziegler, H., Osmund, C. B., Stichler, W. and Trimborn, P. (1976). *Planta*, **128**, 85

Zingaro, R. A. and Cooper, W. C. (eds) (1974). "Selenium", D. Van Nostrand, Florence, Ky.

Zipkin, I. (1973). "Biological Mineralization", p. 487 (Zipkin, I., ed), John Wiley and Sons, London, New York

Zitko, V. (1975). *Sci. Total Envir.* **4**, 185

Zitko, V., Aiken, D. E., Tibbo, S. N., Besch, K. W. T. and Anderson, J. M. (1970). *J. Fish. Res. Bd. Canad.* **27**, 21

Zobell, C. E. (1963). "Organic Geochemistry", p. 543 (Berger, I. A., ed), Pergamon Press, Ltd., Oxford, New York

Zoller, W. H., Gladney, E. S. and Duce, R. A. (1974). *Science*, **183**, 198

Zoller, W. H. and Gordon, G. E. (1970). *Analyt. Chem.* **42**, 257

Zook, B. C., Carpenter, J. L. and Roberts, R. M. (1972). *J. Vet. Res.* **33**, 899

Zunino, H. and Martin, J. P. (1977). *Soil Sci.* **123**, 65

Zyakun, A. M., Bondar, V. A. and Nozhevnikova, A. N. (1976). *Dokl. Akad. Nauk. SSSR*, **227**, 497

Addendum

Arunachalam, J., Gangadharan, S. and Yegnasabramanian, S. (1979). "Nuclear activation in the Life Sciences, 1978", p. 499, I.A.E.A., Vienna

Beath, O. A., Draize, J. H. and Gilbert, C. S. (1935). *Bull. Wyo. Agric. Exp. Sta.* **206**

Beauford, W., Barber, J. and Barringer, A. R. (1977). *Science* **195**, 571

Buat-Ménard, P. and Arnold, M. (1978). *Geophys. Res. Lett*, **5**, 245

Cadle, R. D. (1975). *J. Geophys. Res.* **80**, 1650

Cornelis, R., Mees, L., Hoste, J., Ryckebush, J., Versieck, J. and Barbier, F. (1979). "Nuclear Activation in the Life Sciences, 1978", p. 165, I.A.E.A., Vienna

Edmonds, J. S., Francesconi, K. A., Cannon, J. R., Raston, C. L., Skelton, B. W. and White, A. H. (1977). *Tetrahedron Lett.* **18**, 1543

FAO Statistics (1975). Annual Fertilizer Review, F.A.O., Rome

Harold, F. M. (1977). *Annu. Rev. Microbiol.* **31**, 181

Humble, G. D. and Raschke, K. (1971). *Plant Physiol.* **48**, 447

Kelly, M. (1965). *Nature, Lond.* **208**, 809

Lantzy, R. J. and Mackenzie, F. T. (1979). *Geochim. Cosmochim. Acta* **43**, 511

Nriagu, J. O. (ed) (1978). "Biogeochemistry of Lead in the Environment", Thomond, Limerick

Pourbaix, M. (1973). "Atlas d'Equilibres Electrochimiques", Gauthier-Villars, Paris

Renan, M. J., Drennan, B. D., Keddy, R. J. and Sellschop, J. P. F. (1979). "Nuclear Activation in the Life Sciences, 1978", p. 479, I.A.E.A., Vienna

Sabbioni, E., Marafante, E., Pietra, R., Goetz, L., Girardi, F. and Orvini, E. (1979). "Nuclear Activation in the Life Sciences, 1978", p. 179, I.A.E.A., Vienna

Schiegl, W. E. and Vogel, J. C. (1970). *Earth Planet. Sci. Lett.* **7**, 307

Schmidt, U. and Huber, F. (1976). *Nature, Lond.* **259**, 157

Schwarz, K. and Spallholz, R. E. (1978). "Cadmium 77", Metal Bulletin Ltd., London

Shacklette, H. T., Boerngen, J. G. and Tumer, R. L. (1971). *U.S. Geol. Surv. Circ.* 644

Ure, A. M., Bacon, J. R., Berrow, M. L. and Watt, J. J. (1979). *Geoderma* **22**, 1

Ward, P. F. V., Hall, R. J. and Peters, R. A. (1964). *Nature, Lond.* **201**, 611

Subject Index